Magic, Science, Technology, and Literature

Jarmila Mildorf / Hans Ulrich Seeber / Martin Windisch (Eds)

Second Edition

Kultur und Technik

Schriftenreihe des Internationalen Zentrums für Kultur- und Technikforschung (IZKT) der Universität Stuttgart

herausgegeben von:
Helmut Bott, Franz Brümmer, Gerd de Bruyn, Georg Maag, Walter Göbel, Wolfram Pyta, Ortwin Renn, Werner Sobek

Band 03

LIT

Magic, Science, Technology, and Literature

Jarmila Mildorf / Hans Ulrich Seeber / Martin Windisch (Eds)

Second Edition

LIT

Bibliografische Information der Deutschen Bibliothek

Die Deutsche Bibliothek verzeichnet diese Publikation in der Deutschen Nationalbibliografie; detaillierte bibliografische Daten sind im Internet über http://dnb.ddb.de abrufbar

ISBN 3-8258-9311-1

Second Edition 2014

Grevener Str./Fresnostr. 2, 48159 Münster
Tel.: 0251 62 03 20 Fax.: 0251 23 19 72
e-Mail.: lit@lit-verlag.de http://www.lit-verlag.de
Umschlag und Satz nach Entwürfen von 2Schwertfeger

Contents

Scientific Perspectives

Media Differences

Preface to the Second Edition

It is rare for a collection of essays to go into a second print run. The fact that the present volume now sees its second edition attests to the fact that the subject matters covered in these essays – magic, science, technology, and literature – are undiminished in their significance and topicality. The first edition, which is reprinted in its original form here, emerged from an international symposium entitled "Magic, Science, Technology, and Literature," which we held at the Universität Stuttgart in 2005. Eight years after their first publication, the wide-ranging and stimulating contributions have lost nothing of their productivity, salience and creativity in dealing with concepts of magic at the cultural interface of literature, science and technology. Some of the contributions have meanwhile been extended to larger book-length studies. We refer interested readers to Andreas Höfele's book *Stage, Stake, and Scaffold: Humans and Animals in Shakespeare's Theatre* (Oxford University Press, 2011), Roger Luckhurst's *The Mummy's Curse: The True Story of a Dark Fantasy* (Oxford University Press, 2012), and to Hans Ulrich Seeber's *Literarische Faszination in England um 1900* (Universitätsverlag Winter, 2012). On a more light-hearted note, Elmar Schenkel published a collection of short stories entitled *Der magische Globus: rätselhafte, verrückte und magische Geschichten* (AKV Edition Hamouda, 2012). The growing number of academic studies relating to magic, science, technology, and literature – also and especially with regard to more popular cultural artefacts – suggests that these topics are likely to stay with us for some time. We hope that the present volume will continue to interest readers and to provide inspiration for further research.

While we of course thank all those who were involved in the production of the first edition we want to give special thanks here to Elke Uhl and Nora Heinzelmann at the *Internationales Zentrum für Kultur- und Technikforschung* (IZKT) Stuttgart, who brought the second edition on its way.

Paderborn / Stuttgart, February 2014

Preface to the First Edition

This volume emerged from an international symposium entitled "Magic, Science, Technology, and Literature," which we held at the Universität Stuttgart from 27 to 29 January 2005. The papers delivered at this symposium and the subsequent discussions arising from them soon confirmed that we had entered a rich and fascinating field of inquiry. The present book with its wide-ranging and stimulating contributions attests to the productivity and undiminished salience of concepts of magic at the cultural interface of literature, science and technology. We hope that the volume will be a source of 'enchantment' to interested readers much in the same way as it excited its editors while they compiled the chapters.

As with any other book, this publication owes a large debt of gratitude to numerous people whom we would like to thank here. First and foremost, we are grateful for the financial and logistical support we received from the *Internationales Zentrum für Kultur- und Technikforschung* at the Universität Stuttgart both during the symposium and in the preparation of this book. Financial assistance was also kindly granted by the *Vereinigung von Freunden der Universität Stuttgart*. Furthermore we would like to thank John Fowler and Jennifer Pyroth for translating and proofreading some of the manuscripts, and Miriam Höß, who spent many hours formatting and streamlining the chapters. Benjamin Wille's and Jörg Schwertfeger's help with the layout and design of the book also has to be mentioned here. Last but not least we would like to thank all our contributors, whose work made editing this book such a pleasurable task.

Introduction

Hans Ulrich Seeber / Jarmila Mildorf / Martin Windisch

Magic, Science, Technology, and Literature: A Dynamic Relationship

Dissociating science from magic is part of that process which Max Weber described as *Entzauberung*, demystification. And indeed, modern practitioners of the art of magic do not wish to be called magicians in the sense of sorcerers. Knowing the tricks and the techniques of the trade, they do not believe in magic as the manipulation of supernatural forces. Instead of, like shamans or black magicians, conjuring up spiritual forces or the dead, they practise the art of deception, of creating illusions.

However, this rational approach to the matter is not the only one in our culture, and not even the most typical one. Instead, the ongoing process of rationalisation and abstraction has produced, in dialectical fashion, its very opposite. In view of the complexities of modern science and technology the classic ideal of scientific realism has been abandoned in the popular genre of fantasy, which enjoys, as the success stories of Rowling's *Harry Potter* and of Tolkien's *The Lord of the Rings* testify, extraordinary popularity. Magic motifs such as magic rings, magic wands, magic stones or magic mirrors reappear in that literature, seemingly without offending the reader's sense of reality. However, whereas the appeal of popular fantasy literature may often be reductive or even regressive, Rushdie's mixture of realism and fantasy pays full attention to the complexities of cultural divisions and conflicting mentalities. Magic realism in novels such as Rushdie's *Midnight's Children* exploits the newly-won imaginative freedom for purposes of historiographic metafiction. He also comments metaphorically on the perception that modern technology is viewed as a form of magic not only by members of a traditionally magic culture.

The language of literary criticism is by now prepared to acknowledge with approval not only magic effects but also the presence in a contemporary literary text of magic and romantic elements. When Niall Williams published his postmodern combination of neo-romanticism and neo-realism, *Four Letters of Love* (1997), *The Times* greeted it with the following words: "A breathtaking affirmation of magic, miracles and the power of human love. Read it."[1] *Four Letters of Love* also alludes to that well-known analogy between poetic and magic effects, which has time and again been emphasised by critics and poets: "'*Dine hunc ardorem mentibus addunt, Euryale, an sua cuique deus fit dira cupidos.*' They were sounds so soft and full in my mouth that the very saying of them was a kind of magic, a kind of disappearing, upwards and outwards beyond the breaking of the surf and the distant shining of the lights on the islands" (*Four Letters of Love* 132).

How then do we account for the striking re-emergence of magic in our culture, literature and films in particular? As already indicated, the first reason may be, apart from modernism's and postmodernism's rejection of classical realism, simple escapism or regression. The complexities of modern science and technology are such that simple, magical, animistic answers become irresistibly attractive. But even so, the seeming incompatibility of magic with orthodox science is really beside the point. Magic is used as a convention, as a medium to explore basic questions of human life such as the quest for and the maintenance of power. The second, more serious reason has to do with developments inside modern science and technology themselves which provoked the language of magic. In his famous address to the Parapsychological Society of Edinburgh Arthur Koestler[2] stressed the astonishing parallels between magic ideas and the insights of modern science. How does one explain the telekinetic power of a star on a planet? The answer is of course gravitation, but the early observers, including Galileo and Newton, rejected this explanation as occult humbug, namely the assumption that a star or planet affects a distant other planet without being connected with it by some sort of medium.[3] However, the concept of a medium is in itself an occult idea. In the light of quantum physics the notion of contagious or sympathetic magic does not seem to be incompatible with modern insights into the relation between parts and the whole any more. Again, according to Koestler and the British physicist Paul Davies, modern quantum physics has opened a Pandora's box, creating a veritable surrealist panorama or version of mysticism. Physics and parapsychology are not so far apart, nor are physics and religion in the age of postmaterialist physics. The third reason has to do with modern man's horizon of expectations. The structure of wishing itself seems to have changed, since science and technology seem to be able, very much like magic, to meet and to realise every conceivable wish: we fly, we communicate instantly with people at the other end of the globe, we listen to voices long dead, as if we were conversing with the spirits of our ancestors, and we watch pictures sent to us from the moon, ugly faces are miraculously transformed into beautiful ones. It is such developments which encouraged authors to recycle magic motifs. Ernst Jünger even interprets the development of science and technology as having reached their 'magic phase' ("*magisches Stadium*");[4] similarly, Arthur C. Clarke states in his Third Law: "Any sufficiently advanced technology is indistinguishable from magic."[5]

The dynamics of the relationship between magic, science, technology, and literature are extraordinarily complex. Clear demarcation lines in terms of binary oppositions are theoretically possible, but difficult in practice. Historically, the acceptance of magic as a mode of knowledge became untenable as soon as Galileo demonstrated that the Aristotelian dualism between *physis* and *techne* was not valid, that both were in fact subject to the laws of nature and the notion of secret knowledge quite unsound (cf. Schiemann). The close connection between magic and the developments of science are too well-known to labour at length. In the

Renaissance the attempts by the magus to practise natural magic, i.e. to exploit natural forces with the help of 'occult' knowledge and procedures paved the way for modern science. Thus alchemists employed by princes – indeed, Count Wolfgang II of Hohenlohe was himself a famous alchemist around 1600 – to discover a short cut to gold and power made crucial contributions to the emergence of chemistry. Their visions of transformations and metamorphoses were taken up by writers and poets to illustrate the poetic process.[6] Even the icons of the modernising process such as Bacon and Sir Isaac Newton were entangled in the traditions of magic.[7] Only recently more of Sir Isaac Newton's occult studies in alchemy, which were written in a coded language and were thought to be lost, have been rediscovered in the archives of the Royal Society (*Frankfurter Allgemeine Zeitung*, 4 July 2005). What is exciting about the manuscript is the fact that Newton does not merely reproduce the opinions of contemporaries but makes a contribution of his own to the controversial topic. In psychology, it is clear that in the nineteenth century the modern discipline grew out of studies in mesmerism and occult fascination. Hypnotism and suggestion, which were widely discussed around 1900[8] and had a considerable impact on literary theory and literary practice, were the scientific equivalents of mesmerism and occult fascination.

Again, the interrelation between literature on the one hand and magic, science and technology on the other are quite intriguing. The verbal artist, who uses the *technology* of writing and printing and employs *techniques* of manipulating language for aesthetic effects, is both, as already Shakespeare (*The Tempest*) knew, a magician and a technician. His art of conjuring up reality with the help of words (an art which is reminiscent of occult necromancy and of incantation) and of enchanting his recipients by fascination or suggestion remind one of the classical role of the magician. In the performative art of the theatre the rhetorical device of *demonstratio ad oculos* is translated by quasi-magical procedures into visible actions. Frances Yates, without actually considering the performative dimension of *The Tempest* (see Döring, this volume), has characterised the relationship between magic and the playwright in the following manner: "Though Shakespeare never wielded a wand, nor thought of himself as a magus, he is a magician, master of spell-binding use of words, of poetry as magic. This was the art in which he was supreme and which Prospero symbolizes."[9] The role of the artist as magus was rediscovered around 1900, as Robert Stockhammer showed (*Zaubertexte*), when artists reacted against the traditions of realism and scientism. Science Fiction is really at the other end of the spectrum since its orthodox practitioners attempt to stay within the limits of accepted scientific discourse, although it is clear that poetic myths provide crucial subtexts for the production of the key texts, Wells' *The Time Machine*, for example.

The abstractions of modern science and technology have given poetic language and literature a new lease of life. A process of disenchantment and re-enchantment was triggered (neoclassicism/romanticism/realism/surrealism), the beginnings of which can be traced back to the Renaissance. The occult concepts of *harmonia mundi* and of the synthesising power of the imagination had far-reaching implications for the theory and practice of poetry. Furthermore, faced with impenetrable abstractions, the popularisers of modern science had to resort to metaphoric speech to be able to talk about the topic at all. Crucial truths of quantum theory are beyond the grasp of everyday human experience and the human imagination, *Vor-stellungskraft*. As soon as a scientist wishes to popularise and make accessible scientific insights to the general public his language can easily resemble that of a prose poem. We all know that the second law of thermodynamics makes entropy and the final end of the universe inevitable. The scientist Edward Harrison describes the gradual disintegration of the universe in the following metaphoric manner: "The stars begin to fade like guttering candles and are snuffed out one by one. Out in the depths of space the great celestial cities, the galaxies, cluttered with the memorabilia of ages, are gradually dying. Tens of billions of years pass in the growing darkness. Occasional flickers of light pierce the fall of cosmic night, and spurts of activity delay the sentence of a universe condemned to become a galactic graveyard."[10] Furthermore, instead of merely illustrating scientific abstractions and technological innovations or presenting a vision of future events, metaphors are also used by scientists as models providing theoretical explanations. Analogy is not only a slippery tool of the theoretical imagination, but also a heuristic instrument.

Clearly the topic of the conference addressed crucial issues, particularly those pertaining to the relationship between magic, science and technology on the one hand and language on the other. Most papers read at the conference focused on magic. However, it was clear that in a modern context magic is only relevant because it cannot be separated from developments in science and technology and from the question of public acceptance. Insofar as Western literature and art are concerned, its presence is usually an index not of truth claims but of imaginative freedom.

The chapters in this volume together illuminate the interface between magic, science, technology, and literature throughout the centuries from antiquity to the present day. Since many of the chapters branch out into various epochs and cross disciplinary boundaries, we grouped them together under four major headings which highlight points of convergence among the various contributions. Needless to say that such classifications are approximations at best and that one could have identified other group similarities. The four parts of the volume are meant as rough guides for readers to help them find their way around in what emerges as a rather diverse and multi-faceted area of research. We attached the following labels to the four parts: *Magical Renaissance, Cultural Transformations, Scientific Perspectives*, and *Media Differences*. The first group of chapters focuses on magic in the dramatic literature and pictorial art of the Renaissance period. They explore not only magic and performance *in* the Renaissance but also the Renaissance *as* magic, where traditional forms of magic were revived to entertain but also started to be contested. The second set of chapters spans the time from antiquity to postmodernity and encompasses magic from classical metamorphoses to postcolonial magical realism. The contributions in this part investigate transformations of time and space as well as the translatability of cultures in literature and beyond. One common feature of the second and the third part of this volume is their wide historical scope. Thus, the chapters in the third part also deal with magic from antiquity to the present day, albeit with a stronger emphasis on the history of science. This history, the chapters demonstrate, includes viewpoints which insist on the incompatibility of scientific and magical discourses on the one hand but also allow for the expansion of, for example, literary discourse through scientific input on the other. This cross-fertilisation is shown to add new dimensions to both scientific and non-scientific cultural domains. The volume is then rounded off by a fourth group of chapters which cross generic and media boundaries by considering discourses of enchantment in literature, film and music. The three chapters in this final part highlight the ways in which magic resurfaces in our postmodern world thanks to the enhancement of sensual experience through new technologies. The contributions can thus be said to develop a 'technology of effects' in the sense that they explore the mutual influence of technological development and aesthetic effect. The following chapter outline provides brief summaries of the individual contributions.

Magical Renaissance

Andreas Höfele opens his chapter about "Raising Tempests: Religion, Science, and the Magic of Theatre" with a description of a genre of texts called *Wetterpredigt* (weather sermon). While these weather sermons instituted God as the source of bad weather, popular belief had it that thunderstorms were caused by witches, sorcerers and the devil himself. This conflation of Christian belief with popular demonology can also be found in Shakespeare's *The Tempest*, where Prospero's "quasi-divine power" (28), exhibited in his creation of the storm, constitutes an act of hubris as well as "a masterstroke of theatrical illusion" (29). Höfele discusses the close connection between magic, science and the theatre by pointing towards the theatricality of natural experiments, which may seem wonderful to those not initiated to their causes. Furthermore, the theatre in the Renaissance period functioned as a discursive link between the disparate discourses of meteorologies, religion and science. Moral tales at the time construed "a God-as-Playmaker, the creator of spectacular moral drama literally or metaphorically conceived as thunderstorms" (34). In *The Tempest*, however, it is Prospero himself who creates theatre and storm and ultimately brings about his own downfall. Thus, Höfele concludes, the play is deeply troubling as it no longer assures one of a transcendental reality.

Tobias Döring also focuses on the relationship between magic and theatre in his chapter entitled "Magic, Necromancy, and Performance: Uses of Renaissance Knowledge in Marlowe's *Doctor Faustus*." He highlights the significance of books for conceptions of magic, which also assume an important role in Marlowe's play since Faustus is presented as "a reader, translator and interpreter" (39). However, the power of books is quickly superseded by the performative power of the theatre, both in *Dr Faustus* and in the Renaissance period at large. Döring first delineates to what extent magic was restricted to a script in books, also in Marlowe's play where Faustus aims at gaining magical powers from reading magical books, and he then offers an alternative reading by depicting the metatheatrical allusions in the play that correspond with magical effects. Döring investigates a number of scenes with regard to their rhetorical performativity, and he contends that rhetorical devices such as prosopopeia and other tropes are essentially like magic in that they evoke the illusion of presence despite the absence of real or literal concepts. "Marlowe's plays explore and show how to do magic things with theatre" (50), Döring contends, and ultimately theatrical performance makes authors lose control of their scripts as scripts are put to use for the theatre and thereby undergo re-interpretation.

Stephan Laqué discusses in his chapter entitled "The Force of Attraction in *A Midsummer Night's Dream*" the laws of gravity in Shakespeare's play. As Laqué points out at the beginning of his chapter, it was not before Newton's introduction of the term 'gravity' that people began to distinguish among forces of attraction such as gravitation, magnetism and love. In Shakespearian times these various forces were subsumed by analogy under one powerful force, whose source was regarded as divine or magical. Explanations concerning the force of attraction have to be considered against the background of a general development in scientific thought from the animistic view, which held that all objects possess a soul, to the mechanistic view, which assumes that the cosmos is nothing more than quantifiable matter and can be explained as such. Giordano Bruno's works on magic and on bonding serve as analytical frameworks for Laqué's discussion of the magic exercised by Puck and Oberon, the changing affiliations among the young lovers, and Bottom's "fear of bonding" (65). Laqué argues that the ending of the play constitues a "wedding of irreconcilable elements" which "creates the dynamic bond of an intellectual *coincidentia oppositorum*" (67). This demonstrates the possibility still given at Shakespeare's time to combine the animistic view with mechanical rationality.

Following the lead of Sigmund Freud's reading of *Macbeth*, Martin Windisch focuses on the visual representation of 'mysteries of state' in *Macbeth* and *King Richard II* in his chapter on "Pictorial Magic and the Staging of the Body Politic." Windisch first discusses the relationship between verbal and visual imagery in drama with regard to its effect on the audience. He highlights the function of mirrors and other optical equipment, such as anamorphic devices, and he considers processes of mystification accompanying them. In his interpretation of *Macbeth*, emphasis is put on the problem of the royal succession culminating in the tormenting sight of the witches' show of eight kings and magic glass. Windisch reads this against the background of Kantorowicz's *The King's Two Bodies* and also relates the Gorgon-like appearance of Banquo's ghost to the royal funeral effigies exhibiting the *dignitas* of the monarchical superbody. Magic and monstrosity appear, as Windisch contends, to be dialectically intertwined in the famous 'perspective scene' of *Richard II* with its dark implications of Queen Isabel's engendering an heir. He finally connects the continuous challenge to the legitimacy of Henry IV, partly caused by problematically rudimentary funeral rites given to Richard, with the Falstaff figure, who is said to embody hauntingly lifelike mimicry effigies causing conspiratorial unrest.

Cultural Transformations

The first chapter of the second part, Susanne Rupp's "Of Men and Pigs: Reconciling Epic and Magic in Alexander Pope's Translation of the *Odyssey*" takes the reader to the eighteenth century. Rupp discusses the ways in which eighteenth-century readers sometimes criticised Homer's works for being too fanciful and 'magical' and how Pope attempted to refute such criticism by justifying Homer's use of magical features in his epics, particularly in the *Odyssey*. Rupp unravels the argumentative strategies Pope adopts in order to achieve this goal. Thus he, for example, historicises the writing of the epics by stating that Homer had to appeal to his Phæacian audience and that Homer wrote as a historian committed to the ultimate truths inherent in his stories, and he also reads some of the more magical episodes such as the turning of Odysseus' men into pigs in allegorical terms. Another justification Rupp identifies in Pope's commentary on Homer is the didactic purpose behind magical features, which are meant to embellish and thus to make more pallatable the moral messages of the stories.

In his chapter on "*The Dunciad* and the City: Pope's *Heterotopia*," Brean Hammond argues the point that Alexander Pope's *Dunciad* is probably "the first literary work to raise the question of the *production* of space in recognizably modern form" (111). Taking Jeri Johnson's considerations of James Joyce's treatment of cityscapes as a starting point, Hammond explores to what extent *The Dunciad* represents London in its material reality but also invests it through symbols and allusions with additional layers of political, ethical and metaphysical meanings, e.g. through the enmeshment of Virgil's cultural geography in *The Aeneid* with Pope's own. While the poem is shown to become a contested space which is used to exert cultural control, the *Dunciad* also "*stages* cultural contest" (113), as Hammond argues, referring back to his earlier work on Pope's writing against mainstream cultural and historical beliefs and concepts. Hammond takes Henri Lefebvre's spatial typology as a backdrop for his discussion of Pope's satirical appropriation of London's history after the Great Fire and of its "almost-phantasmagoric mutations" (117) in *The Dunciad*. Pope's presentation of this particular cityscape is compared with Foucault's concept of "heterotopias," i.e. real spaces which "have a mirror-like relationship to utopias" (120).

"The Mummy's Curse: A Genealogy" by Roger Luckhurst explores the mummy's curse as part of a larger cultural formation, which the author labels the 'Egyptian Gothic.' Luckhurst traces the mummy curse back to possible origins and he also shows how memories and 'subjugated knowledges' (in Foucault's sense) of the curse proffer new ways of approaching official historical renditions of the Egyptian Question. Luckhurst poses the question where anecdotes and rumours surrounding the mummy curse in the late nineteenth and early twentieth century came from since Ancient Egyptian culture did not really have a concept of the curse. His explanations point in two directions, one of them literary-historical, the other politico-historical. One resource, Luckhurst argues, is the revival of the late Victorian Gothic, which was able to accommodate the mummy curse. Secondly, the mummy curse's underlying representation of Egypt as a mysterious land of the dead also "works exactly to obscure the other blindingly obvious element of this conjuncture: the revolt of *living* Egyptians" (133). In other words, Luckhurst identifies one of the major influences on people's imaginings of the mummy curse in Britain's conflict with Egyptian nationalists and Islamic revolutionaries at the time, which led to a "complex cross-cultural translation" (134) of discourses.

Walter Göbel, in his chapter on "Magic Pastoral: Tina McElroy Ansa," discusses McElroy Ansa's work in the context of the development of the African American novel from naturalism to modernism, postmodernism and magic realism. While much of African American writing has been marked by antagonism and notions of martyrdom, Göbel discerns in McElroy Ansa's novels *Baby of the Family* and *You Know Better* a combination of black magic and domestic idyll. This new genre, Göbel argues, emphasises the spiritual dimension of the African American experience and subverts the enlightenment discourse of the dichotomy of reason and imagination. Magic is here liberated from Christian suspicions of magic as blasphemous and heretical, and it is reconfigured as being central to a good education and as a part of everyday life experience and understanding. While *Baby of the Family* "can be interpreted as a re-writing of the classical story of African American infant martyrdom" (144), *You Know Better* "anticipates a black female utopia" (146). Ultimately, Göbel contends, this combination of pastoral elements with a positively connotated kind of magic confers a fairy-tale quality on McElroy Ansa's work and thus gives rise to a new mode of expression for the African American novel.

Scientific Perspectives

The third part of the volume is opened by Gregor Schiemann's chapter on "Physics and Magic: The Disenchanting of Nature," in which he explores the gradual distancing between magic and science from ancient times to the present day. Schiemann selects two historical moments in order to illustrate this development: Aristotle's discussion of *physis* in opposition to *techne*, and Galileo's recourse to the experimental method. While *physis* refers to all things natural, *techne* denotes humans' technical skills to modify what can be found in nature. The relationship between the two is one of imitation and completion. Schiemann uses this distinction as a background to a discussion of alchemical practices, which often referred back to Aristotle's work. Alchemy also regarded *techne* as technical skill, but one that could be used to ameliorate nature, "to replace the impure by the purer form" (158). However, alchemists still often regarded their art as white or natural magic. With the rise of technology, a shift towards the empirical method could be observed. Schiemann demonstrates how Galileo, in contrast to some of his contemporaries, already lacked an interest in magic and instead concentrated on the explanation and verification of natural phenomena through scientific experiments. Schiemann concludes his outline by referring to the ways in which the very belief in magic has now come under closer scientifc scrutiny whereby the "relationship between scientific and magical interpretations of nature is turned upside down" (161).

Robert Stockhammer explores in his contribution on "The Techno-Magician: A Fascination Around 1900," the attraction that magic held for people at the turn of the century. He points out the difficulties in defining 'magic,' and he tries to circumvent this problem by stating first what magic is *not*. Thus it is not the contrary of psychology, nor the contrary of technology. Nor is it absolutely opposed to science around 1900. What emerges, Stockhammer argues, is a class of empirically working scientists such as Edison and Charcot who can be regarded as "techno-magicians." Stockhammer presents Philippe-Auguste Villiers's novel *L'Ève Future* as a literary example of the interrelatedness of science, technology, magic, and literature at the time, and he identifies three different approaches to literature with regard to magic: firstly, the reconfiguration of magic around 1900 can be considered literary in the sense that it relied heavily on written texts; secondly, literature in a more restricted sense can be regarded as a site for restoring magic; thirdly, narratives are important for creating an image of the magician in people's minds. Villiers's novel is shown to portray a multimedia system which leaves room for media in both the technological sense of new apparatuses as well as spiritual media, a concept which not even Edison himself rejected categorically.

Elmar Schenkel commences his chapter entitled "Ghostly Geometry: The Fourth Dimension in Literature" with various definitions and explanations of the term "fourth dimension." While the term was only rediscovered in the 1970s and 1980s due to changed concepts of space in the wake of the electronic revolution, it already played a significant role around 1900 for similar reasons. Schenkel demonstrates to what extent the fourth dimension was bound up with non-Euclidean mathematics and was largely the result of scholars' attempt to measure the planet. Literary texts of the time also dealt with the 'fourth dimension.' While in Edwin A. Abbott's novel *Flatland* and in George MacDonald's *Lilith* the fourth dimension has a spiritual function and gives rise to philosophical deliberations, H. G. Wells uses the concept as an "eyeopening" (184) fictional tool. As Schenkel notes, Wells "turns scientific ideas into means of cognition" (184) by offering thought experiments in his fiction, and he is the first author to use 4D notions in both spatial and temporal terms. Joseph Conrad and Ford Madox Ford's joint novel *The Inheritors* furthermore addresses the question of the translatability of reality by taking into account the psychic quality of the fourth dimension. Thus, Schenkel argues, the fourth dimension constituted an "adventure in consciousness" (187) around 1900 and, probably for the first time, a science fiction concept influenced narrative method.

In "Disenchanting the World, Enchanting the Reader: The Negotiation of Biochemistry in Sinclair Lewis' *Arrowsmith*," Jarmila Mildorf discusses Sinclair Lewis' novel *Arrowsmith*, which was written at a time when the media had already started to contribute to an increase in lay knowledge of medical problems and biochemistry through a range of newspaper articles, health advice literature, discussion forums, etc. Mildorf shows how the novel explores the moral and ethical dilemmas involved in walking the tightrope between scientific practice and society's expectations of science's healing powers and to what extent it is informed by but also recontextualises and questions contemporary discourses on biochemistry. Images of magic and of magicians are shown to function as central tools for making scientific discourse accessible to and potentially fascinating for lay readers. Mildorf demonstrates how features more endemic to literary texts, such as affective styles, figural perspectivisation and satire, are used to negotiate the 'magic' world of medicine and biochemistry within the novel's "artistic spaces of freedom" (191). Readers are thus not only invited to learn more about the world of science, Mildorf argues, but also to sympathise with scientists and to become aware of their own (mis)conceptions.

Dirk Vanderbeke points out right at the beginning of his chapter entitled "'Science Is Magic That Works': The Return of Magic in Literature on Science" that magic has been discredited over the last centuries because it had to compete with science in terms of a true/false dichotomy. Nevertheless, authors such as James George Frazer, Agrippa von Nettesheim and even Nobel Prize winner Niels Bohr have pointed out that magic and science share basic presuppositions and that indeed science can be rephrased as 'magic that works.' Magic and science can at the same time be regarded as different, but also as converging, approaches to knowledge. Vanderbeke corroborates this contention by focusing on the problem of gravity, which he compares with the concept of 'contagious magic,' and he demonstrates how physicists today again use magic imagery and rhetoric in their scientific debates and how they contemplate phenomena which would formerly not have been taken seriously. Vanderbeke then discusses literary texts by Jeanette Winterson, Jim Dodge, Peter Ackroyd, Tom Stoppard, Thomas Pynchon and Umberto Eco, which, as he maintains, have not "been slow to grasp the narrative potential of the scientific imagery" (216).

Media Differences

The final part of this volume starts with Hans Ulrich Seeber's chapter on "Magic and Literary Fascination," in which he explores the connection between writers' tendency to draw upon magic as an analogical image for the mysterious power of literary texts and the concept of fascination. Seeber argues that an implicit and as yet undertheorised "aesthetics of fascination" has emerged since the Romantic period in authors such as Samuel Taylor Coleridge, Matthew Arnold, Friedrich Nietzsche, Joseph Conrad and James Reeves. Seeber demonstrates that poetry, for example, is said to have a mysterious suggestive power which changes readers' perceptions and engages them emotionally. Fascination thus becomes almost spiritual or religious. Seeber traces the aesthetics of fascination back to Edmund Burke's seminal discussion of the sublime, which encompasses astonishment and a degree of horror, and he provides a detailed definition of the term 'fascination,' which also takes into account its treatment in the history of ideas. Seeber then discusses what he calls 'remagicisation' in Oscar Wilde's *The Picture of Dorian Gray,* Lord Dunsany's *The King of Elfland's Daughter* and Salman Rushdie's *Midnight's Children.* He points out that fascination is not only engendered by fantastic elements but can also be experienced in mimetic-illusive texts written in the realist tradition.

In "Magic and the Film: The Enchantments of a Modern Medium," Johann N. Schmidt provides a survey of the affinities between magic and film techniques throughout the history of the film. Schmidt argues that the dialectic between the visible and the invisible lies at the heart of magic, and he demonstrates how, for this reason, early scientific presentations with their special effects were similar to magic tricks. The image, the film's central feature, mediates between reality and illusion in that it displays a likeness with the thing presented but is also of a ghost-like transparency. Schmidt further illuminates to what extent the development of the film, alongside other new technologies, coincided with a mechanisation of production, which made endless repetitions possible. Today, Schmidt maintains, video and DVD have additionally changed the temporality of reception, which is complemented by new possibilities of transforming the image. The history of the film, however, reaches even further back if one takes into account the magic lantern and the camera obscura, which both comprised narrative structure and were used for spectacular shows. Schmidt's review of pioneers of the cinema such as

Stellan Rye, Josef von Sternberg and Friedrich Wilhelm Murnau, who all revelled in the "manipulation" of the visual medium, demonstrates that enchantment and technophilia "are still just two sides of the same coin in the creation of a poetic universe out of the scientific spirit" (245).

The final chapter of this volume, "A Magic Lantern of Sound? The Aeolian Harp between Cheap Showmanship and Spiritual Mystery, Here Shown as a Catalyst within the History of Music" rounds off the discussion of magic elements by leading the reader into the world of music. Walter Windisch-Laube traces back the history of the Aeolian harp as both a musical instrument and as a symbol from the seventeenth century to the present day, and he highlights the ways in which the Aeolian harp has often been presented as a divine instrument producing 'magical' sounds. Windisch-Laube offers concrete musical examples, and he illuminates the processes whereby technological development and the instrument's symbolism have mutually influenced and propelled each other as "the aspiration to find or even produce the absolute, pure sound has become an urgent issue" (252). Thus, the Aeolian harp has thrived until today. Even modern jazz has made use of this instrument in order to enchant and delight the audience with exotic and unusual tones.

As can be seen in this brief outline, the various contributions responded in rather diverse ways to the topic of the symposium, providing general theoretical frameworks on the one hand and literary, cinematic, and musical case studies on the other. What all contributions have in common is that they place their object of discussion in a wider socio-historical and cultural context, thus mapping out a complex field of artistic, scientific, philosophical, religious, historical and social discourses in which magic constantly resurfaces and recedes. Readers are invited to explore this vast field by following the various paths opened up in the different chapters.

Notes

[1] Niall Williams, *Four Letters of Love* cover.

[2] Arthur Koestler, "Physik und Synchronizität," in Bauer, *Psi und Psyche* 111-120.

[3] Ibid. 111.

[4] Quoted in Martin Schwonke, *Vom Staatsroman zur Science Fiction* 134.

[5] Quoted by Dirk Vanderbeke in the present volume, page 211.

[6] Cf. Elmar Schenkel, *Die Elixiere der Schrift*.

[7] Cf. Frances A. Yates, "Francis Bacon 'Under the Shadow of Jehova's Wings'" and Richard S. Westfall, "Newton and Alchemy."

[8] Cf. Robert Stockhammer, *Zaubertexte*; Pamela Thurschwell, *Literature, Technology and Magical Thinking* 1880-1920; and Seeber, this volume.

[9] Frances Yates, *The Occult Philosophy in the Elizabethan Age* 190.

[10] Quoted in Paul Davies, *God and the New Physics* 201.

Works Cited

Bauer, Eberhard, ed. *Psi und Psyche: Neue Forschungen zur Parapsychologie: Festschrift für Hans Bender.* Stuttgart: Deutsche Verlagsanstalt, 1974.

Davies, Paul. *God and the New Physics.* London: Dent, 1983.

Schenkel, Elmar. *Die Elixiere der Schrift: Literatur und Alchemie.* Eggingen: Edition Isele, 2003.

Schwonke, Martin. *Vom Staatsroman zur Science Fiction: Eine Untersuchung über Geschichte und Funktion der naturwissenschaftlich-technischen Utopie.* Göttinger Abhandlungen zur Soziologie 2. Stuttgart: Enke, 1957.

Stockhammer, Robert. *Zaubertexte: Die Wiederkehr der Magie und die Literatur 1880-1945.* Berlin: Akademie, 2000.

Thurschwell, Pamela. *Literature, Technology, and Magical Thinking 1880-1920.* Cambridge: Cambridge UP, 2001.

Westfall, Richard S. "Newton and Alchemy," in *Occult and Scientific Mentalities in the Renaissance.* Ed. Brian Vickers. Cambridge: Cambridge UP, 1984. 315-335.

Williams, Niall. *Four Letters of Love.* London: Picador, 1997.

Yates, Frances A. "Francis Bacon 'Under the Shadow of Jehova's Wings'," in *The Rosicrucian Enlightenment.* London: Routledge, 2002.

---. *The Occult Philosophy in the Elizabethan Age.* London: Routledge, 1979.

Magical Renaissance

Raising Tempests

Religion, Science, and the Magic of Theatre

Andreas Höfele

I

On the second of July 1505, some four miles north of Erfurt, a young traveller walking towards the town was overtaken by a thunderstorm. A bolt of lightning striking close by, he cried out in mortal fear, "Help me, holy Saint Anne, and I shall become a monk!" And, true to his vow, the pious young man joined the Augustinians a fortnight later. All the biographical evidence suggests that the storm only precipitated a decision that had been long in the making, that it was only the last straw after months, if not years, of anguished soul-searching. Tempting as it may be, it would be pointless to speculate on the different course of European history if the weather had taken a fair turn for Martin Luther that day. More to my present purpose is the belief in thunderstorms as signs of divine wrath, which this famous episode exemplifies. This belief survived well past the Reformation. Indeed, it even increased its powerful hold on people's – inevitably bad – conscience. God's thunderbolt was thought to serve as a terrifying wake-up call rousing Christians to the realisation of their sins.

Lest they prove deaf to the call, this message was also trumpeted at them from the pulpit and in a multitude of tracts enjoining believers to contrition and repentance. In Germany, *Wetterpredigt* (weather sermon) became an established term for a homiletic genre thriving throughout the sixteenth and seventeenth centuries. Although the term is lacking in English there was no shortage of such texts in Britain during the same period. *Fearfvl Newes of Thunder and Lightning* promises the title of a tract published in 1606, in which we read how "the force of Gods terrible voice, had shaken, rifted, and rent "the tower of a church in Gloucestershire." "So powerfull and provident is Almighty God, that he striketh and leaveth unstriken where he will, and whom he will [...]. And although this Thundring voyce of Almighty God, did no where shewe so terrible an effect, as on and in the Church, yet without the Church, in the Field and in the houses neere, an extraordinary presence of his power, not without his mercie, was sensibly to bee perceived by diverse, whom it cast, one hither, another thither, and some downe, but hurt neither man, woman, child, nor any living creature els [...]."[1] "[B]ut all these aforesaid grievous calamities," explains the author of "A Briefe Relation [...] of the late great Tempeste" in 1613, "we must all confesse and acknowledge, to be the true tokens that our mercifull God is highly offended with us."[2] A perfect specimen of a weather sermon is provided, as late as 1695, by the notorious Salem witch-hunter Cotton Mather. Here is the title in full: "Brontologia sacra: The voice of the glorious God in the thunder explained and applyed in a sermon uttered by a minister of the Gospel in a lecture unto an assembly of Christians abroad, at the very same time when the thunder was by the permission and providence of God falling upon his own house at home.: Whereunto are added some reflections formed on the Lords-Day following by the voices of thunders, upon the great things which the great God is now doing in the world.: A discourse useful for all men at all times, but especially intended for an entertainment in the hours of thunder."[3]

Weather tracts and sermons frequently engaged in fierce polemics against Catholic ways of providing shelter from the storm, such as the use of sacred objects, candles, incense, holy water and church bells. But another, perhaps even stronger adversary they had to contend with was the entrenched popular belief that thunderstorms were not caused by direct divine intervention but by inferior powers: the devil, his demonic retinue or indeed human evildoers in cahoots with them, witches and sorcerers. For Reginald Scot, in his grand rebuttal of the efficacy of witchcraft, popish superstition and the belief in witchery go hand in glove. "The fables of Witchcraft," he protests, "have taken so fast hold and deep root in the heart of man, that [...] a clap of thunder, or a gale of wind is no sooner heard, but either they run to ring the bels, or crie out to burne witches."[4]

Scot is categorical in his rejection of any intermediate agency in the causation of storm and tempest. Other early modern writers were more accommodating. Many a Continental weather sermon assumes a chain of command in which God, the devil and the witch, as it were, collaborate: the Devil is authorised to execute God's wrath but, being the Devil, is always eager to exceed his brief; the witch, the devil's duped accomplice, trades her or his soul for the mere illusion of power.

As a German tract from 1599 explains: When the Devil, a spirit of the air, from long experience of natural causes, scents a heavy storm brewing, he calls upon his sworn strumpets and miscreants to cook up a potion and to turn the pot over, making mirth and delight for themselves and bringing down terror and heartache on others.[5] Don't believe, is the author's message, that witches can really command the weather, but burn them anyway for their evil intentions. Conflating Christian belief with popular demonology, weather sermons also drew upon Aristotelian cosmology. Thunderstorms are sublunary phenomena generated in the three layers of the elementary sphere. The middle layer is where "aerae potestates, or, air-spirits" abide, those "evil spirits under the sky" that St. Paul mentions in his letter to the Ephesians.[6] Making weather, concludes a late-seventeenth century *Thesaurus Consiliorum*, is God's work and not man's, be he ever so high-born, much less an old woman's. God sometimes permits His hangman, the Devil, to raise a harmful wind and thunder in a particular place. This the Devil can do, being a spirit of the air and having control over the inconstant and fickle meteors. But he can never do more than God permits and ordains.[7]

II

Turning from weather tracts and sermons to Shakespeare's *Tempest*, one cannot but notice correspondences. Opening with, according to the Folio stage direction, "A tempestuous noise of thunder and lightning heard,"[8] the play orchestrates a thunderstorm at sea with all the sight and sound effects Jacobean stagecraft was capable of. Royal authority is nullified by *force majeure*. "What care these roarers for the name of king?" the boatswain snaps when councillor Gonzalo demands that he show deference. A moment of carnivalesque topsy-turvydom, to be sure, but in this tempestuous setting it sounds a distinctly apocalyptic note.

Christian tradition has always counted thunder and lightning among the omens of Judgment Day, and the repeated talk of hanging in this scene, for all its gallows humour, points in this direction. The very bravado of that talk contrasts sharply with the behaviour of the King and Prince Ferdinand, who are, we hear, "at prayers" (*The Tempest 1.1.53*) below deck. Like those hardened sinners that weather sermons never fail to put up as cautionary examples, the two villains of the play, Antonio and Sebastian, are deaf to the voice of thunder. Not so the King. To him the roaring of the tempest becomes the voice of his own guilty conscience:

O, it is monstrous, monstrous!
Methought the billows spoke and told me of it,
The winds did sing it to me; and the thunder,
That deep and dreadful organ-pipe, pronounced
The name of Prosper: it did bass my trespass.
Therefore my son i' th' ooze is bedded; and
I'll seek him deeper than e'er plummet sounded,
And with him there lie mudded.
(*The Tempest* 3.3.95-102)

The King's expression of remorse in Act 3 follows hard upon another blast of thunder and lightning, the fitting prelude to Ariel's pronouncement of doom:

Thunder and lightning.
Enter Ariel, like a harpy, claps, his wings upon the table, and with a quaint device the banquet vanishes.
ARIEL
You are three men of sin, whom Destiny,
That hath to instrument this lower world
And what is in't, the never-surfeited sea
Hath caused to belch up you, and on this island,
Where man does not inhabit - you 'mongst men
Being most unfit to live. [...]

You fools! I and my fellows
Are ministers of fate [...]
Thee of thy son, Alonso,
They have bereft; and do pronounce by me
Ling'ring perdition, worse than any death
[...]
whose wraths to guard you from,
[...]
is nothing but heart's sorrow,
And a clear life ensuing.
He vanishes in thunder.
(*The Tempest* 3.3.53-82)

Although the episode of the vanishing banquet is based on Virgil[9] and although Ariel appears in the guise of a harpy, this pagan monster pronounces an unmistakably Christian sentence, leading the culprits from sin to despair.to a future of "nothing but heart's sorrow" and bleak, unending penance in "a clear life ensuing" (*The Tempest* 3.3.81-82). No wonder then that the King associates the voice of thunder with the sound of a church instrument, "[t]hat deep and dreadful organpipe" (*The Tempest* 3.3.98).

Ariel's speech exhibits the same fusion of Christian and Aristotelian conceptions that we find in many weather sermons. It also shows a similar division of labour between "Destiny," the higher power, and its "ministers" who stir up the elements of "this lower world," Destiny's "instrument." Ariel, it may be noted, plays the part here that weather sermons often allocate to the Devil. This is not to suggest that he ought to be conceived of as altogether satanic; it is only perhaps another good reason for not picturing him as some bland Mendelssohn fairy doing inane things with flowers. Prospero, on the other hand, is unmistakably playing God. We hardly need the weather sermons to know this. It is abundantly clear from the text itself and is confirmed by its much-explored references to Neoplatonic conceptions of magic.[10] But while Neoplatonic readings tend to idealise Prospero, reference to the Christian weather tracts cited here not only enhances his Godlike status but also makes that status deeply ambivalent.

If by your art, my dearest father, you have
Put the wild waters in this roar, allay them.
(*The Tempest* 1.2.1-2)

Prospero's quasi-divine power is sprung on the audience with the metatheatrical *dénouement* of the storm scene. No sooner has the ship gone down with all hands and passengers than we learn that the catastrophe was just a show. The uproar of unruly nature was in fact ruled by art; art in the still largely undifferentiated

Renaissance sense encompassing pursuits from all areas of human ingenuity.

Miranda, whose name means 'she who is to be admired,' is shown here, as so often in the play, as *mirans*, 'she who is amazed.' Notwithstanding her suspicion that her father may be the author of the spectacle, she is fully taken in by it, articulating the ideal audience response of terror and pity:

O, I have suffered
With those that I saw suffer – a brave vessel
(Who had no doubt some noble creature in her)
Dashed all to pieces. O, the cry did knock
Against my very heart.
(*The Tempest* 1.2.5-9)

This is followed by her helpless wish that the lost might have been saved, the first indirect measurement of Prospero's well-nigh illimitable power:

Had I been any god of power, I would
Have sunk the sea within the earth or ere
It should the good ship so have swallowed ...
(*The Tempest* 1.2.10-12)

Prospero is this very "god of power." Together with the curt command that she not waste her feelings on mock tragedy, he "switches off" the tempest, so to speak:

Be collected;
No more amazement. Tell your piteous heart
There's no harm done.
(*The Tempest* 1.2.13-15)

The surprise twist in the second scene, revealing the tempest as nothing more than a play within the play, establishes the much-discussed intertwining of theatre and magic so characteristic of the work.[11] A masterful demonstration of magic powers, Prospero's storm is also a masterstroke of theatrical illusion, of mutually potentiating magic and theatre. It confirms the quasi-divinity that Sir Philip Sidney bestows on the poet as a Godlike creator. "The Greeks called him 'a poet', which name hath, as the most excellent, gone through other languages. It cometh of this *poiein*, which is 'to make'; wherein I know not whether by luck or wisdom, we Englishmen have met with the Greeks in calling him 'a maker' [...] Neither let it be deemed too saucy a comparison to balance the highest point of man's wit with the efficacy of Nature; but rather give right honour to the heavenly Maker of that maker, who having made man to His own likeness, set him beyond and over all the works of that second nature: which in nothing showeth so much as in Poetry, when with the force of a divine breath he bringeth things forth far surpassing her doings […]"[12] Sidney describes the working of poetic invention as a kind of creative alchemy turning nature's brazen world into a golden one of the poet's own making. But when doing so, the poet is careful not to be "too saucy," presumptuous, irreverent. The poet may exceed the perfection of postlapsarian earthly nature, but poetic power does not compete with the creative power of God, nor does it diminish it. On the contrary, it glorifies that power, for God is the poet's creator, "the Maker of that maker."

The tempest that Prospero unleashes goes further. On the one hand, it is a permissible exercise of theatrical mimesis according to Sidney. On the other hand, however, it is also an intervention in the workings of nature and as such not merely the image of divine execution, but rather the appropriation of it. This would be – and I intentionally say 'would be' – a classical hubris motif. But Shakespeare's *Tempest* lacks an essential element. Hubris, as the Greeks saw it, is not hubris if it goes unpunished. Likewise pride, Christian *superbia*, always comes before the fall and only becomes manifest in the fall. The gods are not amused when mortals mess with their prerogatives. Prospero does this and gets away with it. *The Tempest* is not a tragedy of the *de casibus* type.

Yet the connection between magic and hubris was a commonplace of the period. The exemplary case, Christopher Marlowe's overly curious Doctor Faustus, is associated in the prologue of the tragedy, very much in the customary emblematic way, with the fall of Icarus:

Till swol'n with cunning of a self-conceit,
His waxen wings did mount above his reach,
And melting, heavens conspired his overthrow.
For falling to a devilish exercise,
And glutted now with learning's golden gifts,
He surfeits upon cursed necromancy.
(*Doctor Faustus*, Prologue 20-25)[13]

The same nexus between magic and hubris is maintained in King James' *Daemonologie*, published in 1597. What the royal author has to say about the typical career of a sorcerer reads like a moralising paraphrase of Marlowe's play: "[…] mounting from degree to degree, vpon the slipperie and vncertaine scale of curiositie; they are at last entised, that where lawfull artes or sciences failes, to satisfie their restles mindes, even to seeke to that black and vnlawfull science *Magie*. Where […] they blindlie glorie of themselves, as if they had by their quicknes of ingine, made a conquest of *Plutoes* dominion, and were become Emperours over the Stygian habitacles. Where, in the meane time (miserable wretches) they are become in verie deede, bond-slaues to their mortall enemie: and their knowledge, for all that they presume thereof, is nothing increased, except in knowing evill, and the horrors of Hell for punishment thereof, as *Adams* was by the eating of the forbidden tree. "[14] "A sound Magician is a mighty god" (*Doctor Faustus* A-text; i 93),[15] says Faustus in his initial enthusiasm, anticipating Miranda's "god of power." But the power Faustus strives for eludes him like a fata morgana. Although the master of a world empire in his mind, he becomes a slave to the power he thinks he can employ as the instrument of his wishes. The pact with the devil does not guarantee his omnipotence but rather his impotence; "omnimpotence" as one critic put it.[16] Faustus expects magic to offset the difference between thought and deed by making everything that can be imagined achievable. But, quite unexpectedly, he must accept that in fact Lucifer is even able to censor his thoughts.[17]

III

What distinguishes Prospero from Faustus? It is the tempest. This is not the only difference, of course, but a sufficiently crucial one to begin with. What for Faustus was only ambition, namely "[to] raise the winde, or rende the clouds" (*Doctor Faustus* A-text; i 89) – Prospero actually performs. His curiosity bears the fruits that are withheld from Faustus: knowledge and power.[18] How can this be? One could answer – and this answer has frequently been given – because Prospero's magic is totally different from that of Faustus.[19] Prospero does not practice devilish black magic, according to Pico della Mirandola "the most deceptive of all the arts," but rather the hermetic Neoplatonist "white" magic that Pico praises as the "most perfect highest wisdom."[20] "Prospero," Frances Yates maintained in 1975, "is poles apart from the witch Sycorax and her evil son."[21]

But subsequent critics have pointed out that the play deconstructs such tidy polarisation, most resonantly when Prospero enumerates his feats of magic in Act 5:

I have bedimmed
The noontide sun, called forth the mutinous winds,
And 'twixt the green sea and the azured vault
Set roaring war; to the dread-rattling thunder
Have I given fire and rifted Jove's stout oak
With his own bolt: the strong-based promontory
Have I made shake, and by the spurs plucked up
The pine and cedar; graves at my command
Have waked their sleepers, ope'd and let 'em forth
By my so potent art.
(*The Tempest* 5.1.41-50)

Borrowed from Ovid's evil sorceress Medea (in Arthur Golding's English translation) this passage clearly transgresses the limits of what may pass as permissible white magic. Its classical provenance notwithstanding, the speech lands Prospero in the none-too-comfortable vicinity of Early Modern witch hunting. Control over the elements had been an attribute of the saints, but after the eclipse of the medieval Church such power became witchery.[22] Subsumed under the heading "Witcherie" in a legal commentary of 1594,[23] the power "to cause haile and tempests" becomes a really perilous asset. It is thus only fitting that Reginald Scot should open his bold challenge to the abuses of witch hunting, *The Discoverie of Witchcraft* (1584), as Shakespeare does his play, with the magical causation of thunder, lightning and tempest. Both Shakespeare and Scot skip over the petty charms and spells of lowly sorcery[24] and begin with the greatest of all magic tricks. Nothing could have demonstrated Prospero's power more convincingly than his command of natural forces. Conversely for Scot, the more such power exceeds human capabilities, the more evident its fraudulence. Scot writes: "[...] But certeinlie, it is neither a witch, nor divell, but a glorious God that maketh the thunder. I have read in the scriptures, that God maketh the blustering tempests and whirlewinds: and I find that it is the Lord that altogether dealeth with them, and that they blowe according to his will. [...] But little thinke our witchmongers, that the Lord commandeth the clouds above, or openeth the doores of heaven, as *David* affirmeth; or that the Lord goeth forth in tempests and stormes, as the Prophet *Nahum* reporteth: but rather that witches and conjurers are then about their businesse."[25] In the light of Scot, there are two sides to Prospero, both equally reprehensible. On the one hand he stands convicted as a charlatan, on the other as a usurper of the Biblical "Lord [who] goeth forth in the tempests and storms."

It is likely that Shakespeare was familiar with *The Discoverie of Witchcraft*, but at the time of the writing of *The Tempest* (1610/1611) the dominant authority on witchcraft was not the sceptical Scot but the King himself, James I, with his 1597 *Daemonologie, in Form of a Dialogue*. Unlike Scot, whose "damnable opinions"[26] he sharply rejected, ordering the public burning of the book upon his ascension to the English throne, James harboured not the slightest doubt that witches and wizards can conjure tempests, although here, too, the devil is but the ape of God and can muster only inferior imitations of the real thing, storms of lesser magnitude and duration.[27] His Majesty is the surer of himself for having, he assumes, firsthand knowledge. The King's unquestioning belief in witches stemmed from an experience that he had described six years before the *Daemonologie* in a publication called *Newes from Scotland*. In it, he reports on the apprehension, interrogation, torture and condemnation of a whole ring of witches, along with its warlock, a certain Dr. Fian, a schoolmaster. The King himself presided over much of the proceedings and "in respect of the strangenes of these matters, tooke great delight to be present at their examinations."[28] The curiosity which James damns in his *Daemonologie* seems to have taken a strong hold on him here. The confessions of the alleged witch Agnis Sampson were, to be sure, so absurd that the King could not bring himself to believe her, whereupon, the report states, she took him aside and whispered to him "the verye woordes which passed between the Kings

Maiestie and his Queene at Vpslo in Norway the first night of their mariage."[29] Convinced, the King was then prepared to believe what Agnis Sampson had to say about "a christened cat" that she and her sister witches had cast into the Firth of Forth at Leith on Edinburgh harbour: "[T]his doone, there did arise such a tempest in the Sea, as a greater hath not beene seene: which tempest was the cause of the perrishing of a Boate or vessel coming ouer from the towne of Brunt Iland to the towne of Lieth, wherein was sundrye Iewelles and riche giftes, which should haue been presented to the now Queen of Scotland, at her Maiesties comming to Lieth. Againe it is confessed, that the said christened Cat was the cause that the Kinges Maiesties Ship at his comming forth of Denmarke, had a contrary winde to the rest of his Ships, then being in his companye, which thing was most strange and true, as the KIngs Maiestie acknowledgeth, for when the rest of the Shippes had a fair and good winde, then was the wind contrarye and altogither against his Maiestie: and further the saide witche declared, that his Maiestie had never safely come from the Sea, if his faith had not preuailed aboue their ententions."[30]

The actual events – the marriage of James I and Anne of Denmark in Norway, the passage of the royal couple to Scotland in April 1590 – blend here with that magical world in which, two decades later, Shakespeare has a different wedding party suffer shipwreck on its return from Tunis to Naples. Given that Shakespeare's *Tempest* was performed twice at court, we may imagine an interesting encounter of art with reality: Duke Prospero meeting King James, the magician, the demonologist.

IV

And perhaps – though admittedly this is pure speculation – in the person of the King's attorney general, Sir Francis Bacon, the New Philosophy made up a third party to this encounter. If Prospero's raising of tempests was witchcraft for the King, it may have been science for Bacon.[31] The first edition of his *New Atlantis* (1627) concludes with a list of practical results to be derived from future inquiry into the secrets of nature. Besides such items as the "prolongation of life" and the "[m]aking of new species," the list, entitled "Magnalia Naturae," also includes "[i]mpressions of the air, and raising of tempests."[32] Given the epistemic and technological parameters of his day, Bacon's list reads like pure fantasy or indeed like pure magic. Bacon's empiricist new philosophy points away from, and beyond, Prospero's Neoplatonic magic. But Shakespeare's *Tempest* encapsulates an historical moment where the two have not yet parted company, where they are still conceivable as variations of a single theme.

Magic, Sir Walter Ralegh suggests, may just be a name for what the ignorant do not understand: "the connection of natural agents [...] wrought by a wise man to the bringing forth of such effects as are wonderful to those that know not their causes."[33] This formula covers a wide range of early modern activities which have since become strictly separated. It also registers a connection between the study of nature and theatre. The "bringing forth of such effects as are wonderful" to the beholder is an integral part of early modern science. In his essay *Les Météores*, Descartes gives instructions for producing artificial rainbows.[34] The purpose of the exercise is to give an exemplary and – considering the Biblical symbolism of the rainbow – highly strategic demonstration that all celestial phenomena are brought about purely by natural causes, "so that there is no reason to marvel at anything that shows itself there [i.e. in the sky]."[35] But at the same time as it produces disenchantment, the experiment also generates re-enchantment by way of theatrical illusion. In order to secure their admiration, Descartes writes, the audience should be kept at a distance from the machinery pumping the water into the air, so that they cannot "discover the artifice."

Theatre is at the centre of Bacon's scientific utopia. Salomon's House, described as "the very eye of this kingdom," i.e. New Atlantis,[36] is not just a place for research and development, but also, crucially, one for creating spectacles. "We have also great and spacious houses, where we imitate and demonstrate meteors – as snow, hail, rain, some artificial rains of bodies and not of water, thunders, lightnings; also generations of bodies in air – as frogs, flies, and divers others."[37] Theatricality is an integral part of experimental science as envisaged by Bacon and his contemporaries. If the printed book is a key medium of the age and indispensable for the advancement of learning, then so is the theatre. The Theatre of the World, no less than the Book of Nature, is a powerful trope the Renaissance inherited from its predecessors and adapted to the demands of a rapidly increasing and diversifying body of knowledge. Well into the eighteenth century, 'theatre' became a favourite title for books claiming to offer insight and clarification, an authoritative survey of a field or subject, supported by reliable evidence. Theatre, etymologically a place made for watching – a 'very eye,' we might say, of the world – provided a highly plausible epistemological model for an intellectual culture privileging the visual.

This model was just as useful for moral instruction. Thomas Beard's *The Theatre of God's Judgements* published in 1597 is a case in point. This was a collection of cautionary tales ranging from antiquity to the present, in which the Almighty is shown to let no sinner escape his severe punishment. Beard is particularly harsh on play actors and playwrights and has no doubt that God detests them too, witness his graphic if largely spurious account of the death of Christopher Marlowe on which he dwells with unconcealed glee. Yet for all his anti-theatrical fervour Beard uses the term 'theatre' to enhance the prestige of his book, quite unaware, it seems, of any contradiction. Two decades after Beard's original publication, Edmund Rudierd published an abridged version of it under the title *The Thunderbolt of God's Wrath*.[38] Taken together, the two titles construe a God-as-Playmaker, the creator of spectacular moral drama literally or metaphorically conceived as thunderstorms. Theatre emerges as the bridge that allows disparate discourses, the disparate meteorologies, of religion and science to meet.

Which brings us back to Shakespeare's *Tempest*. The contexts sketched here, I would suggest, argue against taking the theatricality of Prospero's magic as an index merely of flimsiness, vanity or mere 'hocus-pocus' as Barbara Mowat argued in an influential essay.[39] The realisation that it is all just a show does not in itself, and not all at once, deflate the seriousness of Prospero's God-game, to borrow a term from John Fowles' *The Magus*. The deflation occurs, not in the second scene of act 1, but in the first of act 4, and when it occurs it is not confined to Prospero's magic only, nor to the 'magic' of theatrical allusion either, but, expanding outward beyond the borders of the play world, it encompasses no less than simply everything under the heavens:

The cloud-capped towers, the gorgeous palaces,
The solemn temples, the great globe itself,
Yea, all which it inherit, shall dissolve,
And, like this insubstantial pageant faded,
Leave not a rack behind.
(*The Tempest* 4.1.152-156)

I said earlier that Prospero's hubris can hardly be defined as such because it lacks the crucial element of the fall by which the magnitude of its presumption may be measured. This needs, however, some qualification. To be sure, there is no gaping hell in *The Tempest*, no Mephistophilis coming to fetch the magus in the end. But a downward movement there nonetheless is, running counter to the successful dynastic marriage scheme. Prospero's epilogue hovering indecisively between the play world and the world of the audience sounds a curiously despondent note. The power is gone, the god-like magus reduced to his own strength, "[w]hich is most faint" (*The Tempest* 5.1.321).

The descent that has brought Prospero to this depth can be traced in his abjuration speech earlier in act 5. Beginning with the noontide sun and ending with drowning – if only the drowning of "my book" –, the speech re-inscribes the trajectory of Icarus' fall. But Prospero's fall is self-induced, initiated by *a mise en abîme* which is not confined to artistic procedures only but which plunges "the great globe itself" into the void. If the epistemological authority of the theatre-as-medium gives credit to Prospero's magic art, the ontological downside of a world-as-theatre begets a state of existential vertigo. Unlike Calderon's *El gran teatro del mundo*, Shakespeare's *Tempest* offers no reassurance that the vanity of mere appearances is securely anchored in a transcendental reality. Drawing together the discrepant worlds of post-Reformation homiletics and the secular utopia of science, the play is not quite at home in either. Caught between the nemesis of the weather sermons and the promise of a New Atlantis, Prospero escapes the fear of the one without gaining the hope of the other. If we are "such stuff / As dreams are made on" (*The Tempest* 4.1.156f.), these dreams are deeply troubled.

Notes

[1] P. S., *Fearefvl Newes of Thunder and Lightning, with the terrible effects thereof, which Almighty God sent on a place called Oluestone in the county of Glocester the 28. of Nouember last C 3.*

[2] "A Briefe Relation of the great losses both by Sea and Land, by reason of the late great Tempests," in *The last terrible Tempestious windes and weather.*

[3] Cotton Mather, "Brontologia Sacra: The Voice of the Glorious God in the Thunder, etc." (1695).

[4] Reginald Scot, *The Discoverie of Witchcraft* 25-26.

[5] "Wann der Teuffel / als ein Geist auch in den Lüften / auß langwirigem Brauch vnd Erfahrung der Natürlichen Vrsachen abnimpt vnnd spürt / daß ein schwer vnd gefahrlich Wetter ohne das entstehen vnnd werden will / so mahnet er seine arme geschworene Metzlin / die Vnholden / auff / vnnd richtet sie an / sie sollen diß oder jenes nemmen / anfahnen kochen / vnnd darnach das Häfelein umbstürzen / so werden sie jhnen selbs ein Frewd vnd Lust / anderen Leuten aber Schrecken / vnnd groß Hertzeleid machen" (Johann Schopff, *Wetter Glöcklin* (1599) 72ff., quoted in Heinz Dieter Kittsteiner, *Gewissen und Geschichte: Studien zur Entstehung des moralischen Bewußtseins* 34).

[6] Ephesians 6:12.

[7] Georg Dedeken, *Thesaurus Consiliorum et Decisionum Volumen Secundum* (1671) 340. Quoted in Kittsteiner 35.

[8] William Shakespeare, *The Tempest*, ed. Stephen Orgel. All references are to this edition.

[9] *Aeneid* III 225 et seq.

[10] See Brian Vickers, *Occult and Scientific Mentalities in the Renaissance;* John S. Mebane, *Renaissance Magic and the Return of the Golden Age;* Barbara Howard Traister, *Heavenly Necromancers: The Magician in English Renaissance Drama.*

[11] See, for example, Karol Berger, "Prospero's Art;" Andrew V. Ettin, "Magic into Art;" Alvin B. Kernan, *The Playwright as Magician*; Andreas Höfele, "Wissen und Macht: Der Gelehrte als Magier im Drama der Shakespearezeit."

[12] Sir Philip Sidney, *An Apology for Poetry, or, The Defence of Poesy* 99 and 101.

[13] Christopher Marlowe, *The Complete Plays* 265.

[14] King James the First, *Daemonologie* (1597) 10-11.

[15] Christopher Marlowe, *Doctor Faustus* 1604-1616. Parallel Texts.

[16] Constance Brown Kuriyama, *Hammer or Anvil: Psychological Patterns in Christopher Marlowe's Plays.*

[17] "LU[CIFER]: Thou shouldst not thinke of God, thinke of the deuil, / [...] / FAU: Nor will I henceforth: pardon me in this, / And Faustus vowes neuer to looke to heauen, / [...]" (*Doctor Faustus* A-text; vi 722-725).

[18] Cf. Andreas Höfele, "Wissen und Macht: Der Gelehrte als Magier im Drama der Shakespearezeit."

[19] Frank Kermode's introduction to the Arden *Tempest* made this view canonic for a time.

[20] Giovanni Pico della Mirandola, *Über die Würde des Menschen* (1486) 50.

[21] Frances Yates, "Magic in Shakespeare's Last Plays" 17.

[22] Cf. Keith Thomas, *Religion and the Decline of Magic.*

[23] William West, *Three Treatises, Of the second part of Symboleographie.* Kurt Tetzeli von Rosador, *Magie im elisabethanischen Drama* 36-37, cites West at length in his excellent introduction to the Elizabethan definition of magic.

[24] Scot returns to these later in the book, with the result that his work, entirely contrary to his intent, became a kind of how-to-do-it guide to witchcraft.
[25] Scot, *The Discoverie of Witchcraft* 25-26.
[26] King James the First, *Daemonologie* 46.
[27] "They can rayse stormes and tempestes in the aire [...]" which are recognisable by the "suddaine and violent raising thereof, together with the short induring of the same" (ibid. 46).
[28] King James the First, *Newes From Scotland* 14.
[29] Ibid. 15.
[30] Ibid. 17.
[31] One must, of course, be careful not to give that term too modern, indeed too 'scientific' a meaning.
[32] *The Works of Francis Bacon*, vol. III. 167.
[33] Walter Ralegh, *History of the World*, ch. 11, section 2 202.
[34] René Descartes, "De L'Arc-En-Ciel," *Les Météores* 749-761.
[35] "en telle sorte qu'on n'ait plus occasion d'admirer rien de ce qui s'y voit" Descartes, *Les Météores* 720.
[36] *The Works of Francis Bacon* 137.
[37] Ibid. 158.
[38] Edmund Rudierd, *The Thvnderbolt of Gods Wrath Against Hard-Hearted and Stiffe-Necked Sinners* (1618).
[39] Barbara Mowat, "Prospero, Agrippa, and Hocus Pocus."

Works Cited

Bacon, Francis. *The Works of Francis Bacon*, vol. III. Ed. James Spedding, Robert Leslie Ellis, and Douglas Denon Heath. London: Longman, 1859.

Beard, Thomas. *The Theatre of God's Judgements*. London: Islip, 1597.

Berger, Karol. "Prospero's Art." *Shakespeare Studies* 10 (1977): 211-239.

Descartes, René. *Œuvres Philosophiques*, vol 1, Les Météores. Ed. F. Alquié. Paris: Éditions Garnier Frères, 1972.

Ettin, Andrew V. "Magic into Art." *Texas Studies in Literature and Language* 19 (1977): 268-293.

Heninger, Simeon Kahn. *A Handbook of Renaissance Meteorology*. Durham, North Carolina: Duke UP, 1960.

Höfele, Andreas. "Wissen und Macht: Der Gelehrte als Magier im Drama der Shakespearezeit." *Poetica* 26 (1994): 88-104.

Kermode, Frank. "Introduction," in William Shakespeare, *The Tempest*. Ed. Frank Kermode. The Arden Shakespeare. London: Methuen, 1954.

Kernan, Alvin B. *The Playwright as Magician*. New Haven: Yale UP, 1979.

King James the First. *Daemonologie* (1597), *Newes from Scotland* (1591). Ed. G.B. Harrison. London: John Lane, The Bodley Head; New York: E. P. Dutton, 1924.

Kittsteiner, Heinz Dieter. *Gewissen und Geschichte: Studien zur Entstehung des moralischen Bewußtseins*. Heidelberg: Manutius, 1990.

Kuriyama, Constance Brown. *Hammer or Anvil: Psychological Patterns in Christopher Marlowe's Plays*. New Brunswick: Rutgers UP, 1980.

Marlowe, Christopher. *Doctor Faustus*. 1604-1616. Parallel Texts. Ed. Walter Wilson Greg. Oxford: Oxford UP, 1950.

---. *The Complete Plays*. Ed. John Barry Steane. Harmondsworth: Penguin, 1986.

Mather, Cotton. "Brontologia Sacra: The Voice of the Glorious God in the Thunder, etc." London, 1695. Reprinted in Cotton Mather, *Magnalia Christi Americana*. London, 1702. Hartford, 2 vols, 1820, 1853.

Mebane, John S. *Renaissance Magic and the Return of the Golden Age: The Occult Tradition in Marlowe, Jonson and Shakespeare*. Lincoln, Nebraska: U of Nebraska P, 1989.

Mowat, Barbara. "Prospero, Agrippa, and Hocus Pocus." *ELR* 11 (1981): 281-303.

P. S. *Fearefvl Newes of Thunder and Lightning, with the terrible effects thereof, which Almighty God sent on a place called Oluestone in the county of Glocester the 28. of Nouember last*. London: Frauncis Burton, 1606.

Pico della Mirandola, Giovanni. *Über die Würde des Menschen* (1486). Trans. Werner Rüssel. Zurich: Manesse, 1988.

Ralegh, Sir Walter. *History of the World*. London: Burre, 1614.

Rudierd, Edmund. *The Thvnderbolt of Gods Wrath Against Hard-Hearted and Stiffe-Necked Sinners*. London: Thomas Panier, 1618.

Scot, Reginald. *The Discoverie of Witchcraft*. Arundel: Centaur Press, 1964.

Shakespeare, William. *The Tempest*. Ed. Stephen Orgel. Oxford: Oxford UP, 1987.

Sidney, Sir Philip. *An Apology for Poetry, or, The Defence of Poesy*. Ed. Geoffrey Shephard. London. Edinburgh: Nelson, 1965.

Tetzeli von Rosador, Kurt. *Magie im elisabethanischen Drama*. Braunschweig: Georg Westermann, 1970.

The last terrible Tempestious windes and weather. London: Hunt, 1613.

Thomas, Keith. *Religion and the Decline of Magic*. London: Weidenfeld and Nicolson, 1973.

Traister, Barbara Howard. *Heavenly Necromancers: The Magician in English Renaissance Drama*. Columbia, Missouri: U of Missouri P, 1984.

Vickers, Brian, ed. *Occult and Scientific Mentalities in the Renaissance*. Cambridge: Cambridge UP, 1984.

West, William. *Three Treatises, Of the second part of Symboleographie*. London 1594.

Yates, Frances. *Shakespeare's Last Plays*. London: Routledge, 1975.

---. "Magic in Shakespeare's Last Plays. – On *The Tempest*." *Encounter* 44 (1975): 14-22.

Magic, Necromancy, and Performance

Uses of Renaissance Knowledge in Marlowe's *Doctor Faustus*

Tobias Döring

Magic Books, Magic Performances

For Renaissance contemporaries, just as for most of us today, magic is primarily a matter of books. "Remember/ First to possess his books": Caliban's reminder to the rebels in *The Tempest* (3.2.86-87) has rarely been forgotten by anyone engaging with, let alone attempting to resist, the art and power of magicians. Whatever attitude towards their questionable working one might take, books and codices are widely seen as their true mark and serve as emblems of their intellectual standing. Traditionally, the written medium thus serves to distinguish magic practices also along the lines of gender, reserving the scholarly attributes and greater status for male practitioners, as opposed to witches and their oral charms. From John Dee's library to Prospero's books, European iconography and cultural imaginations have long focussed on the written medium in which magic learning is perpetuated or produced and where, supposedly, the secret arts reside. "What has thou got there? A book? Why, thou canst not tell ne'er a word on't": dabbling students and apprentices in magic, like the clowns in *Doctor Faustus*,[1] usually begin by picking up their master's books, trying to make sense of all the wonderfully foreign ciphers, circles, signs or forms set down therein. Yet the stories of the great master magicians, too, are regularly told as stories of illicit reading. "Oh Faustus, lay that damned book aside" (*Doctor Faustus* 1.1.69): from the Good Angel's first appearance and first warning until the Doctor's final cry and desperate promise, when the clock strikes midnight and the devils come to fetch his soul, "I'll burn my books" (*Doctor Faustus* 5.2.200) – the fortunes and failures of this damned scholar are epitomised in nothing so much as in books.

Even the theatrical version by which this figure gained such popularity constantly foregrounds the pleasures and perils of reading. Christopher Marlowe's *Tragical History* opens with the familiar scene of the dissatisfied German scholar at his desk – "And this the man that in his study sits" (Prologue 28) – so that we first encounter Faustus as a reader, translator and interpreter, powerfully presented to us, in Clare Harraway's apt phrase, as a "consumer of print."[2] Books are to Faustus and his kind what the apple is to Adam: a catalyst for greater knowledge as for the fall from grace. We may be so familiar with this well-established cultural connection between magic and the book, prevalent at least since the Renaissance, that we rarely pause to think about its implications. This, then, is the project of this paper: it sets out to take another look at Marlowe's drama in order to explore the potential relevance of the media in which magic features here and through which its consequences are deliberated or determined. In particular, my paper suggests to reconsider the role of the printed book in relation to the new medium that rose to unprecedented prominence in Elizabethan England as the leading cultural medium of the era: stage performance in the playhouse. While *Doctor Faustus* will provide the main example for my argument, its scope should reach beyond this single case and, hopefully, allow for some broader view to be established at the end.

The cultural debates on magic and magicians, which raged throughout the early modern period, were indeed staged and conducted as a battle of books. A brief look at the two most prominent Elizabethan writers in the field immediately bears this out. Both Reginald Scot and King James VI present their arguments in ways that foreground books self-consciously as the major marks of knowledge and contention. Scot, the protestant sceptic who writes against the popular belief in charms and occult practices, prefaces his *Discouerie of Witchcraft* (1584) with a comprehensive list of some two hundred names giving us "the forren authors vsed in this Booke;"[3] and throughout his entire text he adds careful notes in the margin to refer us to his printed sources, i.e. the books he draws on. King James VI, who wrote his *Daemonologie* (1597) as a counter-blast and refutation of Scot's work, casts his argument in the dramatic form of a dialogue; but his two interlocutors are constantly concerned with the correct interpretation of Scriptures and the discussion of books: "marke the words of the text," as Epistemon expostulates with Philomanthes early on.[4] And the crucial passage, in which Epistemon describes the scholar's dangerous progress into higher and forbidden spheres of knowledge – "mounting from degree to degree, vpon the slipperie and vncertaine scale of curiositie"[5] – seems to invoke this progress rather like climbing up a ladder towards ever higher book shelves. In fact, the entire passage reads like a gloss, or echo, to the Chorus in Act III of Marlowe's play describing the advance of learned Faustus: "To find the secrets of astronomy,/ Graven in the *book* of Jove's high firmament,/ Did mount him up to scale Olympus top" (*Doctor Faustus* 3.1.2-4, emphasis added). The cosmological learning that Faustus desires figures here in the traditional image of a book – clearly recalling his repeated library orders to Mephostophilis to satisfy his deepest longings: "Yet fain would I have a book wherein I might behold all spells and incantations," "Now would I have a book where I might see all characters and planets of the heavens," "Nay, let me have one book more" (*Doctor Faustus* 1.5.168-176). Whether as metaphor or medium of Renaissance knowledge, books are what magicians live by.

This notion may have consequences, too, for our critical engagements with the role of magic. For example, in her groundbreaking study, published twenty-five years ago, Frances Yates described *The Occult Philosophy of the Elizabethan Age* as a system of hermetic knowledge combined with Neoplatonic Cabbalism, whose interest for the Queen and key figures associated with her court at certain points, like Spenser, Dee or Sidney, she discussed with great speculative power. In this conceptual framework, Marlowe's use of magic in his plays is characterised in purely negative terms, as a vicious and satirical attack on the great tradition of hermetic thinking. According to Yates' historical account, his view of conjuring is nothing but reactionary "propaganda" that belongs to the contemporary "revolution from Renaissance magic" and targets, among others, Spenser's splendid vision of "white magic, imperial reform, and Christian Cabala."[6] I suspect that Marlowe features in this way in Yates' story because she treats magic entirely as a matter of books. Viewing occultism as a system of knowledge in the history of ideas that she so interestingly reconstructs, she reads Marlowe's plays, too, just like books and sees only what they say, as opposed to what they do – or might have done – when performed on stage. The epilogue to *Doctor Faustus*, for instance, is cited by her as conclusive evidence for the final condemnation of the magus who practiced "more than heavenly power permits" (*Doctor Faustus* 5.3.27).[7] Thus, she clearly disregards the possibility that the actual spectacle which has preceded this conclusion and which the playhouse audience has watched may have done a great deal more – or, at least, a great many other things – than the orthodox wisdom of the Chorus would admit. Beginning with the Prologue, in fact, there have throughout been strong suggestions that the narrative presented by the Chorus, closely following the familiar story told in the English *Faust-Book*, should not be taken as reliable and does not necessarily provide appropriate judgements for the dramatic scenes it frames; at several points the on-stage interaction rather blurs or goes beyond the moral boundaries reiterated, or re-cited, by the bookish and conventional Chorus.

Against this background, and against such a reduction of magic to a script in books, I would like to examine *uses* of Renaissance knowledge on and for the stage. I propose to look at magic, not as a hermetic corpus of ideas – however diverse, heterogeneous or variable these may be[8] – but as a set of practices, as ways of interacting or performing, as modes of physical and verbal exchange which can also change the way things are. In short, my focus lies on the performative dimension of this issue and should help to find out how to do things with books. Whenever a play like *Doctor Faustus* stages magic rituals and practices, I argue, it does so to explore the possibilities of its own medium: the theatre. Especially the scenes of necromantic conjuring also conjure up the peculiar conditions in the playhouse where historical protagonists – a German professor, an Emperor, a classic military hero – appear as shadows of the dead and so establish and, at the same time, test the impact of the theatre on the contemporary cultural field, a field defined in rival bids for power. It has been argued that the scholar's tricks of magic in the course of his career do not live up to the overreaching fantasies of power that propel him at the outset.[9] Indeed, we never see Faustus in the play command any emperors or kings or whatever else may move between the quiet poles, as he deliriously raves about when he first turns to the "heavenly" necromantic books on his desk (*Doctor Faustus* 1.1.49). And yet, it is the process of theatrical performance by which such fantasies may still be brought to life. For spectators, at any rate, the stage itself appears to be the very world "of profit and delight, / Of power, of honour, of omnipotence" which has been promised to the studious artisan (*Doctor Faustus* 1.1.52-54): the playhouse space is metadramatically invested here with a kind of magic efficacy that may be programmatic, I suggest, for the cultural work of early modern theatre.

This approach proceeds with an understanding of literary texts and, in particular, of drama based on notions of performance and performativity ultimately deriving from Austin's speech act theory.[10] Such notions have, over recent years, gained considerable attention and many uses in the area of literary and cultural studies, which cannot be reviewed here,[11] but which all shift our critical emphasis from texts to acts, from products to processes and from codes or structures to modes and changing strategies. I am well aware that this 'performative turn,' as it is sometimes referred to, has also affected Marlowe studies for quite some time, so that few readers would nowadays subscribe to the static, de-theatricalised view of his playtexts, just illustrated with reference to Yates. In fact, one of the most recent critical engagements with *Doctor Faustus* accounts for the whole drama entirely in terms of role-play, spectacle and theatricality – at the expense almost of its theological or philosophical aspects. Thomas Healy, in an elegant and thoroughly absorbing reading, describes the play as "a performance that has as its main endeavour the staging of theatrical opportunities," a series of scenes all preoccupied "with creating theatre, with organizing performances;" to him, such a self-conscious, perhaps self-gratifying interest in performance may come to seem the whole play's "ultimate rationale," while any clarification of its metaphysical speculation is constantly being deferred.[12]

But such a view, attractive as it clearly is, may be in danger of neglecting the scriptural and scripted interests which *Doctor Faustus*, with its frequent foregrounding of books, also pursues. I would like to suggest, therefore, that Marlowe's performances of magic should best be situated in a field mutually defined by scripts and acts. It is the particular use, or appropriation, of conventional writing from which performative acts generally derive their power – just as, according to speech act theory, the illocutionary act functions by means of the convention which a speaker cites.[13] This interplay of script and performance, or convention and realisation, or model and parody, is the crucial point for any speech act to take place – a point appreciated not least by the aspiring magician. Cornelius, Faustus's tutor in the occult arts who is himself modelled on Cornelius Agrippa, says to his companion: "first let him know the *words* of art, / And then, all other *ceremonies* learned,/ Faustus may try his cunning by himself" (*Doctor Faustus* 1.1.157-159, emphasis added), and so suggests that any magic performance, too, must resort to some existing verbal formula, the "words of art" that are taught first.

Taking our cue from these instructions, we should look more closely at the words in Marlowe's playscript and then raise questions about what kind of ceremonies or stage performances they may have prompted or provoked.

Was this the face that launched a thousand ships,
And burnt the topless towers of Ilium?
Sweet Helen, make me immortal with a kiss.
Her lips suck forth my soul: see where it flies.
Come, Helen, come, give me my soul again.
Here will I dwell, for heaven is in those lips,
And all is dross that is not Helena.
I will be Paris, and for love of thee
Instead of Troy shall Wittenberg be sacked,
And I will combat with weak Menelaus,
And wear thy colours on my plumed crest.
Yea, I will wound Achilles in the heel,
And then return to Helen for a kiss.
(*Doctor Faustus* 5.1.97-109)

These, surely, are the most famous lines of the entire play, marking the climax of its promise to perform whatever pleasing shapes we like. The actual figure to whom they are addressed may just be a devil in drag, summoned by Faustus's command and Mephostophilis' compliance. But this figure is presented and rapturously addressed here as Helen of Troy, the icon of beauty and desire, so that the figure also serves as focus to explore the cultural status of performance. Again, one might be rather too familiar with these lines, so often cited and rehearsed, to consider what they actually imply. How are we to gauge the effect and value of the apparition they invoke? What is Faustus saying, or perhaps doing, with this wondrous speech?

In the critical debate, there is no agreement on this simple question. On the one hand, we could argue that the Helen-figure just personifies and fulfils Faustus's longing for debased sensual pleasure. The moment of her summoning is introduced with Faustus pleading to Mephostophilis, his "good servant," that he may "glut the longing" of his "heart's desire" (*Doctor Faustus* 5.2.88-89), a plea recalling the earlier pageant of the Seven Deadly Sins, Gluttony among them, which marked the beginning of their master–servant contract (*Doctor Faustus* 2.1.113-174). Indeed, when Faustus speaks to "Sweet Helen" he implores her to consummate their encounter "with a kiss," a hope for physical contact between the magus and the figure of his art that might be seen – in Yates' sense – as part of Marlowe's propaganda against occult philosophy, which he stages only in degrading form, ridiculing secret wisdom as debauchery. But there are other ways to see this scene. Addressing Helen, Faustus dreams himself into the role of Paris and, by implication, Hector, while he wants to give his local habitation a new name, making Wittenberg into another Troy, a dream of scene changing and literary re-enactment, rather like the usual practice in a London playhouse where conventional speech acts would locate the scene, variously, in Troy, in Carthage or in Germany. Healy therefore notes that Faustus "does not want to possess Helen for simple sexual gratification;" she mainly "stimulates his excitement about role-playing" and feeds "his imagination for theatre."[14] In this reading, the speech expresses something of an actor's ambition to take on ever more and more exciting roles before his show will soon be over – as if Faustus, like Bottom in *A Midsummer Night's Dream*, were saying here, "let me play Paris, or the lion, too!" But I think it is equally important to acknowledge the specific scripts on which the stage performer draws here, for we must surely say of Faustus and his kiss what Juliet says of Romeo on another, perhaps not quite dissimilar, occasion: that he kisses by the book.

Throughout the entire play, of course, Marlowe is dramatising the source version of his story that he found in the English *Faust-Book*, the existing moral narrative of the doctor's *Damnable Life* translated from the German. So it may be significant that his dramatisation differs from its sourcebook precisely at this point. In Chapter 55 of the *Faust-Book*, we read that Faustus harbours "a great desire to lie with fair Helena of Greece," and the narrator is as explicit in his judgement as in revealing the protagonist's true motives: "to the end that this miserable Faustus might fill the lust of his flesh." Helen, duly summoned by Mephostophilis, is made into "his common concubine and bedfellow, for she was so beautiful and delightful a piece that he could not be one hour from her."[15] This traditional version of the story still retains some traces of Faustus's much earlier, gnostic ancestor called Simon Magus, the biblical figure of Simon of Samaria, whose story is told in Acts (chapter 8); according to gnostic tradition, Simon is said to have travelled around with a sexual companion identified as Helen of Troy.[16] In the sixteenth-century protestant *Faust-Book*, however, the homiletic or polemic purpose leaves no scope for hermetic re-interpretations of this relationship: their alliance is carnal and condemnable, its motivation miserable lust, its product "a man child;" at Faustus's end, concubine and son "vanished away both together."[17]

Significantly, then, in Marlowe's drama Helen does *not* figure as bedfellow nor mother of a child. Here, her stage apparition remains disembodied, distanced, truly *questionable* in both senses of this word and is physically touched – if at all – only for one short, uncertain, fleeting and precarious moment. John Mebane, in fact, reads Faustus's wish to be "immortal in a kiss" as an allusion to the moment of divine ravishment and mystical union referred to by the Cabbalists as "the death of the kiss," a commonplace in the occult tradition which he finds "unmistakable" in Faustus's speech.[18] But I think we can best appreciate this moment of performative engagement when we consider the specific uses and the urgent questioning to which such occult knowledge is subjected here. "Was this the face that launched a thousand ships?": like a bemused spectator in the playhouse, Marlowe's magician questions the kind of identification theatrical impersonators always claim when presenting a familiar figure in the here and now for all spectators to behold, and yet with an awareness of the distance, or difference, from their own, their actual sphere. This ambiguity is inscribed into the variable emphasis of Faustus's language: "Was *this* the face?" provides the necessary deictic signal to anchor the apparition in the here and now of actual experience. "*Was* this the face?" – as opposed to the more plausible verb form "*Is* this the face?" – removes the figure here observed into a realm of history, of cultural memory and textuality, from which the ongoing performance both differs and, at the same time, clearly takes its shape.

In such a metadramatic perspective, powerfully established through the theatrical references in his lines, Faustus's climactic speech should not be taken as posing a rhetorical question, but as a serious and sceptical exploration into the wondrous working of a performance where the present figure both is and is not an embodiment of the past. In this view, the stage present-ing, or presence-ing, of the desired Helen always remains double – she is real and a role, she is a mask and a myth, she is a woman acted by a devil acted by a boy, she is here and not here, long dead and still alive – because performance always keeps a consciousness of difference in the bodily identification between the magic and the mythic persona. Conjurors, Reginald Scot writes in his treatise, "raise vp what bodies they list, though they were dead, buried, and rotten long before."[19] This is the magic art of necromancy, which we are also witnessing in some way in the Helen scene. But when Marlowe's dramatic art shows us such necromantic acts, it also foregrounds the conditions of its own theatricality, challenging us to interrogate its effects. We are, it should be noted, not the only watchers and witnesses to this scene. During Faustus's speech to Helen, the Old Man enters silently and so offers an onstage model for the audience in his role as critical observer. Soon after, in the A-text of the play-script, he provides a choric comment on Faustus's deplorable behaviour. Yet as spectators in the playhouse, we are neither urged to share his moral condemnation of the Helen show nor to share Faustus's fascination with it. Both attitudes are staged here simultaneously and side by side: another case of doubleness on and for the stage. So perhaps we are justified to generalise this case and take it as an indication of the peculiar position of the early modern theatre in the embattled political arena. For all its involvement in the controversial issues of religion, of faith and doubt and government, in which not just *Doctor Faustus* but surely all of Marlowe's plays and many other playwrights of his time engage, the popular theatre remains curiously uninvolved. As an institution of the incipient entertainment industry, it only ever stages, never states, doctrinal positions; deniability lies in the nature of performance.

This reading might gain plausibility when we place the scene discussed into some of its wider context. The Helen apparition, I would briefly like to show, also serves as an occasion to explore the power and potential danger emanating from words, i.e. from verbal utterances realised in performance. This is Helen's second appearance; her first entry occurred at the beginning of the scene when Faustus, prompted by the scholars' bidding and helped by Mephostophilis, already conjures up "that peerless dame of Greece" (5.1.22). At this point he gives the audience of raptured students clear instructions before the figure can appear: "Be silent, then, for danger is in words" (5.1.26). His injunction offers us a vital clue, I think, to situate his subsequent questioning of magic figures, at Helen's re-appearance, into the appropriate cultural field: the Renaissance debates about rhetoric, the "danger in words." Rhetoric surely is the oldest and still cogent account of how to do things with words, and we can see the Helen scenes, whatever else they signify or do, as an onstage inquiry into the possible effects of verbal acts. The spectators' passionate desire to behold, in person, the figure they have hitherto just read as a name in books translated or debated as part of their classical learning, requires what Debora Shuger's study of Renaissance rhetoric has called "the legitimation of passionate discourse." Marlowe's acts of stage conjuring thus demonstrate *in actu* the kind of emotional persuasion which "turns the heart toward spiritual reality by fulfilling, not subverting, man's need for the sensible and corporeal." In this way, Shuger continues, "it gives invisible truth a local habitation and a name through metaphor, symbol, prosopopeia and all the figures that create drama, vividness and force."[20] Shuger is not at all concerned with theatre or actual drama, but her language suggests clearly how stage plays can perform a realisation of the kind of verbal force she is describing – just what we are witnessing, I submit, in Faustus's necromantic acts. Among the rhetorical figures Shuger cites, the central figure to bring on the dead is certainly prosopopeia, or personification, the rhetorical move to give to what is past, or absent, a new face or mask. As amply testified in *Doctor Faustus*, the stage is the site where such a summoning takes place and where its power can be questioned: "Was this the face that launched a thousand ships?" Prosopopeia, Greenblatt argues in the context of his *Hamlet* reading, is the "rhetorical device that lies behind all haunting."[21] I would add to this and argue that prosopopeia is also the figure, not behind, but within necromancy, i.e. in the kind of magical performance offering, as Puttenham says of the rhetorical figure, "counterfeit representations" which "appear they were true before our eyes though they were not present."[22] Stage performance operates through such counterfeits: words in motion, true appearances not present, rhetorical figures acted out.

I emphasise the need to take a questioning or critical attitude, mindful of deniability, because the play itself seems to invite spectators to take a sceptical view of the entertainment it provides. To pursue this issue and its consequences for the uses of Renaissance knowledge, a closer look at two other scenes of conjuring in the play will help to situate the magic arts, as shown and practised in *Doctor Faustus*, in relation to the monarchy and the church, i.e. the two other institutions of sacred charisma and, as such, potential rivals to the theatre in early modern culture.[23] The Helen scene recalls the earlier necromantic show performed at the Emperor's court, answering the special request made by Charles V to "behold that famous conqueror, / Great Alexander, and his paramour, / In their true shapes" (*Doctor Faustus* 4.2.28-30). At this point, Faustus's standing and activity correspond very closely to an Elizabethan theatre entrepreneur and head of a travelling company, eager to accommodate a noble patron's wishes and perform to his liking. The magic apparition that follows is described in the stage directions just like a conventional dumb show, complete with sonnet, trumpets and salute to the patron, so that the whole scene serves to demonstrate the powerful seductions of spectators through the effects of sight. The Emperor is even trying to embrace the figures: his thoughts, he says, are "so ravished / With sight" (*Doctor Faustus* 4.2.56-57) that he momentarily forgets himself. But Faustus, as the play-maker, holds him back reminding him "these are but shadows" (*Doctor Faustus* 4.2.55) – a keyword which recurs throughout the text to speak about the miracles that magic can perform. The word gives a key also for the performative reading of *Faustus* and its magic that I am suggesting here. Yates, to one of Faustus's earlier references to "shadows," gives as a gloss that *shadows* "probably" means "magic images," as in Giordano Bruno's book *De umbris idearum*.[24] But this learned explanation, paradoxically, overlooks the most immediate and relevant use of the term *shadows* in Elizabethan language: to identify actors as well as the characters they play. Magic, above all, lies in the medium and profession of the stage.

However, the most interesting point in the Emperor scene is the split audience it sets up to investigate the seductions of sight even in those who claim to resist it. Such a sceptical, openly critical position towards magic performance is featured on stage in the figure of Benvolio, a courtier who refuses to suspend his disbelief and has no use for wonders. In a series of asides, punctuating the dialogue between Charles and Faustus, he voices his distrust of the magician and his reputed power in no uncertain terms: "Blood, he speaks terribly! But for all that, I do not greatly believe him. He looks as like a conjuror as the Pope to a coster-monger" (*Doctor Faustus* 4.2.24-26). Again, it is significant that the Benvolio episode is not part of the relevant chapter in the *Faust-Book* (Chapter 29) and only occurs in the stage-version. Here, it functions as a counter-performance to the magician's conjuring, a negation of his impact and a resolute denial of the specular forces he claims to command. Right away Benvolio declares: I "can see nothing" (*Doctor Faustus* 4.2.42). These disrespectful remarks rather sound as if this courtier had read Reginald Scot, the sceptic who describes in his *Discouerie* how he himself tried out the so-called secret rites, spoke the appropriate words from magic books, went through the motions of the occult arts, performed all the required moves and gestures – and saw nothing happen.[25] Scot thus tries to "discover" the frauds of magic in the spirit of experimental testing. Marlowe's court scene, on the other hand, in a grand gesture of theatrical self-assertion, goes on to show how the sceptic is subdued. The magician punishes Benvolio by endowing him with a pair of horns, a precise mythological reference that is at once spelled out, lest anybody miss it: "you are straight resolved / In bold Acteon's shape to turn a stag" (4.2.95-96). The horns are the essential features from the Acteon myth that is here magically re-enacted, i.e. the story of the hunter who watched Diana naked, was therefore turned into a stag and torn apart by his own hounds. This well known story provides the central figure for Elizabethans to manifest the dangers of spectatorship, the ravishment of passionate viewing, and so warn against what Philip Sidney calls the "sweet violence" of theatre: to move, affect and change even those who think they can resist it.[26]

That such instructions about the power of theatre and speech acts – manifest in the Emperor's ravishment just as in Benvolio's horns – are included in this play-within-the-play suggests a message also to contemporary spectators, who were certainly aware that the control over, just as the control in performance was a difficult and vexed issue. This can also be gauged from the third example I would like to consider, which is the initial scene of conjuring in the play, when Faustus for the first time tries out the newly-gained secret knowledge and puts his occult reading to a performance test. Here is the text of the play-script:

Faustus, begin thine incantations
And try if devils will obey thy hest [...].
Then fear not, Faustus, to be resolute
And try the utmost magic can perform.
Thunder.
Sint mihi dei acherontis propitii, valeat numen triplex Jehovae,
ignei areii, aquatani spiritus salvete: orientis princeps Belzebub,
inferni ardentis monarcha et demigorgon, propitiamus vos, ut
appareat, et surgat Mephostophilis (Dragon) quod tumeraris: per
Jehovam, gehennam, et consecratam aquam quam nunc spargo;
signumque crucis quod nunc facio; et per vota nostra ipse nunc
surgat nobis dicatus Mephostophilis.
Enter a Devil.
I charge thee to return and change thy shape.
Thou art too ugly to attend on me.
Go, and return an old Franciscan friar:
That holy shape becomes a devil best.
Exit Devil.
I see there's virtue in my heavenly words.
Who would not be proficient in this art?
(*Doctor Faustus* 1.3.5-28)

A major point to note in this performance of a magic incantation is that it includes, as the Latin text reveals, the use of holy water (*consecratam aquam quam nunc spargo*) and the holy sign of crossing (*signum crucis quod nunc facio*), i.e. Catholic elements which are employed here to call up the devil. Should we consider this a demonstration of dangerous blasphemy or righteous attitudes?

It is an oft-debated question what exactly different members in an Elizabethan audience, mixed of very different backgrounds and persuasions, might have seen in such a blatant display of the verbal and the ritual features associated with the outlawed faith.[27] Faustus's subsequent response raises the same question: when he congratulates himself on his success and praises the "virtue" in his "heavenly words" he is evidently claiming ritual efficacy for his acts and speech – another clearly Catholic position. Paradoxically, it is left to the devil to refute these popish points and voice the truly protestant position of voluntary ascent by claiming that he has come, not obeying Faustus's conjuring, but of his "own accord" (*Doctor Faustus* 1.3.44). For many contemporaries, such an orthodox claim made by the devil must have had the same unnerving quality as, for us, the paradox of the Cretan liar: a performative contradiction, which ever way you turn it. While protestant polemicists like Phillip Stubbes routinely railed against the playhouses in terms of their associations with the spectacle and superstitions of Catholic worshipping,[28] Marlowe's magic stage performance works as if to return such claims, investing scenes of conjuring not just with metadramatic power, but with distinctly Catholic remembrances – as well as with a sense of unease and suspicion about who, ultimately, may control them. How strong and real this sense must have been is recorded in the notorious anecdote about the *Faustus* performance by certain players in Exeter who, all of a sudden, noticed with dismay that one devil too many had appeared on stage – at which point their performance was abruptly ended and the whole night spent in prayer.[29]

Magical Theatricality and the "Omnimpotence" of Authors

In different ways, then, these three scenes of conjuring all exemplify the uses of Renaissance magic as a theatrical technique, while also exemplifying the distinction, noted at the outset, between magic as a body of knowledge codified in books and magic as a *modus operandi*, i.e. as bodies in performance. In Marlowe's play, the forms and fortunes of Renaissance occult thinking like the hermetic figures of tradition are dramatically enacted – and often metadramatically displayed – so that they become figures in and of performance. As such, they powerfully assert and, at the same time, question the wordly status of the stage on which these performances are watched and potentially validated by spectators. This analysis, I would like to suggest, holds true for this playwright's other dramatic works as well and, quite possibly, for many of his colleagues at the time. In some way or other, all of Marlowe's stage plays stake their claims on magical theatricality: in the two *Tamburlaine* plays these are the scenes of spectacular violence and charismatic power in the title hero drawing all other figures into his performative orbit; in *The Jew of Malta* this happens through the Machiavellian villain's campaign of deceit and revenge, which draws on the whole repertoire of theatrical techniques and which ends, literally, on the scaffold of the stage;[30] in *Edward II* there are the transgressive spectacles of desire that Gaveston stages to draw the "pliant" king which way he pleases (1.1.53), using the same term here as Faustus when he first raises the devil (1.3.29). In sum, all of Marlowe's plays explore and show how to do magic things with theatre.

This analysis bears also on our more general views of magic and its career throughout Renaissance culture. For a long time, it has been generally accepted to think about this question, according to Keith Thomas' classic study, in terms of the "decline" of magic.[31] But in the view suggested by the works of Marlowe we may rather see the cultural presence and the place of magic in the early modern world to be continued, though displaced – or transformed – into the new practices of theatre and so to be transformed from arcane knowledge and hermetic books into popular spectacle. In one of the most famous comments on the new medium of the professional stage and its public effects, one of Marlowe's colleagues and contemporaries praised the playhouse for presenting "our forefathers' valiant acts, that have lien long buried in rusty brass and wormeaten books," whereby all kinds of historical heroes are here "raised from the grave of oblivion and brought to plead their aged honours in open presence."[32] Thomas Nash is speaking here about English history plays, but several of his formulations – "buried in books," "raised from the grave" – exactly imply the kind of magical and necromantic acts we have been witnessing in Marlowe's *Faustus*. The theatre, then, is principally seen as using its performative techniques to raise the dead and let them act in front of our eyes so effectively that even "ciphers" – in the Prologue's words from Shakespeare's *King Henry V* – work on our "imaginary forces." In this way, the "wooden O" of the playhouse turns into "this circle" of magic conjuring, identified by Faustus (*Doctor Faustus* 1.3.8), the space of performance in which we, too, can be charmed.

This is a cultural space where verbal acts may become manifest and rhetorical figures are enacted. Fifty years ago, Harry Levin published his influential Marlowe reading entitled *The Overreacher* to characterise the stature and ambition of his heroes. Levin's study is still used and sometimes cited in early modern research, but it is not often remembered that its title, in fact, derives from a rhetorical figure, the figure of hyperbole, which in Elizabethan translation, through Puttenham's work, was known under this term. Levin was probably the first to look at Marlowe's stage as "a vehicle for hyperbole," a place of performance that takes conventional figures of speech literally: "operating visually as well as vocally, it converts symbols into properties," as Levin puts it: "students no longer satisfied to read about Helen of Troy, must behold her in her habit as she lived."[33] As a vehicle for this kind of satisfaction, derived from turning letters into life, Marlowe's stage serves to perform old rhetorical devices, like hyperbole or prosopopeia, and newly test their impact on spectators – a vehicle to mobilise traditional knowledge for the purposes of playing. For this reason, Marlowe's plays do not work to support orthodoxies nor to enforce traditional morals and beliefs – even the familiar metaphysical figures which *Doctor Faustus* takes over from the Morality Plays, the Good Angel and Evil Angel, reappear here more like citations, remnants of an old tradition that no longer holds. In the arena of performance they, too, are subject to playful, sceptical or parodistic uses.

When the Prologue says "we must now perform / The form of Faustus' fortune, good or bad" (7-8), this curious phrase already suggests the sense of obligation ("must perform"), as if citing the familiar figure from the *Faust-Book* where the moral judgement is laid down, while also opening a new alternative ("good or bad"), as if to open up the sourcebook to different, possibly heterodox, meanings. It is therefore significant that the play opens with the extended scene of book reading and rereading, an academic exercise in textual interpretation which many Elizabethan spectators would undoubtedly have recognised as a scene of *mis*reading. As Faustus cites and translates variously from Aristotle, Galen, Justinian and – most notoriously – Jerome's Bible, he twists and turns and falsifies the lines, misattributes quotations, cuts them short and changes their established meaning. At the same point, then, where the traditional authority of book learning is staged, the performance turns it into an actual reader's authority *over* books to use them at his will and liking for his own, perhaps wrong, purposes. Thus, the first scene immediately dramatises the interplay of script and interpretation, or norm and realisation, or perhaps model and parody, which I referred to earlier and which always sets the pattern for and in performance. In a historical period of transition, where the principle of *sola scriptura* was one of the leading notions of the new religion, this opening must have been a troubling scene, showing not just how an individual mind is misled into sin or error, but how the very process of reading is principally open to such erroneous interpretation. As the play therefore begins by asking what kind of certain knowledge we can gain from books, it also launches a process of questioning the traditional certainties on which cultural authority used to be based.

In recent years, it has often been argued and demonstrated that the early modern period was a time when the figure of authorship was first carved out and socially established, with the commercial rise of print and books.[34] But in the opening scene of *Doctor Faustus* we observe how this figure, from the very start of its career, is accompanied by a shadow and uncomfortable counterpart: the reader, who makes the authorial text his own by making it the subject of interpretation – thereby turning it into a script for his own performance. In the broader view, which may become discernible here, such a process cannot just be seen when magic acts are placed on stage; it also affects the books about magic on which we draw until this day when trying to reconstruct occult practices or systems of knowledge. Scot's *Discouerie*, for instance, was written in the clear attempt to do away with all the superstitions about witchcraft, spells and incantations and the like. But it contains so many pages cataloguing them that the book, paradoxically, rather helps to preserve the knowledge it wants to destroy. Even though the author repeatedly rejects all popish hocus-pocus, his own comprehensive presentation of the various charms, gestures, tricks and tokens, in fact, may give us fairly detailed instructions in the occult arts. At certain points Scot inserts chapters like the following entitled "All the former practices breeflie confuted,"[35] yet this speech act cannot erase nor disavow what has been shown before – just as the Epilogue to *Doctor Faustus* cannot cancel the power of theatrical performance that has gone before. All uses of a script eventually slip from authorial control.

Faustus's own fantasies of power, which are never realised in the play except in magical theatricality, have been characterised as expressions of his "omnimpotence,"[36] a brilliant term to capture the twin aspects of this stage magician: his powerlessness as well as his power. However, in the performative perspective I have been trying to suggest, this term also captures the twin aspects of the author figure: his omnipotence to create worlds and his impotence to control them. Nothing illustrates this better than the final line of Marlowe's text – a line never acted in performance nor meant as a stage direction. It is a line clearly added to the written work, to be appreciated only as the termination of a book. It reads: *Terminat hora diem, Terminat Author opus.* In David Riggs' recent reading, this motto "asserts Marlowe's ultimate authority over his work," because the "overlapping cycles of nature and art summon up the shade of the author, Christopher Marlowe, completing his labour at midnight, when Doctor Faustus's last exit coincides with the final strokes of Marlowe's pen."[37] And yet, I think, this motto only shows the author acting as an overreacher, making a heroic but hyperbolic claim to closure. For it need no ghost come from the grave to tell us that this grand self-assertion is quite futile. The very text of Marlowe's *Doctor Faustus* is notorious for its varying versions, its open-endedness and openness to a long process of revision and rewriting, not to mention reinterpretation and restaging, so that this book can never be terminated with a single stroke of pen. Instead, like magical theatricality, it lives and survives ever changing in performance, as its words and figures continue to be conjured on the stage.

Notes

[1] *Doctor Faustus* 2.2.11-12. All quotations from Marlowe's plays follow J. B. Steane's Penguin edition; references to act, scene and line numbers given in the text. Stean's version offers a speculative conflation of the A- and B-text, which effectively make up two rather different plays. More recently, most editors regard the shorter A-text to be closer to Marlowe's original – however we construe this problematic term.

[2] Clare Harraway, *Re-citing Marlowe: Approaches to the Drama* 31.

[3] Reginald Scot, *Discouerie of Witchcraft* preface without pagination.

[4] James I, *The Workes* 95.

[5] Ibid. 99.

[6] Frances A. Yates, *The Occult Philosophy in the Elizabethan Age* 124-125.

[7] Ibid. 119.

[8] Alexandra Lembert, *The Heritage of Hermes: Alchemy in Contemporary British Literature* 135.

[9] For instance, Kurt Tetzeli von Rosador, "The Power of Magic: From *Endimion to The Tempest*" 7.

[10] See Stephen C. Levinson, *Pragmatics*, and Sandy Petrey, *Speech Acts and Literary Theory*.

[11] For recent critical surveys and theoretical accounts, see Uwe Wirth, *Performanz: Zwischen Sprachphilosophie und Kulturwissenschaften*, Marvin Carlson, *Performance: A Critical Introduction, and Erika Fischer-Lichte, Ästhetik des Performativen.*

[12] Thomas Healy, "Doctor Faustus" 188-189.

[13] This is the central issue of a famous debate between leading speech act theorists Austin and Searle, and poststructuralist critic Jacques Derrida; the scope and theoretical significance of this debate are usefully summarised by Jonathan Culler, *On Deconstruction: Theory and Criticism after Structuralism* 115-120.

[14] Healy, "Doctor Faustus" 188.

[15] Vivien Thomas and William Tydeman, *Christopher Marlowe: The Plays and their Sources* 230.

[16] See A. D. Nutall, *The Alternative Trinity: Gnostic Heresy in Marlowe, Milton, and Blake* 42.

[17] Thomas and Tydeman, *Christopher Marlowe: The Plays and their Sources* 230.

[18] John M. Mebane, *Renaissance Magic and the Return of the Golden Age: The Occult Tradition and Marlowe, Jonson and Shakespeare* 27.

[19] Scot, *Discouerie of Witchcraft* 377.

[20] Debora K. Shuger, *Sacred Rhetoric: The Christian Grand Style in the English Renaissance* 138.

[21] Stephen Greenblatt, *Hamlet in Purgatory* 251.

[22] Puttenham quoted in Greenblatt, *Hamlet in Purgatory* 312.

[23] This thesis was, famously, formulated by Stephen Greenblatt in *Shakespearean Negotiations: The Circulation of Social Energy in Renaissance England*; it also forms the framework for Tetzeli's discussion of "The Power of Magic," to which my reading is throughout indebted.

[24] Yates, *The Occult Philosophy in the Elizabethan Age* 203; this gloss refers to the following lines: "I'll be as cunning as Agrippa was, / Whose shadow made all Europe honour him" (1.1.116-117).

[25] See, for example, Book XV, Chapter VII, "An experiment of the dead."

[26] In *The Defence of Poesy*, Sidney retells a story from Plutarch about a tyrant "that was not ashamed to make matters for tragedies, yet could not resist the sweet violence of tragedy", Sidney, *Major Works* 230.

[27] See Gareth Roberts, "Marlowe and the metaphysics of magicians" 69.

[28] Phillip Stubbes, *The Theater of the Popes Monarchy* (1584).

[29] See, for example, the account quoted in David John Palmer, "Magic and Poetry in *Doctor Faustus*" 195.

[30] Darryll Grantly, "'What meanes this shew?': Theatricalism, Camp and Subversion in *Doctor Faustus* and *The Jew of Malta*", and Sara Munson Deats and Lisa S. Starks, "'So Neatly Plotted, and so well Perform'd': Villain as Playwright in Marlowe's *The Jew of Malta*".

[31] See Keith Thomas, *Religion and the Decline of Magic: Studies in Popular Beliefs in Sixteenth and Seventeenth-Century England.*

[32] Thomas Nash, *Pierce Penniless his Supplication to the Devil, [etc.]* (1592) 64.

[33] Harry Levin, *The Overreacher: A Study of Christopher Marlowe* 24.

[34] See Douglas A. Brooks, *From Playhouse to Printing House: Drama and Authorship in Early Modern England*; Lukas Erne, *Shakespeare as Literary Dramatist*, for the first time makes the important and challeng-

ing claim that William Shakespeare, too, was aspiring to the status of literary authorship.

[35] Scot, *Discouerie of Witchcraft* 430.

[36] I owe this term, as well as other thoughts that have informed my reading, to Andreas Höfele, "Wissen und Macht: Der Gelehrte als Magier im Drama der Shakespearezeit," who in turn owes this term to Constance Brown Kuriyama, *Hammer or Anvil: Psychological Patterns in Christopher Marlowe's Plays.*

[37] David Riggs, *The World of Christopher Marlowe* 246.

Works Cited

Brooks, Douglas A. *From Playhouse to Printing House: Drama and Authorship in Early Modern England.* Cambridge: Cambridge UP, 2000.

Carlson, Marvin. *Performance: A Critical Introduction.* London: Routledge, 2003.

Culler, Jonathan. *On Deconstruction: Theory and Criticism after Structuralism.* London: Routledge and Kegan Paul, 1983.

Deats, Sara Munson, and Lisa S. Starks. "'So Neatly Plotted, and So Well Perform'd': Villain as Playwright in Marlowe's *The Jew of Malta.*" *Theatre Journal* 44.3 (1992): 375-389.

Erne, Lukas. *Shakespeare as Literary Dramatist.* Cambridge: Cambridge UP, 2003.

Fischer-Lichte, Erika. *Ästhetik des Performativen.* Frankfurt am Main: Suhrkamp, 2004.

Grantley, Darryll. "'What meanes this shew?': Theatricalism, Camp and Subversion in *Doctor Faustus* and The Jew of Malta," in *Christopher Marlowe and English Renaissance Culture.* Ed. Daryll Grantley and Peter Roberts. Hants: Scolar Press, 1996. 224-238.

Greenblatt, Stephen. *Shakespearean Negotiations: The Circulation of Social Energy in Renaissance England.* Oxford: Clarendon Press, 1988.

---. *Hamlet in Purgatory.* Princeton, NJ: Princeton UP, 2001.

Harraway, Clare. *Re-citing Marlowe: Approaches to the Drama.* Aldershot: Ashgate, 2000.

Healy, Thomas. "Doctor Faustus," in *The Cambridge Companion to Christoper Marlowe.* Ed. Patrick Cheney. Cambridge: Cambridge UP, 2004. 174-192.

Höfele, Andreas. "Wissen und Macht. Der Gelehrte als Magier im Drama der Shakespearezeit." *Poetica* 26.1–2 (1994): 88-104.

Kuriyama, Constance Brown. *Hammer or Anvil: Psychological Patterns in Christopher Marlowe's Plays.* New Brunswick: Rutgers UP, 1980.

James I. *The Workes* (1616). Facsimile reprint. Hildesheim, New York: Olms, 1971.

Lembert, Alexandra. *The Heritage of Hermes: Alchemy in Contemporary British Literature.* Glienicke: Galda and Wilch, 2004.

Levin, Harry. *The Overreacher: A Study of Christopher Marlowe.* Cambridge, MA: Harvard UP, 1952.

Levinson, Stephen C. *Pragmatics.* Cambridge: Cambridge UP, 1983.

Marlowe, Christopher. *The Complete Plays.* Ed. J. B. Steane (1969). Harmondsworth: Penguin, 1986.

Mebane, John S. *Renaissance Magic and the Return of the Golden Age: The Occult Tradition and Marlowe, Jonson, and Shakespeare.* Lincoln: U of Nebraska P, 1989.

Nash, Thomas. *Pierce Penniless his Supplication to the Devil, Summer's Last Will and Testament, The Terrors of the Night, The Unfortunate Traveller, and Selected Writings.* Ed. Stanley Wells. London: Edward Arnold, 1964.

Nuttall, A. D. *The Alternative Trinity: Gnostic Heresy in Marlowe, Milton, and Blake.* Oxford: Clarendon Press, 1998.

Palmer, David John. "Magic and Poetry in *Doctor Faustus*" (1964), in *Marlowe: Doctor Faustus: A Casebook.* Ed. John Jump. Houndmills: Macmillan, 1969. 188-203.

Petrey, Sandy. *Speech Acts and Literary Theory.* London: Routledge, 1990.

Riggs, David. *The World of Christopher Marlowe.* London: Faber, 2004.

Roberts, Gareth, 2000. "Marlowe and the metaphysics of magicians," in *Constructing Christopher Marlowe.* Ed. James Alan Downie and J.Tim Parnell. Cambridge: Cambridge UP. 55-73.

Scot, Reginald. *Discouerie of Witchcraft* (1584). Facsimile reprint. Theatrum Orbis Terrarum, Amsterdam: Da Capo Press, 1971.

Shakespeare, William. *The Norton Shakespeare.* Based on the Oxford Edition. Ed. Stephen Greenblatt, Walter E. Cohen, Jean E. Howard, and Katherine Eisaman Maus. New York: Norton, 1997.

Sidney, Philip. *A Critical Edition of the Major Works.* Ed. Katherine Duncan-Jones. Oxford: Oxford UP, 1989.

Shuger, Debora K. *Sacred Rhetoric: The Christian Grand Style in the English Renaissance.* Princeton, NJ: Princeton UP, 1988.

Stubbes, Phillip. *The Theater of the Popes Monarchy* (etc). London: Thomas Dawson, 1584.

Tetzeli von Rosador, Kurt. "The Power of Magic: From *Endimion to The Tempest.*" *Shakespeare Survey* 43 (1991): 1-13.

Thomas, Keith. *Religion and the Decline of Magic: Studies in Popular Beliefs in Sixteenth and Seventeenth Century England.* London: Weidenfels and Nicholson, 1971.

Thomas, Vivien, and William Tydeman, eds. *Christopher Marlowe: The Plays and their Sources.* London: Routledge, 1994.

Wirth, Uwe, ed. *Performanz: Zwischen Sprachphilosophie und Kulturwissenschaften.* Frankfurt am Main: Suhrkamp, 2002.

Yates, Frances A. *The Occult Philosophy in the Elizabethan Age.* London: Routledge and Kegan Paul, 1979.

The Force of Attraction in *A Midsummer Night's Dream*

Stephan Laqué

In 1665 an apple caught the attention of Sir Isaac Newton as he was resting in his orchard at Woolsthorpe Manor in Lincolnshire. Legend has it that as the apple fell off the tree, Newton was struck both by the apple and by the idea that the attraction which was causing the fruit to fall to the ground must be the same force which determines the motions of the moon by holding it in its orbit.[1] The enormous reach of this moving force was established and in turn made quantifiable in Newton's general law of gravitation. An epochal step no doubt, but one which suggested command over a phenomenon whose nature or quality remained a mystery. In this paper, I want to trace this mysterious force and some of the ways in which it has been handled and conceptualised in Shakespeare's *A Midsummer Night's Dream*.

At the time *A Midsummer Night's Dream* was first seen on stage, Newton's birth was still more than half a century away and the term 'gravity' had not been coined. The introduction of the term marks a process of differentiation between forces which had formerly been regarded as indistinct. Well into the seventeenth century, they were seen as a single force of attraction which included the physical forces of gravitation and of magnetism and the emotional force of love. In the third book of *Paradise Lost*, Milton charts the cosmos which Satan enviously eyes and in a phrase describing the sun's position all three aspects are brought together. The constellations, Milton writes, "Turn swiftly their various motions, or are turn'd / By his [the sun's] Magnetic beam, that gently warms / The Universe."[2] Since this conglomerate force was quite rightly perceived as a fundamental principle governing all existence, the inscrutable mystery surrounding its source caused it to be viewed as either divine or magical.

To a conception of the cosmos which relied to a great extent on relations and analogies, forces of such pervasiveness which attract and link were, of course, not just unproblematic safeguards of stability, but also the origin of mutability. Quoting Giambattista della Porta's example of the mutual attraction which metal chains develop under the influence of a magnetic stone, Foucault in *The Order of Things* describes this incalculable force as "a game of sympathies" whose reach and direction are unpredictable and which threatens to rob distinct entities of their individuality.[3] Thus, the force of attraction had fundamental metaphysical and ontological implications far beyond its tangible effects in the physical world.

Whether perceived as predominantly constructive or detrimental, the capacity to attract and to bind was a unique fascination and challenge to poets, philosophers and scientists during the centuries which saw the emergence of the new sciences with reactions ranging from humble and uninquisitive awe to confident theories of how these forces might be manipulated by magic. The different approaches reflect that general development of the world picture which has since Keith Thomas' seminal study come to be known as "The Decline of Magic;" a decline which marked the gradual shift from an animistic view of the cosmos to a mechanistic world picture.[4]

The animistic view held that all objects possess a soul, i.e. an internal motivating force which is the cause of their potential to change and to interact with other bodies. Aristotle refers to Thales' view "that everything is full of gods,"[5] that everything is endowed with a divine soul. According to Aristotle's account in *De Anima*, Thales had been led to this assumption by the contemplation of the effect a magnet has on iron: "Thales, too, to judge from what is recorded of his views, seems to suppose that the stone is in a sense the cause of movement, since he says that a stone has a soul because it causes movement to iron."[6] Here, at the beginning of western philosophy, the force which moves and attracts was the epitome of vitality, of the soul and living origin of everything. Its mysterious capacity and its pervasiveness gave it a divine aura beyond human control. In his essay *On Magic* of 1589 Giordano Bruno posits that "There is no body which is completely devoid of spirit and intelligence."[7] This animistic view occasions a mutability both intriguing and disturbing. "[B]ecause of the active principle and spirit or universal soul," Bruno writes, "nothing is so incomplete, defective or imperfect, or, according to common opinion, so completely insignificant that it could not become the source of great events."[8] According to Bruno, objects owe this extraordinary potential to a firm and organic union between physical and mental properties and forces: If matter has a soul, it will act and show effects according to quasi-mental impulse and taste and it will, on the other hand, be open to the same forms of manipulation as the human mind. Hence, when Hamlet on his way to his seat at the performance of "The Murder of Gonzago" turns from his mother to Ophelia saying "No, good mother, here's metal more attractive"[9] he is not just using an image to illustrate his preference, but he is also pointing out a well-established parallel: The mutual attraction between him and Ophelia is based on the very same force by which metal is attracted to a magnet.

Around the beginning of the seventeenth century, the process of dissociation and abstraction that would lead to the mechanistic universe of René Descartes was well under way. For the idea of describing the cosmos in terms of an artificial machine, the authority of Plato's *Timaios* was gratefully quoted. In order to elucidate the composition of the world out of the four elements, Plato has Timaios use the image of *tò tû kósmu sôma*, "the body of the world."[10] In the 4th century, Chalcidius had rendered this passage in his commentary on the *Timaios* as "this wonderful and visible machine."[11] Within Plato's philosophy, this 'machine' was a living organism and throughout the middle ages the *machina mundi* was seen as suffused by the *anima mundi* which determined its motion. However, as the recognition of the Aristotelian tenet of an internal motivating force (*entelechy* in his terminology) waned, natural philosophy began to regard the machine of the cosmos as a purely mechanical system made up of inanimate components and exclusively moved by outside forces. Plato's introduction of the Demiurge,[12] the divine but markedly anthropomorphic artisan-creator of the cosmos, was taken to support this view. Increasingly, then, the cosmos came to be seen as an artful but ultimately dead construction analysable in terms of quantifiable matter and space without consideration of internal dispositions or forces of attraction.

A Midsummer Night's Dream is very much a play about attraction and binding, about inscrutable forces and strange links. The wedding of Theseus and Hippolyta acts as a frame both for the action and for this context in the play. Theseus is the Duke of Athens, of the ancient fountainhead of reason. By virtue of its patron goddess and her foundation of the first law-court on the Areopagus, Athens is also the home of wisdom and law. Quite appropriately then, the play starts off with a court-scene which intrudes upon the wedding preparations of Theseus and Hippolyta. Theseus is called upon to bring Hermia to her senses who resists her father's choice of Demetrius for her husband. Egeus cites "the ancient privilege of Athens" (*A Midsummer Night´s Dream* 1.1.41)[13] to kill his recalcitrant daughter. Athenian law demands that Hermia either bend to her father's wish or suffer death. As a third alternative, Theseus offers her the retreat to a cloister, but in general, he grants Egeus the right to enforce his design against the professed mutual attraction between his daughter and Lysander. Judging from Theseus' reference to his wife's reaction to this state of affairs, Hippolyta obviously does not approve: "Come, my Hippolyta; what cheer, my love?" (*A Midsummer Night's Dream* 1.1.122). Hippolyta's dejection would be understandable, since she has experienced a similarly forced match at the hands of her groom. As Theseus recounts:

Hippolyta, I woo'd thee with my sword,
And won thy love doing thee injuries;
But I will wed thee in another key,
With pomp, with triumph, and with revelling.
(*A Midsummer Night's Dream* 1.1.16-19)

Theseus announces a new approach to the matrimonial link he is about to establish. Paradoxically, he had tried to form a bond with the violently severing action of his sword. Now, in order to bring about the ultimate bond of marriage, he intends to respect rite and ceremony and to allow the irrational forces of human nature to take their course "with triumph and with revelling."

Oberon, who is in many ways Theseus' counterpart in the 'green world' of the forest, is equally in need of self-repositioning. Even as he is about to trick his Queen Titania into relinquishing the custody of her Indian boy to him, he explains to Puck that "we are spirits of a different sort" (*A Midsummer Night's Dream* 3.2.388) thus drawing a line between themselves and those "Damned spirits all, /

That in cross-ways and floods have burial" (*A Midsummer Night's Dream* 3.2.382-383). Considering the confusion and potential damage which Puck and Oberon are causing, this is a timely reminder that the magic which they represent is not to be confused with the influence of the occult. Rather, they represent a form of magic which is based on an intimate knowledge of nature both human and botanical, which they exploit in order to direct forces of attraction and to establish bonds.

Using the extract of the flower "love-in-idleness," Puck and Oberon work on the disposition of their objects by distorting their perception. According to the second definition which Giordano Bruno gives in his treatise *On magic*, "'magician' refers to someone who does wondrous things merely by manipulating active and passive powers, as occurs in chemistry, medicine and such fields; this is commonly called 'natural magic'."[14] The underlying magical dynamics on which this manipulation can build are precisely the inner forces of attraction. Thus Bruno's fourth definition according to which "magic refers to what happens as a result of attraction and repulsion between things, for example, the pushes, motions and attractions due to magnets and such things, when all these actions are due [...] to the spirit or soul existing in things. This is called 'natural magic' in the proper sense."[15] Oberon and Puck are practitioners of this "natural magic" which they use to create and confound bonds. This makes them indeed "spirits of a different sort" quite removed from the magic of demons and talismans. Puck is a semi-material servant who needs to be instructed in the proper use of the different flowers by Oberon. He therefore hardly comes across as a transcendentally powerful demonic spirit (elemental or otherwise).[16] Brunian terminology of natural magic, however, offers a most fitting appellation for Puck and Oberon working together: they are a "bonding agent" enlisting "the hand which binds." Bruno defines the "bonding agent" in terms of an internal force: "A body does not have any feeling of its own, but only because of a certain force which resides in it and which emanates from it. This force is called, metaphorically, the 'hand which binds', and it is oriented and adapted to bonding in multiple ways."[17] For natural magic in the Brunian sense to work, this "certain force" needs to be complemented by knowledge about the object of the bond, since, according to Bruno, "he who knows how to bind needs to have an understanding of all things, or at least of the nature, inclination, habits, uses and purposes of the particular things that he is to bind."[18] Whereas Puck is baffled by human folly – "Lord, what fools these mortals be!" (*A Midsummer Night's Dream* 3.2.115) he bemusedly exclaims –, we learn from Oberon's all-too-human quarrelling with Titania just how intimately familiar he is with the whole range of human inclinations and dispositions. The line separating Puck and Oberon from occult practices is therefore not drawn along the ethical categories of black vs. white magic. They are of a totally different kind in that their magical influence is based on an understanding of the dynamics of attraction. Seen as a representation of this dynamics, Puck and Oberon are prime examples of the much quoted description of poetry which Theseus offers in the last act:

[A]s imagination bodies forth
The forms of things unknown, the poet's pen
Turns them to shapes, and gives to airy nothing
A local habitation and a name.
(*A Midsummer Night's Dream* 5.1.15-17)

In the figures of the Fairy King and his servant, the power and mutability of the forces which regulate life by attraction and bonds are "bodied forth."[19]

In 1589 Bruno introduced his theory of bonds towards the end of his essay *On Magic* and also dedicated a whole treatise entitled *A general account of bonding* to this force. As Hilary Gatti has noted, this work has "some scientific pretense"[20] in its constant – though one must add, rarely conclusive – recourse to experience and observation. It is, however, not an anticipation of modern empirical science, but a work of speculative psychology which is firmly rooted in an animistic theory of attraction: "Everything which is bound has an awareness in some sense, and in the nature of that awareness, one finds a certain type of knowledge and of appetite, just as a magnet attracts or repels different kinds of things."[21] Because "a bonding power is not simple or reducible to only one thing, but is composite [and] variable in nature,"[22] what emerges from these processes of attraction is not a fixed set of coherences, but a dynamic web of relations that is open to change and manipulation: "no bonds are eternal. Rather, things alternate between being bonded and escaping from a bond, or they transfer from one type of bond to another. This is a natural occurrence"[23] – a natural occurrence which is hardly apt to dispel the mystery surrounding the force which attracts and binds.

From the outset, the four lovers in Shakespeare's *A Midsummer Night's Dream* are aware of the inexplicable and powerful forces they are subject to in their affiliations. When Hermia addresses the Duke to plead against her father, she cites these forces as an excuse for her temerity: "I know not by what power I am made bold" (*A Midsummer Night's Dream* 1.1.59) and after Lysander has revealed to her his plans to elope, she includes within her elaborate invocation of mythology an undefined oath "By that which knitteth souls and prospers loves" (*A Midsummer Night's Dream* 1.1.172). The very inscrutability and uncontrollable enormity of the forces working on her are a source of her determination. To Helena, on the other hand, understanding the attraction which binds Demetrius to her rival Hermia is of crucial importance. She entreats Hermia:

Your eyes are lode-stars, and your tongue's sweet air
More tuneable than lark to shepherd's ear,
[...]
O, teach me how you look, and with what art
You sway the motion of Demetrius' heart.
(*A Midsummer Night's Dream* 1.1.183-193)

Again, amorous attraction is associated with the physical and astronomical power of magnetism. Helena uses the same parallel again when she accuses Demetrius of his indifference:

You draw me, you hard-hearted adamant –
But yet you draw not iron, for my heart
Is true as steel. Leave you your power to draw,
And I shall have no power to follow you.
(*A Midsummer Night's Dream* 2.1.195-198)

Both Helena and Demetrius are baffled by this attraction which is plainly against reason: "Do I entice you? Do I speak you fair?" (*A Midsummer Night's Dream* 2.1.199) Demetrius exasperatedly asks and Helena sums up the irrationality of her behaviour: "I am your spaniel" (*A Midsummer Night's Dream* 2.1.203). As Helena hopes that either she will become more 'magnetic' to Demetrius or that he will loosen his pull on her, attraction emerges as a force of nature which demands to be investigated and controlled.

Naturally, one might say, the lovers are prepared to accept the force which binds. Bruno holds that apart from the binding agent "a passive power or disposition in a subject or patient"[24] is an indispensable prerequisite for the establishing of bonds. This makes the lovers ideal objects for the working of natural magic, since, as Bruno states, "all bonds are either reduced to the bond of love, depend on the bond of love or are based on the bond of love."[25] Puck and Oberon exploit this disposition by manipulating the eyes with the love-in-idleness juice. Bruno takes the eyes to be the essential gateway for the creation of bonds: "The senses are the entrances through which bonds are cast. And vision is the most important of them all."[26] It is this capacity of the eyes which Oberon seeks to employ. As he explains, the love-in-idleness juice

on sleeping eyelids laid,
Will make or man or woman madly dote
Upon the next live creature that it sees.
(*A Midsummer Night's Dream* 2.1.170-172)

This is not a simple case of distorted perception, of misguided rational assessment, but a more direct mental interference. The play is not concerned with rational correspondences or causal links, but with bonds. In keeping with Bruno's observation that "fantasy and opinion bind more things than does reason,"[27] the play does not suggest that reason might be a remedy for love's tribulations – reason would simply not have been a sufficiently powerful binding force.[28] The bonding agent "acts on the imagination of someone, asleep or awake, to produce internal images [...] these matters are cases of mania or melancholy, and are called 'the dreams of the wakeful';"[29] a paradox which nicely captures the transitory state of dynamic bonding in which Puck and Oberon exercise their power.

The play amply demonstrates that Oberon's seemingly straightforward strategy is open to all sorts of confusion and Bruno grants that in the operation of bonding, the agents of attraction tend to go wrong. In his *General Account of Bonding*, he dedicates a chapter to "The blindness of the bonding agent"[30] and he goes on to describe this blindness as a trait inherent in all bonds, but especially so in the bond of love. Love cannot be adjusted or manipulated by bonds of reason: "[W]hat use is it to appeal to analogies, similarities, family traits and other such meaningless words when we see a person who hates nothing more than another person who is his genial companion, while at the same time and without reason, he also loves that person more than anything else?"[31] The person Bruno describes here is very much reminiscent of Lysander and Demetrius in whom love and hatred are very closely related as well. In the play as in Bruno, the bonds of love are too strong for the ordering influence of the bonds of reason. They therefore have to be addressed directly by the creation of equally powerful bonds – a task for which Oberon takes over from Theseus, whose legal endeavours in Act I have proved ineffectual.

Apart from the lovers, a second group of people is affected by the pranks of Puck: the mechanicals who intend to honour the wedding of Theseus and Hippolyta with the performance of a play. In the forest they hope for rehearsals which will not be "dogged by company" (*A Midsummer Night's Dream* 1.2.96f.) – a vain hope as the play will show. They have left town too, but unlike the lovers, they are not trying to escape from Athenian law and they therefore refer to the forest as "the palace wood" (*A Midsummer Night's Dream* 1.2.94), a place outside of Athens which is hidden but not legally or essentially distinct from the Athens of Theseus. Obviously, the mechanicals are not in search of (marital) bonds and they generally fall under the category of objects which Bruno classifies as hard to bind: "Of all the things which bind, certainly more of them bind humans than brute animals, and more of them bind those who have an active character than those who are dull witted."[32] It would be hard to deny a certain element of dullness in Quince's troupe of amateur actors, but their imperviousness to bonds and attractions is not exclusively a matter of their mental simplicity or their unfavourable disposition in the Brunian sense. I want to argue that the "rude mechanicals" (*A Midsummer Night's Dream* 3.2.9), as Puck calls them, represent a mechanistic view of the world.[33] This perspective, which was to dominate science from the 17th until well into the 20th century, was quickly evolving and gaining recognition at the time *A Midsummer Night's Dream* was written, so it is hardly surprising that its resonance might have found its way into the *dramatis personae*.

As the mechanistic and mathematical concept of the universe was gradually superseding the animistic cosmos, the profession of the scientist became increasingly accessible to people without gentle birth or metaphysical and theological learning. The new scientists came from a variety of professional backgrounds and introduced a practical approach to the search for knowledge which did not depend on learned speculation. As architects, engineers and navigators, they were more interested in applicable rules than in theories of vital forces. In 1581, one of these 'rude mechanicals,' Robert Norman, published an influential treatise entitled *The Newe Attractiue* on the dip in the compass needle which occurs along with its alignment to the north. Norman, who was a mariner, compass maker and navigator by profession, added a preface to his book which was to vindicate his scientific ambition. He writes with regard to book-learning: "I doe verily thinke, that [...] there are in this land divers Mechanitians, that in their severall faculties and professions, have the use of those Artes at their fingers endes, and can apply them to their severall purposes, as effectually and more readily, than those that would most condemne them."[34] There is a certain measure of arrogance in Norman's claim to competence reminiscent of the pushy enthusiasm of Bottom, who eagerly volunteers for every single part in the "Pyramus and Thisbe" play: "Let me play the lion too" (*A Midsummer Night's Dream* 1.2.66). However, the most condescending claim that Norman makes concerns the method he projects: "I mean not to use barely, tedious Conjectures or imaginations: but briefly as I may to passe it over, grounding my Arguements only upon experience, reason and demonstration, which are the grounds of Art."[35] Norman seeks to limit permissible scientific procedure – "the grounds of Art" – to reason and tangible demonstration. In *The Newe Attractive*, measurements and experiments with variously suspended magnetic needles replace the intricate theories about magnetism which Giordano Bruno and many other speculative philosophers were developing at the time. While, to natural philosophy, "tedious Conjectures or imaginations" were perfectly acceptable means of explaining difficult phenomena, Norman dismisses any investigation into the cause of magnetic attraction: "True it is, that God is mighty and marvellous in all his Works; yet he doth not allow us to say more than Truth of them. And truly, his Power is as greatly shewed in the Magnes, as in any Stone that he hath created: And who so shall go about curiously to seek out the efficient Cause of his Properties, I suppose the longer he seeketh, the more he shall marvel, and yet never the nearer his Purpose."[36] Obscure forces within matter are of no interest to the mechanistic view of things and Norman shrouds this suspension of curiosity towards processes beyond his chosen method in a mixture of pragmatism and pious awe. Norman's choice of perspective was certainly a step in the direction of experimental science, but it was also a provocation.

Bottom's approach is equally down-to-earth and informed by an almost grotesque indifference to the forces of attraction. Titania's unwavering advances do not seem to affect him at all and his pertinent insight that "reason and love / keep little company together nowadays" (*A Midsummer Night's Dream* 3.1.138f.) settles the business for him. He is the only Athenian who is granted the privilege of meeting with the supernatural cast of the play, but his dealings with the ethereal Queen of Fairies and her train is exclusively practical and mainly of a culinary nature. As the fairies introduce themselves to him, he considers their usefulness: in the case of Cobweb as wound-dressing, in the case of Peaseblossom and Mustardseed as victuals. Once he gets to enlist their services in Act IV, it is, again, for food and for scratching. The one person to be in contact with the animistic world, he is totally resistant to the wonder and attraction it holds in store. One may assume that, given a similar opportunity, Giordano Bruno would have had somewhat more weighty matters to discuss with the spirits of nature.

Even stronger than Bottom's indifference towards the power of attraction is his fear of bonding which he shares with the other mechanicals in the play and which turns the play they perform before the Duke into a mess of meta-theatrical contortion. While animistic philosophy is informed by a union of physical, psychological and supernatural forces and phenomena, the mechanistic world picture is based on a dissociation of those forces. There is a rigorous delimitation which keeps matter and mind apart and allows scientists to focus exclusively on material objects that are expanded in space. What happens in poetry and especially in the theatre, is precisely a union of the realms of the mental and the physical; again, in the words of Theseus, a "bodying forth" of the immaterial. Having enumerated the narrow objects of astronomers, arithmeticians, lawyers, and so forth, Sir Philip Sidney in his *Apology for Poetry* extols poetry, since "[o]nly the poet, disdaining to be tied to any such subjection, lifted up with the vigour of his own invention, doth grow in effect into another nature." The skill of the poet lies in the "*Idea* or fore-conceit of the work" whose delivering forth "is not wholly imaginative, as we are wont to say by them that build castles in the air; but [...] substantially it worketh."[37] The mental productivity of the playwright transcends the mind and has a "substantial" effect in the stage-world whose illusion is both imaginary and tangible. It is not confined either to mind or to matter, but fuses the two realms. This bonding capacity of poetry is painstakingly blocked by Bottom who introduces his prologues in order to preclude any illusion: "I have a device to make all well. Write me a prologue, and let the prologue seem to say we will do no harm with our swords, and that Pyramus is not killed indeed; and for the more better assurance, tell them that I, Pyramus, am not Pyramus, but Bottom the weaver" (*A Midsummer Night's Dream* 3.1.15-20).

The reason for this procedure is not some metaphysical Angst about threatened identity, but the absurd fear that they might offend the fragile ladies at court.[38] Everyone from the fearful lion down to the decidedly less terrifying wall is made to state his name and to declare that he does not wish to be associated with the role he is playing. The gesture is quite literally 'analytical' in that it neatly severs the bond of illusion which, to adopt Aristotelian terminology, links substance and accident or adjunct. Here, Bottom and Snout are substances while Pyramus and Wall are their adjuncts. By this careful ordering which keeps categories rigorously separated, Bottom and company appear to be making allowance for the critique of poetry of Francis Bacon, one of the protagonists in the mechanisation of the world picture. In the second book of *The Advancement of Learning* Bacon writes that: "Poesy is a part of learning in measure of words for the most part restrained, but in all other points extremely licensed, and doth truly refer to the imagination; which, being not tied to the laws of matter, may at pleasure join that which nature hath severed, and sever that which nature hath joined; and so make unlawful matches and divorces of things."[39] Quince's players are careful not to establish any bonds which might suggest distance from "the laws of matter." They expose the mechanics of stage production in order to demonstrate that the world created on stage is fully explicable in terms of mechanical construction and artificial design and not disposed to create any "unlawful matches."

This dissociation is second nature to Bottom and it is therefore highly appropriate that his 'translation' is purely physical. Even though his ass's head is easily the most striking transformation in the play, it has precious little effect on Bottom, who remains blissfully unaware of his changed appearance. His new head is no more than a cosmetic issue: "I must to the barber's, monsieur, for methinks I am marvellous hairy about the face" (*A Midsummer Night's Dream* 4.1.23-24). However, as soon as he awakes, there is a lingering feeling that something unusual has happened: "I have had a most rare vision. I have had a dream, past the wit of man to say what dream it was. Man is but an ass if he go about to expound this dream" (*A Midsummer Night's Dream* 4.1.203-206). Like Richard Norman, Bottom dismisses any attempt to come to an understanding of his 'vision' as inappropriate and doomed to fail. Unlike the awakened lovers, who follow Theseus out of the forest recounting their dreams, Bottom decides to keep his counsel: "man is but a patched fool if he will offer to say what methought I had" (*A Midsummer Night's Dream* 4.1.208f.). When he meets his fellow actors, he explains his reticence: "Masters, I am to discourse wonders: but ask me not what; for if I tell you, I am not true Athenian" (*A Midsummer Night's Dream* 4.2.29-30). The telling of wonders and the acknowledgement of powers beyond simple material motion obviously run counter to Athenian reason, to being "true Athenian." Instead, Bottom decides to deal with his experience by relegating it to the realm of art: "I will get Peter Quince to write a ballad of this dream: it shall be called 'Bottom's Dream', because it hath no bottom" (*A Midsummer Night's Dream* 4.1.212-215). In Act III, Bottom had rejected the ballad metre for the prologue which mentions the names of the actors and had asked for a more symmetrical metre to be used: "*Quin.* We will have such a prologue; and it shall be written in eight and six. / *Bot.* No, make it two more; let it be written in eight and eight" (*A Midsummer Night's Dream* 3.1.22-25). Any association with the ballad's concern with things numinous and supernatural would have been inappropriate for a text intended to preclude acts of imaginative and 'magical' bonding. In act 4, on the other hand, Bottom chooses the ballad form as a medium to disguise his experience in the animistic cosmos of the fairies as a piece of 'harmless' imaginative writing. His strategic use of the form of the ballad is thus once more indicative of the line he is drawing between the world of the bonds of the imagination and his analytical and mechanistic frame of mind.

Can Shakespeare's *A Midsummer Night's Dream*, then, be read as an imaginative plea for an animistic world view and a remorseless parody of the new sciences? Not quite: The play ends in Athens where it started and we are reminded that Bottom is not the only proper Athenian. Theseus may have bent the law of Athens to support the bonds between the lovers that were forged in the wood, but he remains dismissive about the imaginative qualities of fairyland:

I never may believe
These antique fables, nor these fairy toys.
Lovers and madmen have such seething brains,
Such shaping fantasies, that apprehend
More than cool reason ever comprehends.
(*A Midsummer Night's Dream* 5.1.2-6)

His bride Hippolyta, on the other hand, grants some substantial truth to the sketchy account of the lovers:

But all the story of the night told over,
And all their minds transfigur'd so together,
More wittnesseth than fancy's images,
And grows to something of great constancy.
(*A Midsummer Night's Dream* 5.1.23-26)

Here the exchange is interrupted by the entrance of the lovers and without being reconciled, the two stances come together in the ducal marriage as an "everlasting bond of fellowship" (*A Midsummer Night's Dream* 1.1.85). This wedding of irreconcilable elements creates the dynamic bond of an intellectual *coincidentia oppositorum*, wherein both may keep their distinct qualities, yet complement each other. Which brings me back to Newton's apple: Once it hit the ground, it lay motionless at the physicist's feet. Some force was apparently keeping it from rolling away. Newton went on to call this force *inertia* and described it as a *vis insita* or "internal force" which was – quite animistically – influencing objects from within and rendering them indisposed to change their position.[40] In Newton and at the Athenian court of Theseus, mechanical rationality could still happily coincide with the animistic view, before post-Cartesian rigour ousted this legacy from serious scientific concern. From then until the beginning of the 20th century, Bottom got to play the lion's part, and, one has to hand it to him, he did a very good job.

Notes

[1] This anecdote was first elaborated and made public by Voltaire in his *Essay on Epic Poetry* of 1727. Cf. also D. McKie and G. DeBeer, "Newton's Apple."

[2] John Milton, *Paradise Lost* 3. 582-584.

[3] According to Foucault, this "jeu des *sympathies* [...] joue à l'état libre dans les profondeurs du monde." Its danger lies in its capacity to assimilate: "La sympathie [...] a le dangereux pouvoir d'assimiler, de rendre les choses identiques les unes aux autres, de les mêler, de les faire disparaître en leur individualité [...] comme ces chaînes de métal suspendues par sympathie à l'attirance d'un seul aimant" (Michel Foucault, *Les Mots et les Choses: Une Archéologie des Sciences Humaines* 38f.).

[4] "The essence of the [scientific and philosophical] revolution was the triumph of the mechanical philosophy" (Keith Thomas, *Religion and the Decline of Magic: Studies in Popular Beliefs in Sixteenth and Seventeenth-Century England* 769).

[5] Aristotle, *On the Soul* 411a.

[6] Ibid. 405a.

[7] Giordano Bruno, *On Magic* 125.

[8] Ibid. 111.

[9] William Shakespeare, *Hamlet* 3.2.108.

[10] Plato, *Timaios* 32c1.

[11] "praeclare ista machina mundi visibilis" Plato, *Timaeus: A Calcidio translatus commentarioque instructus* 25. Cf. Karen Gloy, *Das Verständnis der Natur*, vol. 1: *Die Geschichte des Wissenschaftlichen Denkens* 157.

[12] Plato, *Timaios* 28a ff.

[13] William Shakespeare, *A Midsummer Night's Dream*, ed. Harold F. Brooke. All references are to this edition.

[14] Bruno, *On Magic* 105.

[15] Ibid. 105.

[16] For a different interpretation see Bevington: "Puck constantly brings before our eyes a more threatening vision of fairydom than is apparent in Oberon's more regal pronouncements" (David Bevington, "'But We Are Spirits of Another Sort': The Dark Side of Love and Magic in *A Midsummer Night's Dream*" 26).

[17] Giordano Bruno, *A General Account of Bonding* 146.

[18] Ibid. 148.

[19] Shakespeare's adaptation and transformation of fairylore is much more than a device "to suit his comic intentions" (Blount, "Modifications in Occult Folklore as a Comic Device in Shakespeare's *A Midsummernight's Dream*" 12).

[20] Hilary Gatti, *Giordano Bruno and Renaissance Science* 3.

[21] Bruno, *A General Account of Bonding* 157.

[22] Ibid. 147.

[23] Ibid. 159.

[24] Bruno, *On Magic* 132.

[25] Bruno, *A General Account of Bonding* 165.

[26] Ibid. 168.

[27] Bruno, *A General Account of Bonding* 152.

[28] On the critical tradition which holds that the play censures love for its irrationality see Harold Brooks, "Introduction" CXXXf.

[29] Bruno, *On Magic* 139.

[30] Bruno, *A General Account of Bonding* 152f.

[31] Ibid. 152f.

[32] Ibid. 146.

[33] For the opposite view which holds that Bottom and the mechanicals represent an animistic world view at variance with the other Athenians, see Herbert: "Theseus, Lysander, and the rest of Shakespeare's Athenians ridicule in Pyramus' world an extreme animist model with which they have no sympathy" (T. Walter Herbert, *Oberon's Maze'd World* 70).

[34] John Desmond Bernal, *Science in History*, vol. 2, *The Scientific and Industrial Revolutions* 435. Bernal quotes this introduction in full while Hellmann unfortunately chooses to omit it from his edition of the *Newe Attractiue*.

[35] Bernal, *The Scientific and Industrial Revolutions* 435.

[36] Robert Norman, *The Newe Attractiue* 5f.

[37] Sir Philip Sidney, *An Apology for Poetry or The Defence of Poesy* 100f.

[38] The way the players construct the court and its denizens in their heads parodies their residual imagination. As Northrop Frye has brilliantly remarked: "The court ladies belong to the Quince players' fairyland" (Northrop Frye, *Northrop Frye on Shakespeare* 38).

[39] Francis Bacon, *The Advancement of Learning and New Atlantis* 80.

[40] Cf. also the animistic view of magnetism which pervades William Gilbert's experimental study *De Magnete* published in 1600. Gilbert writes that "the magnetic force is animate, or imitates a soul; in many respects it surpasses the human soul while that is united to an organic body" (William Gilbert, *De Magnete* 308).

Works Cited

Aristotle. *On the Soul*. Ed. and trans. Walter S. Hett. London: Heinemann, 1964.

Bacon, Francis. *The Advancement of Learning and New Atlantis*. Ed. Arthur Johnston. Oxford: Clarendon Press, 1974.

Bernal, John Desmond. *Science in History*, vol. 2, *The Scientific and Industrial Revolutions*. 3rd ed. Harmondsworth: Penguin, 1965.

Bevington, David. "'But We Are Spirits of Another Sort': The Dark Side of Love and Magic in *A Midsummer Night's Dream*," in *A Midsummer Night's Dream: Contemporary Critical Essays*. Ed. Richard Dutton. Houndsmills: Macmillan, 1996. 24-37.

Blount, Dale. "Modifications in Occult Folklore as a Comic Device in Shakespeare's *A Midsummer Night's Dream*." *Fifteenth-Century Studies* 9 (1984): 1-17.

Brooks, Harold. "Introduction," in Shakespeare, William. *A Midsummer Night's Dream*. Ed. Harold F. Brooks. The Arden Shakespeare. Walton-on-Thames: Thomas Nelson, 1997.

Bruno, Giordano. *A General Account of Bonding* in Giordano Bruno, *Cause, Principle and Unity, Essays on Magic*. Trans. and ed. Robert de Lucca and Richard J. Blackwell. Cambridge: Cambridge UP, 1998. 143-176.

---. *On Magic* in Giordano Bruno, *Cause, Principle and Unity, Essays on Magic.* Trans. and ed. Robert de Lucca and Richard J. Blackwell. Cambridge: Cambridge UP, 1998. 103-142.

Foucault, Michel. *Les Mots et les Choses: Une Archéologie des Sciences Humaines.* Paris: Gallimard, 1974.

Frye, Northrop. *Northrop Frye on Shakespeare.* Ed. Robert Sandler. New Haven: Yale UP, 1986.

Gatti, Hilary. *Giordano Bruno and Renaissance Science.* Ithaca: Cornell UP, 1999.

Gilbert, William. *De Magnete.* Trans. P. Fleury Mottelay. New York: Dover, 1958.

Gloy, Karen. *Das Verständnis der Natur,* vol. 1: *Die Geschichte des Wissenschaftlichen Denkens.* Munich: Beck, 1995.

Herbert, T. Walter. *Oberon's Maze'd World.* Baton Rouge: Louisiana State UP, 1977.

McKie, D. and G. De Beer. "Newton's Apple." *Notes and Records of the Royal Society* 9 (1952): 46-54 and 333-335.

Milton, John. *Paradise Lost.* Ed. Christopher Ricks. London: Penguin, 1989.

Norman, Robert. *The Newe Attractiue, in Neudrucke von Schriften und Karten über Meteorologie und Erdmagnetismus,* vol 10, *Rara Magnetica.* Ed. Gustav Hellmann. Berlin: Asher, 1898.

Plato. *Timaeus a Calcidio translatus commentarioque instructus.* Ed. Jan Hendrik Waszink. London: Warburg Institute; Leiden: Brill, 1962.

Plato. *Timaios, Sämtliche Werke* vol. 8. Ed. Karlheinz Hülser. Frankfurt am Main: Insel, 1991.

Shakespeare, William. *A Midsummer Night's Dream.* Ed. Harold F. Brooks. The Arden Shakespeare. Walton-on-Thames: Thomas Nelson, 1997.

Shakespeare, William. *Hamlet.* Ed. Harold Jenkins. The Arden Shakespeare. London: Routledge, 1982.

Sidney, Sir Philip. *An Apology for Poetry or The Defence of Poesy.* Ed. Geoffrey Sheperd. London: Nelson, 1965.

Thomas, Keith. *Religion and the Decline of Magic. Studies in Popular Beliefs in Sixteenth- and Seventeenth-Century England.* London: Penguin, 1973.

Pictorial Magic and the Staging of the Body Politic

Martin Windisch

The Verbal and the Visual

In his study of Lady Macbeth as an example of "a person who collapses on reaching success, after striving for it with single-minded energy,"[1] Sigmund Freud offers a description of the magic, spell-binding effect of a drama on its audience: He says, "The dramatist can indeed, during the representation, overwhelm us by his art and paralyse our powers of reflection; but he cannot prevent us from attempting subsequently to grasp its effect by studying its psychological mechanism."[2] This remark involves more than one dimension within an aesthetics of effect, since it leaves open the question to what extent Freud is implying that such an effect only occurs "during the representation," i.e. during a performance, or whether in fact the same effect can be inherent to the text of the drama itself. With regard to drama and the history of its reception, this could be seen as a decisive dividing line, instigating a concentration of scholarly work exclusively on the performing arts, since the magic effect is, at least in most cases, heightened by the representation proper. The shift from the written word to its theatrical performance involves a shift from the verbal to the visual, and yet it also denies this very shift, since the written word has a visual dimension of its own, and performance on stage is as much verbal as it is visual.

The relationship between verbal imagery and visual imagery in Shakespeare's time can be seen as a question of the power of the imagination. The Prologue to Shakespeare's *King Henry V* invites the imaginative work of the audience to create an illusion of the king on stage: "For 'tis your thoughts that now must deck our kings" (Prologue 28). In his magisterial *Shakespearean Negotiations* Stephen Greenblatt highlights the implicit reciprocity of such exchange between audience and actors with a specific emphasis on the analogous structure of representation in the theatre and political representation. Greenblatt also directs our attention to the mirror as optical device and frame of reference for the material exchange implicit in both the collective invention and social emanation at the core of such processes of representation.[3]

During Shakespeare's lifetime the question of the power of the imagination was very much a question of power itself, and of the authority given to the arts. This became even more evident when James I intensified the monarchical desire for representation. Scholarly work on this aspect of the power of representation is abounding, seminal studies such as Stephen Orgel's *The Illusion of Power*, Jonathan Goldberg's *James I and the Politics of Literature*, and Alvin Kernan's *Shakespeare, the King's Playwright* can be taken as exemplary in that respect.[4] Such studies do not, of course, assume a monolithic centre or origin of power. Quite on the contrary, they disrupt the source and origin of power by uncovering something like the ambivalence, or even subversion, of such source and origin. Prime examples are the attribution of prophetic authority and control over fancy and imagination to anti-masquers and witches within the antithetical structure of masques and plays. Ultimately, the desire of securing control over the origin of representation, which is almost synonymous with a desire to embody this origin, comes to a dead end – at least as far as traditional means of representation are concerned.[5] The artistic dilemma of subversion and containment, which has been the focus of new historicist studies for the last thirty years, is still pivotal here.

Illustration 1 and Illustration 2: *Sorcières* (witches' mirrors)

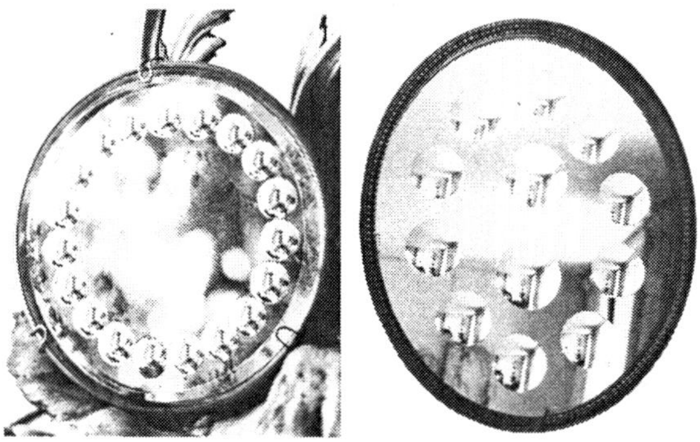

According to Greenblatt's remarks on the material exchange in representation, optical instruments such as mirrors and specific lenses do not only function as metaphors within the work of the imagination but they are made the medium of the very structure at work in such forms of representation – and yet, of course, they are not present on stage, they are not been made use of by either the actors or the theatre-goers in the audience. Taking the example of the Prologue to *King Henry V* once again, an allusion to the so-called witches' mirrors or *sorcières* (Illustrations 1 and 2) is at work, which are mirrors bulged out that actually "Into a thousand parts divide one man, / And make imaginary puissance" (Prologue 24-25). In the same play Shakespeare also alludes to other anamorphic devices, such as turning pictures or perspectives, when the English and the French king have one of their little repartees. The besieged French cities are turned anamorphically into French virgin maidens who are in danger of losing their virginity in the very process of this turn. The anamorphosis brings out the Princess Kate as future wife of King Henry (*King Henry V* 5.2.334-347). In other plays like, e.g., *King Richard II*, *Twelfth Night*, or *Antony and Cleopatra*, anamorphic devices function as structural axes around which the plays move. Orsino, the Duke of Illyria, exclaims in the eye-opening moment of revelation of that precarious cross-dressing of Viola in the disguise of Cesario: "One face, one voice, one habit, and two persons! / A natural perspective, that is, and is not" (*Twelfth Night* 5.1.214-215). And the sea-captain Antonio, a friend of Viola's twin brother Sebastian, adds: "How have you made division of yourself? / An apple cleft in two is not more twin / Than these two creatures. Which is Sebastian?" (*Twelfth Night* 5.1.220-222). With her emphatic exclamation "Most wonderful" (*Twelfth Night* 5.1.223) the Countess Olivia addresses the visual phenomenon with the most resonating and apt term for such magic. A magic which, of course, is at least twofold, since it includes the magical theatrical effects of cross-dressing in the Shakespearean theatre in general. I am here more specifically interested in the pictorial magic resulting from the theatrical illusion of witnessing the effects of optical instruments such as polyoptric lenses converging two (or more) faces, bodies, or pictures, into one, and distorting mirrors, such as convex mirrors, as well as magnifying glasses, prospectives, perspectives, etc. It seems to me that the Shakespearean references to such optical equipment, besides underlining and assisting the poetological self-reflection of such dramatic passages in general, are resonant with a mystification of these devices that can especially be found in the Elizabethan colonial discourse. In a typical example, from Thomas Hariot's *A briefe and true report of the new found land of Virginia* (1588), Hariot puts forth one of the most outspoken critical Elizabethan readings of the European self-assumption of cultural superiority over the American Indians, hinging on the visual effect of the latest European inventions: "Most thinges they sawe with vs, as Mathematicall instruments, sea compasses, the vertue of the loadstone in drawing yron, a perspectiue glasse whereby was shewed manie strange sightes, burning glasses, wildefire woorkes, gunnes, bookes, writing and reading, spring clocks that seeme to goe of themselues, and manie other thinges that wee had, were so straunge vnto them, and so farre exceeded their capacities to comprehend the reason and meanes how they should be made and done, that they thought they were rather the works of gods than of men, or at the leastwise they had bin giuen and taught vs of the gods."[6] These are, in themselves, 'marvellous possessions' which the colonists exhibit for the Indians on Roanoke Island (the first Virginia). The cultural dialects of the reciprocity of marvel and magic in a text like Hariot's at large is a topic going beyond the present paper. What I am trying to hint at here is the analogy and complicity with such strategies apparent in Shakespeare's suggestive use of pictorial magic. Shakespeare's audiences, at least a great many of them, would not have been familiar with optical equipment itself, but they would have had ideas about it from hearsay, probably enhancing its effects to such an extent that the rather vague knowledge of the very existence of optical devices led to a mystification of those devices. Relying on mentality structures based on mystifying processes of that kind, Shakespeare seems to have been well aware of the suggestive potential of the verbal reference to, and constitution of, optical trickery from the realm of the cabinets of curiosity and catoptrical experiments. It is important to note that the Shakespearean theatre had to rely on a minimum of stage props. The theatrical conflation of the verbal and the visual we find in productions today, with magnificent video installations etc., stands in radical contrast to the dominance of language and gestures in Shakespeare's theatre. A contemporaneous window on the illusionary potential of the times can, however, be seen in the splendid forms of magic changes of scenery in the Stuart court masques (as, for example, in *The Masque of Queens* or in *Oberon, The Faerie Prince*) with various forms of special effects (steam, wind, cold, etc.). Evidence for the courtly fascination with the masques cannot only be found in Ben Jonson's annotations for his own edition of his masques, but also in a number of eye-witness accounts, amongst which Bacon's Essay "Of Masques and Triumphs" ranges quite prominently. "It is true, the Alterations of Scenes, so it be quietly, and without Noise, are Things of great Beauty, and Pleasure: For they feed and relieve the Eye, before it be full of the same Object. Let the Scenes abound with Light, specially Coloured and Varied: And let the Masquers, or any other, that are to come down from the Scene, have some Motions, upon the Scene it selfe, before their Comming down: For it drawes the Eye strangely, and makes it with great pleasure, to desire to see that, it cannot perfectly discerne."[7] Concerning the anti-masques Bacon adds a number of further effects: "But chiefly, let the music of them be recreative, and with some strange changes. Some sweet odours, suddenly coming forth, without any drops falling, are, in such company as there is steam and heat, things of great pleasure and refreshment."[8] The latter almost certainly is a reference to Cornelis Drebbel's contribution of special effects to some of the Stuart masques and to his experiments in his laboratory in Eltham. (Amongst Drebbel's inventions are the submarine and an almost perfect form of *perpetuum mobile*.)

Freud, *Macbeth*, and the 'Mysteries of State'

In order to illustrate the precarious and dark side of the staging of the body politic, somehow opposed to the brilliant ceremonies and representations of successful kingship, I will now turn to two Shakespearean scenes, the one taken from *Macbeth*, the other from *King Richard II*. Both scenes are close to the core questions of "mysteries of state" and fully confirm Ernst Kantorowicz's appraisal of Shakespeare's exceptionable role in having eternalised the metaphor of the king's two bodies.[9]

In *The Interpretation of Dreams* Freud had already mentioned Macbeth's key concern with childlessness ("Just as *Hamlet* deals with the relation of a son to his parents, so Macbeth (written at approximately the same period) is concerned with the subject of childlessness").[10] In the aforementioned study of *Some Character-Types met with in Psycho-Analytic Work* (1916) Freud explores this topic further: "It seems to me impossible to come to any decision. Shakespeare's Macbeth is a pièce d'occasion, written for the accession of James, who had hitherto been King of Scotland. The plot was ready-made, and had been handled by other contemporary writers, whose work Shakespeare probably made use of in his customary manner. It offered remarkable analogies to the actual situation. The 'virginal' Elizabeth, of whom it was rumoured that she had never been capable of child-bearing and who had once described herself as 'a barren stock', in an anguished outcry at the news of James's birth, was obliged by this very childlessness of hers to make the Scottish king her successor. And he was the son of the Mary Stuart whose execution she, even though reluctantly, had ordered, and who, in spite of the clouding of their relations by political concerns, was nevertheless of her blood and might be called her guest. The accession of James I was like a demonstration of the curse of unfruitfulness and the blessings of continuous generation. And the action of Shakespeare's Macbeth is based on this same contrast."[11] The deepest speculation on the topics of childlessness, time, and the desire for what Hobbes was to call the *Artificiall Eternity*[12] of the "superbody"[13] of royal dignity can, however, be found in a letter to Lytton Strachey[14] dating from more than a decade later (Dec. 25, 1928), in which Freud investigates the transference of the traumatic effects of this curse of childlessness onto the Macbeth couple, who are said to represent two aspects of Elizabeth: In this concise analysis of the two bodies of *Macbeth*, we are reminded of the emphasis Shakespeare put on the problem of the royal succession in his 'Scottish play.' Alluding to Queen Elizabeth by way of a childless queen and a dramatic echo of the historical exclamation from a few years before (1603), "The queen is dead,"[15] Shakespeare, according to Freud, split the Queen into two, albeit complementary, persons, Macbeth and Lady Macbeth. Queen Elizabeth's occasional hesitancy, evident in her rigour towards Mary Stuart and Essex on the one hand, and her corresponding regret for having them executed on the other, are said to be represented in the Macbeth couple. Taking up Freud's remarks one could, however, shift emphasis and regard the Macbeths as a focal lens making visible the contradictory aspects of the immortality of the body politic and the dilemma of being without an heir. The monstrous murder of Duncan and the agonising line of predecessors and successors visible in the witches' show of eight kings and magic glass are thus juxtaposed. For Macbeth it is a most tormenting and monstrous sight when he becomes aware of his utter powerlessness over and against time:

Why do you show me this? – A fourth? – Start, eyes!
What, will the line stretch out to th'crack of doom?
Another yet? A seventh? I'll see no more.
And yet the eighth appears, who bears a glass
Which shows me many more. And some I see,
That two-fold balls and treble sceptres carry.
Horrible sight! Now I see, 'tis true,
For the blood-boltered Banquo smiles upon me,
And points at them for his.
(*Macbeth* 4.1.115-123)

The line stretching out to eternity is indeed beyond the reach of Macbeth as individual king. It was referred to in the royal funeral effigy on occasion of the demise of a king, and throughout the 16th and 17th centuries it was also represented in several other forms, such as dynastic portraits of the Tudor and Stuart kings, or, e.g., Crispin van de Passe I's both artistically and optically intriguing engraved title page *Christ Triumphing over Sin, Death and Hell* (Illustration 3) in Thomas Scott's *Vox Dei* (presumably Utrecht 1623/24), in which Christ doubly (i.e. through his gaze and in his body) incorporates the Stuart dynasty. In this engraving, Christ holds a large dynastic triangle substituting his own body. In the centre of the triangle which functions as a visual pyramid, James I is represented in the circle of his predecessors and successors. Charles I also appears in full body at the apex of the pyramid. The visual pyramid represents Christ's perspective: In God's prophetic view the members of the dynasty appear as the abstract idea of the ruler in history, or of the body politic. The visual pyramid embodies both, the body of Christ and of history. Double bindings are characteristic of this fascinating cumulation of this visual and dynastic pyramid: Since the rulers represented doubly culminate in Christ's body and Christ's gaze, the focus of "All in All," the engraving invokes the divine origin and legitimacy of secular power ("Touch not mine anointed"). By projecting the contemporary ruler into history, the representation offers a form of counter-balancing synthesis pertaining to *Heilsgeschichte* which Macbeth is not able to envision. The *Artificiall Eternity* so tormenting for Macbeth has, however, found its ultimate pictorial expression in the frontispiece of *Leviathan*, in which the body politic becomes a matter of both, social consent and national memory, in directly answering the specific civil war situation preceding the publication of Hobbes' masterpiece.[16]

Freud's reflection itself had been stimulated by the monument for Elizabeth I in the north aisle of the Lady Chapel in Westminster Abbey: "Als ich vor vielen Jahren vor dem Sarkophag dieser Königin in Westminster stand, schlossen sich bei mir einige Gedanken zusammen [...] Ich meinte, es sei Elisabeth – die kinderlose –, die Shakespeare den Charakter seiner Lady Macbeth eingegeben, von dem er in den historischen Quellen so wenig fand. [...] Eigentlich hatte auch Elisabeth einen Gast, der sich ihr anvertraut (Mary Stuart), töten lassen [...]."[17] For Freud it seems to have been the specific moment of being confronted with the pictorial magic of that tomb which triggered his *Macbeth* analysis. And as if to affirm this effect, art historians' and cultural historians' interest in funeral rites and monuments, and the pictorial magic characteristic of the latter, have increased tremendously within the last fifteen years.[18]

In his magisterial *The King's Two Bodies* Ernst Kantorowicz puts the tormenting and monstrous sight of the line of kings into the context of the fictive person of a corporality by succession.[19] In Roman Polanski's famous film adaptation of 1971, the royal apparitions are holding a mirror each, in which the succeeding figure appears subsequently. But even without the use of such devices, the scene carries an enthralling suggestiveness of reaching out into unfathomable eternity. This is both, a spatial category and a category relating to time, "will the line stretch out to th'crack of doom?," a verse in which the line is, of course, first of all, the hereditary line of kings, but simultaneously a drawn vanishing line with a tormentingly unperceivable endpoint. Macbeth cannot abstain from following this

Illustration 3: Crispin van de Passe I, *Christ Triumphing over Sin, Death and Hell*, engraved title page in Thomas Scott's *Vox Dei*. By Permission of the Folger Shakespeare Library

line, "the monstrous sight" – to adapt Milton's lines here – "Strook [him] with horror [...], but far worse / Urged [him] [...]"[20] to look. But this is not all. The magic glass borne by the eighth apparition adds still something to endlessness and eternity, with all the suggestiveness of early modern occult uses of a shoemaker's bulb.[21]

Such a magic glass carries the air of mystification spoken of before.

The scene, right at the centre of the play, is remarkable also insofar as the information contained in the magic line of kings had been known to Macbeth right from the first encounter with the witches. Had it not been for that, Macbeth would not have arranged for the murder of Banquo and Fleance. Thus the show of kings, quite paradoxically, might seem superfluous, whilst on the other hand it is leading right into the Macbethian dilemma of not seeing his royal dignity and monarchical superbody gifted with *Artificiall Eternity*. As we know from Geoffrey Bullough's fulminant collection of narrative and dramatic sources, the dramatic rendering of the show of kings was possibly influenced by a treatise dating from 1583, by Henry Howard. The treatise's title is *A Defensative against the Poyson of Supposed Prophecies*, and the relevant section reads: "I coulde alledge one notable example of a Conjurer, if that be true, which is reported in the memories of Fraunce, who representing to a Lady of greate calling, all those personnes, as it were in a dumble showe, which should possesse the crown in this our age, caused the king of Navarre, or rather a wicked spirite in his stedde to appeare in the fifth place, to none other ende (as I beleeve) than that she might attempt, the rydding him out of the way, by greater store of indirect devises, whom the destinies reserved to so great and honour."[22] Dating from 1583, i.e. from two years before Henry IV became king of both parts of France and 27 years prior to his assassination, the treatise seems to have been truly prophetic. This, of course, Shakespeare could not know, since *Macbeth* precedes the death of Henri Navarre by four years. If we follow Freud's interpretation, the specific contemporaneous impact of the childless Queen would explain just as much about the pivotal effect the show of kings scene must have had on audiences.

For Shakespeare's audiences the impact of the scary, ghostly appearances was probably more verbal and indirect (through Macbeth's stupefied reaction) than visual and imminent. The audiences' imaginative work was nevertheless concentrating on the visual suggestiveness of what the actor dramatically revealed. Through his eyes, through his power of vision, Macbeth looks into the abyss of an endless desire-structure, in which ambition and envy accelerate each other. Banquo's ghost, and his line of descendants, are the object of this gaze ("Now I see 'tis true; / For the blood-boltered Banquo smiles upon me / And points at them for his"). The scary appearance of this hereditary superbody does also affect our interpretation of the apparition of Banquo's ghost in the banqueting scene (*Macbeth* 3.4), since the undead soul or ghost reappearing from purgatory can also be seen as a very specific figure representing the *dignitas* of Banquo's descendants. Similar to the old Hamlet's ghost in *Hamlet*, Banquo's ghost functions as a royal funeral effigy subverting, or even stronger, denying the legitimacy of Macbeth's kingship.[23] The strongest pictorial manifestation of, and evidence for, this interpretation is Macbeth's allusion to the Gorgon-like aspect of the Banquo-figure: "never shake / Thy gory locks at me!" (*Macbeth* 3.4.50-51), directly linking it to the pictorial magic of the paralysing horrific sight of the assassinated Duncan: "Approach the chamber and destroy your sight / With a new Gorgon" (*Macbeth* 2.3.65-66), as Macduff puts it. This is the utmost paralysis attainable through fascination and the evil eye. When Macduff presents Macbeth's head towards the end of the play, the Medusan head seems to reappear – an impression intensified in Polanski's production, in which the fighting between Macbeth and Macduff reaches its climax with the beheading of Macbeth. His head drops down the stairs and in a brilliant close-up is shown as a Gorgon. The iconography of the scene turns Macduff into a new Perseus. Macbeth's head, Gorgon-like as it is, becomes banned and framed, just as the head of Medusa after having been cut off by Perseus was placed in Athena's aegis.

King Richard II – A Chapter in the Politics of Monstrosity?

primo intuitu monstrosum quid repræsentes

When more than four decades ago Ernst Kantorowicz stated that it was Shakespeare who had eternalised the metaphor of the king's two bodies, he had concentrated, as far as Shakespeare was concerned, on the analysis of *King Richard II*. Meanwhile studies in the field of the representation of power have widened the research on Shakespeare's theatrical representation of the body politic, and turned to other of his plays. Without disparaging Kantorowicz's insights, we can safely say that the topic of the king's two bodies is almost ubiquitous in Shakespeare's work, and that the way in which Richard and Bolingbroke tackle the problem, offers only a few facets of the wide range of paradoxes and dilemmas involved in the concept besides.[24] Furthermore, in his interpretation of *King Richard II*, Kantorowicz does not take into account the anamorphic dynamic introduced in the 'perspective scene' (*King Richard II* 2.2.), which, as Ernest B. Gilman, Allan Shickman, and Christopher Pye have pointed out,[25] offers alternative readings to the play's schematic structure dominated by the creaky turning of a rather medieval wheel of fortune. Bushy's interpretation of the dark foreboding of Queen Isabel has been – amongst other just as probable

Illustration 4: William Scrots, *King Edward VI*, 1546, viewed from the front © National Portrait Gallery, London

sources – referred to William Scrots' 1546 anamorphic portrait of King Edward VI (Illustrations 4 and 5), which in Shakespeare's time was hung, and accessible for visitors, in Whitehall. A German visitor's description dating from 1598, partly quoted in the caption to the present section, insists on the "monstrosity" of the distorted image, before instructing the reader of his *Itinerarium* (first published in Nuremberg, 1612) in how to undo the distortion by looking through a specific hole, which has been reconstructed in the National Portrait Gallery: "A portrait of Edward VI King of England: at first sight a representation of some monstrosity [primo intuitu monstrosum quid repræsentes]; but if it be looked at straight through the hole in the lid or shutter, with which the picture is covered, it is seen in its proper proportions; an ingenious invention of the artist."[26]

Illustration 5: William Scrots, *King Edward VI*, 1546, viewed from the side
© National Portrait Gallery, London

The kind of distortion at work in Shakespeare's 'perspective scene' is, however, much more complex since Bushy's conciliatory explanation includes, or alludes to, fundamental questions of mannerist art theory, such as the origin of pictures from nothingness, the constitution of forms through the casting of shadows, and the ontological status of both shadows and mirror reflections. Of more specific interest in our present context is the rhetorical 'materialisation' of the Queen's sorrows in metaphors of pregnancy and of giving birth, dramatically set against Bushy's attempt to explain away the dark shadows by way of reference to anamorphic art.

[...] Yet again, methinks,
Some unborn sorrow, ripe in Fortune's womb
Is coming towards me, and my inward soul
With nothing trembles. [...]
[...]
'Tis nothing less. Conceit is still derived
From some forefather grief. Mine is not so,
For nothing hath begot my something grief,
Or something hath the nothing that I grieve.
'Tis in reversion that I do possess –
But what it is, that is not yet known what,
I cannot name. 'Tis nameless woe, I wot.
[...]
So, Greene, thou art the midwife to my woe,
And Bolingbroke my sorrow's dismal heir.
Now hath my soul brought forth her prodigy,
And I, a gasping new-delivered mother,
Have woe to woe, sorrow to sorrow joined.
(*King Richard II* 2.2.9-12; 34-40; 62-66)

Since the queen's foreboding concerns the presumed dismal future of her husband as king, the specific identification of the queen's monstrous birth ("prodigy") with Bolingbroke brings into focus the paradox of ascribing a natural birth to the transference of the royal superbody onto Bolingbroke's natural body. As second childless and, as it were, heirless queen to King Richard, the queen's sorrows are of a more tormenting quality – a theme which Shakespeare stages with highly apocalyptic overtones in *Macbeth*. Sorrow and grief themselves stand, however, in sharp contrast to the aloof quality of the 'body politic' which is said to be emotionless, as Plowden points out in some detail: "The King has two Capacities, for he has two Bodies, the one whereof is a Body natural, consisting of natural Members as every other Man has, and in this he is subject to Passions and Death as other Men are; the other is a Body politic, and the Members thereof are his Subjects, and he and his Subjects together compose the Corporation, as Southcote said, and he is incorporated with them, and they with him, and he is the Head, and they are the Members, and he has the sole Government of them; and this Body is not subject to Passions as the other is, nor to Death, for as to this Body the King never dies, and his natural Death is not called in our Law (as Harper said), the Death of the King, but the Demise of the King, not signifying by the Word (Demise) that the Body politic of the King is dead, but that there is a Separation of the two Bodies, and that the Body politic is transferred and conveyed over from the Body natural now dead, or now removed from the Dignity royal, to another Body natural. So that it signifies a Removal of the Body politic of the King of this Realm from one Body natural to another."[27] As is well known, Shakespeare probes the question of how emotionless a king might be, in *King Richard II*. Prior to Richard's dramatic recovery of the emotional apparatus and his increasing awareness of his 'Body natural' throughout his *demise*, Shakespeare highlights the paradox of the corporal incorporality of the 'Body politic' in Queen Isabel's imagery of giving birth to Bolingbroke as to some immaculately conceived heir. When Kantorowicz speaks of the "incarnation" of the "immortal part of kingship"[28] in his interpretation of the passage quoted from Plowden's *Reports* above, the central role attributed to Queen Isabel as a quasi Marian figure ("nothing hath begot my something grief") becomes more vivid. These Christological overtones by no means diminish, but rather intensify the particular monstrosity of the thing conceived and brought forth in place of a 'natural' heir to whom the immaterial superbody would eventually be transferred. The imagery of the perspective scene can thus be said to convey the crucial paradox of the concept of the king's two bodies to the queen's body, which must have been a rather tricky move for Shakespeare writing *King Richard II* in the 1590s.

What Kantorowicz does not really stress in his analysis of *Richard II*, are the two risks at stake for Shakespeare in (firstly) dramatically renewing the memory of King Richard's deposal by way of a forced abdication in a crucial historical situation when the royal succession was not yet settled, and in (secondly) dramatically investigating the "mysteries of state."[29] Shakespeare's precarious demystification of the divine implications of the concept of the king's two bodies in *Richard II*, and his even more radical dissection in *Hamlet*, paradoxically hinge on the most overtly materialist depictions of the decay of the natural bodies of the powerful, a topic which had been of an eminent visual presence in tomb sculpture since the later Middle Ages. The 'double-decker tomb' as analysed by Erwin Panofsky[30] had sharpened the awareness of the split between the two bodies, when the "grisly *transis*" not only uncover the skull and skeleton of the deceased but sometimes grotesquely combine them with animal life by housing worms and toads sculpturally in them. Shakespeare's most prominent allusions to this macabre underside of the otherwise triumphant body of the monarch, are in the grave-digger scene in *Hamlet*, and in King Richard II's reflections on the hollow crown, which contains the uncanny possibility of change from within. The traditional *memento mori*-motif is supplemented by the workings of "villains, vipers damned without redemption! / Dogs easily won to fawn on any man! / Snakes, in my heart-blood warmed, that sting my heart!" (*King Richard II* 3.2.129-131). In stressing the bodily materiality of such lines, I do not want to deny their simultaneous allegorical meaning. Quite to the contrary, the fact of their being closely interconnected points towards the bodily paradox of the royal superbody. As in the case of Queen Isabel's precariously animated foreboding in "Fortune's womb," the animal attack on the king's heart is directed against his mystical superbody, and is yet represented as alien and parasitic life within his natural body. Richard's gloomy talk "of graves, of worms and epitaphs" (*King Richard II* 3.2.145), with the equally affecting imagery of the virtually decaying body,[31] attributes a material quality to the tormenting nothingness of Richard's imaginary role play at Pomfret Castle. Whereas the "being nothing" (*King Richard II* 5.5.41) refers to the part performed within the superbody, and now lost, the graves and worms seem to concern the afterlife of the natural body. And yet, in the corresponding scene in *Hamlet*, the protagonist's little court allegory of "my Lord Such-a-one, that praised my Lord Such-a-one's horse when a meant to beg it" and "my Lady Worm's, chopless, and knocked about the mazard with a sexton's spade. Here's fine revolution and we had the trick to see't" (*Hamlet* 5.1.83-89), might just as well subtly hint at the distorted family situation in the House of Denmark.

At the same time, Shakespeare, in both the History Plays and in *Hamlet*, explores the haunting quality of an effigy which is, as it were, still alive. As more fully and more superbly dramatically developed in *Hamlet* a few years later, the question of whether proper funeral rites were given to his father's predecessor, Shakespeare's *King Henry V* refers to the exhumation and re-interment of King Richard II in 1414, in a passage also discussed by Kantorowicz:[32]

[...] Not today, O Lord,
O not today, think not upon the fault
My father made in compassing the crown.
I Richard's body have interred new,
And on it have bestowed more contrite tears,
Than from it issued forced drops of blood.
(*King Henry V* 4.1.289-294)

Holinshed is quite explicit about King Henry's sense of ritual decorum in giving Richard "all funerall dignitie conuenient for his estate:" "When the king had setled things much to his purpose, he caused the bodie of king Richard to be remooued with all funerall dignitie conuenient for his estate, from Langlie to Westminster, where he was honorablie interred with quéene Anne his first wife, in a solemne toome erected and set vp at the charges of this king. Polychronicon saith, that after the bodie of the dead king was taken vp out of the earth, this new king (happilie tendering the magnifience of a prince, and abhorring obscure buriall) caused the same to be conueied to Westminster in a roiall seat (or chaire of estate) couered all ouer with blacke veluet, & adorned with banners of diuers armes round about. All the horsses likewise (said this author) were apparelled with blacke, and bare sundrie sutes of armes. Manie other solemnities were had at his interrement, according to the qualitie of the age wherein he liued and died."[33] In *Richard II* Shakespeare alludes to the rudimentary funeral rites given to the dead King Richard when the coffin is brought forth with the murdered king ("[Exton:] Great king, within this coffin I present / Thy buried fear. Herein all breathless lies / The mightiest of thy great enemies, / Richard of Bordeaux, by me hither brought" (*Richard II* 5.6.30-33)). Holinshed reports about such an ostentatious display, the body covered with lead, the face left visible immediately after Richard's death ("[...] his bodie was imbalmed, and séered, and couered with lead, all saue the face, to the intent that all men might sée him, and perceiue that he was departed his life [...]"[34]). Whether or not it was known to Shakespeare that historically the plays *Richard II* and *Henry V* centre on a decisive moment in the royal funeral rites, the question of the royal succession was of crucial momentum in the 1590s and it was thus not at all astonishing that Shakespeare gave special emphasis to it. Among the many facets of the figure of Falstaff, whose identity with the Lollard leader Sir John Oldcastle has been stressed in recent studies, is the role play as a royal fake figure. From his reading of Holinshed it seems quite likely that Shakespeare, in putting the divergent material together into the multiple body of Falstaff, tried to incorporate the conspiratorial fake-figures spoken of in the *Chronicles* with reference to myths built around Richard II's living on in hauntingly lifelike (*veristic*) mimicry effigies: "The conspirators being at Brenford, at length perceiued by the lacke of the earle of Rutland, that their enterprise was reuealed to the king, and therevpon determined now openlie with speare and shield to bring that to passe which before they couertly attempted, and so they adorned Maudelen, a man most resembling king Richard, in roiall and princelie vesture, and named him to be king Richard, affirming that by fauour of his kéepers he was escaped out of prison, and so they came forwards in order of warre, to the intent to destroie king Henrie."[35] The figure of Falstaff can thus be said to represent, amongst many other things, a challenge to the legitimacy of Henry IV. The true burial rites given to Richard II and the repudiation of Falstaff towards the end of 2 *King Henry IV* are closely interconnected in terminating those challenges – at least as far as the theatrical staging of the body politic and Shakespeare's intriguing search into the "mysteries of state" are concerned.

Illustrations

For permission to reproduce photographs my thanks are due to the Folger Shakespeare Library, Washington, and the National Portrait Gallery, London.

Illustr. 1 and 2: *Sorcières* (witches' mirrors), i.e. mirrors bulged out. Collection M.-Th. and E. Martin. From

Jurgis Baltrusaitis, *Der Spiegel: Entdeckungen, Täuschungen, Phantasien. [essai sur une légende scientifique. le miroir: rélévations, science-fiction et fallacies.* Paris: Éditions du Seuil, 1978] Trans. Gabriele Ricke and Ronald Voullié. Gießen: Anabas, 1986: Illustr. 20 and 21.

Illustr. 3: Crispin van de Passe I (1565-1637). *Christ Triumphing over Sin, Death and Hell.* Engraved title page on pi2r in Thomas Scott's *Vox Dei.* Presumably Utrecht 1623/24. Folger Shakespeare Library, Washington: STC 22097a copy 1 bd.w. STC 15432. By Permission of the Folger Shakespeare Library.

Illustr. 4: William Scrots. *King Edward VI.* 1546. Oil on panel, anamorphosis, viewed from the front. NPG 1299 (a). © National Portrait Gallery, London.

Illustr. 5: William Scrots. *King Edward VI.* 1546. Oil on panel, anamorphosis, viewed from the side. NPG 1299 (b). © National Portrait Gallery, London.

Notes

[1] Sigmund Freud, *Some Character-Types met with in Psycho-Analytic Work* 318.

[2] Ibid. 323.

[3] Stephen Greenblatt, *Shakespearean Negotiations: The Circulation of Social Energy in Renaissance England* 4 and 8.

[4] Stephen Orgel, *The Illusion of Power: Political Theater in the English Renaissance*; Jonathan Goldberg, *James I and the Politics of Literature: Jonson, Shakespeare, Donne, and their Contemporaries*; Alvin Kernan, *Shakespeare, the King's Playwright: Theater in the Stuart Court, 1603-1613.*

[5] This desire apparently comes to a conclusion in Hobbes' *Leviathan* frontispiece, yet, paradoxically and completely intentionally, by subverting the monarchical origin of such incorporation; cp. Martin Windisch, "Denkbilder der Selbstbehauptung in der Frühen Neuzeit;" Martin Windisch, "'when there is no visible Power to keep them in awe:' Staatstheorie und Bildform bei Thomas Hobbes;" Martin Windisch, "Metapher, Allegorie und Materialität des Körpers als Medien des nationalen Gedächtnisses in der Frühen Neuzeit;" Horst Bredekamp, *Thomas Hobbes Visuelle Strategien: Das Urbild des modernen Staates. Werkillustrationen und Portraits*; and Horst Bredekamp, *Thomas Hobbes Der Leviathan: Das Urbild des modernen Staates und seine Gegenbilder 1651-2001.*

[6] Thomas Hariot, *A briefe and true report of the new found land of Virginia, of the commodities and of the nature and manners of the naturall inhabitants: Discouered bÿ the English Colony there seated by Sir Richard Greinuile Knight in the yeere 1585* 27.

[7] Francis Bacon, "Of Masques and Triumphs" 117.

[8] Ibid. 116.

[9] Ernst H. Kantorowicz, *The King's Two Bodies: A Study in Medieval Political Theology* 26.

[10] Sigmund Freud, *The Interpretation of Dreams* 266.

[11] Freud, *Some Character-Types met with in Psycho-Analytic Work* 320.

[12] Thomas Hobbes, *Leviathan* 135.

[13] Kantorowicz, *The King's Two Bodies: A Study in Medieval Political Theology* 13.

[14] Sigmund Freud, *Briefe 1873-1939* 399-401.

[15] By slightly misquoting *Macbeth* 5.5.16, "The queen, my lord, is dead," Freud reinscribes Seyton's exclamation into the London of 1603: "Wenn in Akt V.5 der Ruf laut wird: The queen is dead, so mag der Londoner jener Zeit gemahnt worden sein, wie wenig lang vorher er die gleiche Nachricht gehört, so daß

ihm die Identifizierung der beiden Königinnen nahe gelegt war" (Freud, *Briefe 1873-1939* 400).

[16] Cp. the references given in note 4.

[17] Freud, *Briefe 1873-1939* 400.

[18] Cp. Nigel Llewellyn, "The Royal Body: Monuments to the Dead, For the Living;" Horst Bredekamp, *Repräsentation und Bildmagie der Renaissance als Formproblem*; Horst Bredekamp and Volker Reinhardt, *Totenkult und Wille zur Macht: Die unruhigen Ruhestätten der Päpste in St. Peter.*

[19] Kantorowicz, *The King's Two Bodies: A Study in Medieval Political Theology* 387. An analogous uncanny show can be found in the procession of the ghosts of Richard's victims in Shakespeare's *King Richard III* 5.3.119-77.

[20] John Milton, *Paradise Lost* 6.862-864.

[21] A historically quite close analogue is Merlin's magic mirror in *The Faerie Queene*, "the glassie globe" (3.2.21.1), "Like to the world itselfe, and seem'd a world of glas" (3.2.19.9).

[22] Geoffrey Bullough, *Narrative and Dramatic Sources of Shakespeare* vol. 7: *Major tragedies: Hamlet, Othello, King Lear, Macbeth* 520-521.

[23] For *Hamlet* compare Windisch, "Denkbilder der Selbstbehauptung in der Frühen Neuzeit" 87-90, and Windisch, "Metapher, Allegorie und Materialität des Körpers als Medien des nationalen Gedächtnisses in der Frühen Neuzeit" 110-113.

[24] For a critical assessment of Kantorowicz's *The King's Two Bodies: A Study in Medieval Political Theology* in the context of the right-wing German intellectualism to which Kantorowicz had affinity before he had to leave Nazi-Germany, and for critical paradoxes in the reception of the book especially during the last three decades, cp. David Norbrook, "The Emperor's new body? *Richard II*, Ernst Kantorowicz, and the politics of Shakespeare criticism," and Wolfgang Ernst and Cornelia Vismann, *Geschichtskörper: Zur Aktualität von Ernst H. Kantorowicz.*

[25] Cp. Ernest B. Gilman, *The Curious Perspective: Literary and Pictorial Wit in the Seventeenth Century* 88-128; Allan Shickman, "'Turning Pictures' in Shakespeare's England;" Allan Shickman, "The 'Perspective Glass' in Shakespeare's *Richard II*;" Christopher Pye, *The Regal Phantasm: Shakespeare and the Politics of Spectacle* 82-105.

[26] Hentzner, *Itinerarium*, Nuremberg, 1612; in the translation of James B. Shaw quoted from James B. Shaw, "The Perspective Picture: A Freak of German Sixteenth-Century Art" 214; 213 for the Latin original.

[27] Plowden, *Reports* 233a, here quoted from Kantorowicz, *The King's Two Bodies: A Study in Medieval Political Theology* 13.

[28] Ibid.

[29] King James I's famous speech of March 21, 1610 is perhaps the most outspoken insistence on the secrecy of "the deepest mysteries of monarchy that belong to the person or state of King and Princes, who are Gods on earth." As is well known, in this speech James replied John Cowell, whom he accused of searching into those *arcana imperii*. Similarly, James speaks of incompetent men who "will freely wade by their writings in the deepest mysteries of monarchy and politick government;" cp. Ernst H. Kantorowicz, "Mysteries of State: An Absolutist Concept and Its Late Medieval Origins," the quotations from James' speech are given on page 383.

[30] Erwin Panofsky, *Tomb Sculpture: Its Changing Aspects from Ancient Egypt to Bernini* 63-64: "The moral significance of these grisly *transis* – often expressed by such inscriptions as "I was what you are; you will be what I am" or "Wretch, why are you proud? You are nothing but ashes and will, like me, be the food of

worms" – is, of course, that of a *memento mori* (a notion eloquently expressed by the term "*representacion de la mort*"), and some of them occur singly rather than in contrastive juxtaposition with a *representacion au vif.* [...] this is also true of one of the earliest recorded instances, the monument of a nobleman named Francis de La Sarra (d. 1362) which can still be seen in the church of La Sarraz some ten miles north of Lausanne. Watched over by members of his family in full dress, he is represented as he was supposed to look after several years in the grave: long worms slither in and out of the body, and the face is covered by toads in such a manner that two of their heads replace the eyes in a macabre anticipation of the effects achieved, more than two hundred years later, by Giuseppe Arcimboldo when he constructed human heads out of sea creatures or vegetables."

[31] "[...] for what can we bequeath / Save our deposed bodies to the ground? [...] And nothing can we call our own but death / And that small model of the barren earth / Which serves as paste and cover to our bones" (*King Richard II* 3.2.145-154).

[32] Kantorowicz, *The King's Two Bodies: A Study in Medieval Political Theology* 26-27.

[33] Raphael Holinshed, *Holinshed's Chronicles of England, Scotland, and Ireland* vol. 3 62.

[34] Holinshed, *Holinshed's Chronicles of England, Scotland, and Ireland* vol. 3 14.

[35] Ibid. 11; cp. ibid. 19, 24, 29, and 43.

Works Cited

Bacon, Francis. "Of Masques and Triumphs. XXXVII," in *The Essayes or Counsels, Civill and Morall.* Ed. with introduction, notes, and commentary by Michael Kiernan. The Oxford Francis Bacon, vol. XV. Oxford: Clarendon Press, 2000. 117-118.

Bredekamp, Horst. *Repräsentation und Bildmagie der Renaissance als Formproblem.* Ed. Heinrich Meier. Carl Friedrich von Siemens Stiftung: Themen Bd. 61, 1995.

---. *Thomas Hobbes Visuelle Strategien. Der Leviathan: Das Urbild des modernen Staates. Werkillustrationen und Portraits.* Berlin: Akademie, 1999.

---. *Thomas Hobbes Der Leviathan: Das Urbild des modernen Staates und seine Gegenbilder · 1651-2001.* Berlin: Akademie, 2003.

Bredekamp, Horst and Volker Reinhardt, eds. *Totenkult und Wille zur Macht: Die unruhigen Ruhestätten der Päpste in St. Peter.* Darmstadt: Wissenschaftliche Buchgesellschaft, 2004.

Bullough, Geoffrey, ed. *Narrative and Dramatic Sources of Shakespeare*, vol. 7: *Major Tragedies: Hamlet, Othello, King Lear, Macbeth.* London: Routledge and Kegan Paul; New York: Columbia UP, 1973.

Ernst, Wolfgang and Cornelia Vismann, eds. *Geschichtskörper: Zur Aktualität von Ernst H. Kantorowicz.* Munich: Fink, 1998.

Freud, Sigmund. *Some Character-Types met with in Psycho-Analytic Work* (1916). *The Standard Edition of the Complete Psychological Works of Sigmund Freud.* Translated from the German under the General Editorship of James Strachey in collaboration with Anna Freud assisted by Alix Strachey and Alan Tyson. Vol. XIV (1914-1916): *On the History of the Psycho-Analytic Movement, Papers on Metypsychology and Other Works.* London: The Hogarth Press, 1957. 309-333.

---. *The Interpretation of Dreams* (1900). *The Standard Edition of the Complete Psychological Works of Sigmund Freud.* Translated from the German under the General Editorship of James Strachey in collaboration with Anna Freud assisted by Alix Strachey and Alan Tyson. Vol. IV (1900): *The*

Interpretation of Dreams (First Part) and Vol. V (1900-1901): *The Interpretation of Dreams (Second Part) and On Dreams.* Reprinted with corrections. London: The Hogarth Press, 1958.

---. *Briefe 1873-1939.* 2nd, enlarged edition, selected and ed. Ernst and Lucie Freud. Frankfurt am Main: S. Fischer, 1968.

Gilman, Ernest B. *The Curious Perspective: Literary and Pictorial Wit in the Seventeenth Century.* New Haven: Yale UP, 1978.

Goldberg, Jonathan. *James I and the Politics of Literature: Jonson, Shakespeare, Donne, and their Contemporaries.* Stanford: Stanford UP, 1989.

Greenblatt, Stephen. *Shakespearean Negotiations: The Circulation of Social Energy in Renaissance England.* Berkeley: U of California P, 1988.

Hariot, Thomas. *A briefe and true report of the new found land of Virginia, of the commodities and of the nature and manners of the naturall inhabitants: Discouered bÿ the English Colony there seated by Sir Richard Greinuile Knight in the yeere 1585.* Frankfurt, 1590. In: *Thomas Hariot's Virginia by Theodore de Bry.* March of America Facsimile Series, Number 15. Ann Arbor: University Microfilms, 1966.

Hobbes, Thomas. *Leviathan.* Ed. Richard Tuck. Cambridge Texts in the History of Political Thought. Cambridge: Cambridge UP, 1991.

Holinshed, Raphael. *Holinshed's Chronicles of England, Scotland, and Ireland.* 6 vols. London, 1808: reprinted New York: AMS, 1965.

Kantorowicz, Ernst H. *The King's Two Bodies: A Study in Medieval Political Theology.* Princeton: Princeton UP, 1957.

---. "Mysteries of State: An Absolutist Concept and Its Late Medieval Origins," in Ernst. H. Kantorowicz, *Selected Studies.* Ed. Michael Cherniavsky and Ralph E. Giesey. Locust Valley: J. J. Augustin, 1965. 381-398.

Kernan, Alvin. *Shakespeare, the King's Playwright: Theater in the Stuart Court, 1603-1613.* New Haven: Yale UP, 1995.

Llewellyn, Nigel. "The Royal Body: Monuments to the Dead, For the Living," in *Renaissance Bodies: The Human Figure in English Culture c. 1540-1660.* Ed. Lucy Gent and Nigel Llewellyn. London: Reaktion Books, 1990. 218-240.

Macbeth. Dir. Roman Polanski. Columbia Pictures, 1971.

Milton, John. *Paradise Lost.* Ed. Alastair Fowler. London: Longman, 1971.

Norbrook, David. "The Emperor's new body? *Richard II*, Ernst Kantorowicz, and the politics of Shakespeare criticism." *Textual Practice* 10.2 (1996): 329-357.

Orgel, Stephen. *The Illusion of Power: Political Theater in the English Renaissance.* Berkeley: U of California P, 1991.

Panofsky, Erwin. *Tomb Sculpture: Its Changing Aspects from Ancient Egypt to Bernini.* Ed. H. W. Janson. London: Thames and Hudson, 1964.

Pye, Christopher. *The Regal Phantasm: Shakespeare and the Politics of Spectacle*, London: Routledge, 1990.

Shakespeare, William. *Hamlet.* Ed. Harold Jenkins. The Arden Shakespeare: Second Series. London: Routledge, 1990.

---. *King Henry V.* Ed. T. W. Craik. The Arden Shakespeare: Third Series. London: Routledge, 1995.

---. *King Richard II*. Ed. Charles R. Forker. The Arden Shakespeare: Third Series. London: Thomson Learning, 2002.

---. *King Richard III*. Ed. Antony Hammond. The Arden Shakespeare: Second Series. London: Routledge, 1990.

---. *Macbeth*. Ed. A. R. Braunmuller. The New Cambridge Shakespeare. Cambridge: Cambridge UP, 1997.

---. *Twelfth Night*. Ed. J. M. Lothian and T. W. Craik. The Arden Shakespeare: Second Series. London: Methuen, 1978.

Shaw, James Byam. "The Perspective Picture: A Freak of German Sixteenth-Century Art." *Apollo* 6 (1927): 208-214.

Shickman, Allan. "'Turning Pictures' in Shakespeare's England." *Art Bulletin* 59 (1977): 67-70.

---. "The 'Perspective Glass' in Shakespeare's *Richard II*." *Studies in English Literature* 18 (1978): 217-228.

Spenser, Edmund. *The Faerie Queene*. Ed. A. C. Hamilton. London: Longman, 1984.

Windisch, Martin. "Denkbilder der Selbstbehauptung in der Frühen Neuzeit." *Mitteilungen 2*. Franfurt am Main: Zentrum zur Erforschung der Frühen Neuzeit, 1994: 55-92.

---. "'when there is no visible Power to keep them in awe:' Staatstheorie und Bildform bei Thomas Hobbes." *Zeitsprünge: Forschungen zur Frühen Neuzeit* 1.1 (1997): 117-165.

---. "Metapher, Allegorie und Materialität des Körpers als Medien des nationalen Gedächtnisses in der Frühen Neuzeit," in *Medien des Gedächtnisses*. Ed. Aleida Assmann, Manfred Weinberg, and Martin Windisch. *Deutsche Vierteljahrsschrift für Literaturwissenschaft und Geistesgeschichte* 72 (1998): Sonderheft. Stuttgart: Metzler, 1998. 90-115.

Cultural Transformations

Of Men and Pigs

Reconciling Epic and Magic in Alexander Pope's Translation of the *Odyssey*

Susanne Rupp

Challenging Homer in the Eighteenth Century

The enlightenment marked – or at least declared – the end of magic and related practices and set out to make them a thing of the past.[1] England saw its last witch trial – resulting in execution – in 1712, and instead of magic, science, epitomised by the achievements of Sir Isaac Newton, loomed large during the first half of the century.[2] The decline of older epistemological paradigms was furthered by representatives of the Augustan Age and their emphasis on rationality, scepticism, and moderation, their insistence on common sense, and their distrust of speculation on supernatural issues.

Despite their strong affiliation with enlightenment values and their empirical outlook, the Augustans nevertheless also looked back into history, insisted on universal and trans-historical values, and modelled their cultural and political practices according to ancient models.[3] Alexander Pope embodied Augustan values to an unprecedented degree. He proclaimed the rule of reason and fashioned himself as a "man of sense."[4] He was a dedicated ancient, affirming the supremacy of classical Greek and Roman literature throughout his life and presenting himself as fighting endless battles against irrationality, stupidity, and "dullness." He was intimately acquainted with the latest developments in natural philosophy and science, though remained sceptical about the moral potential of the disciplines involved and stressed the difference between *scientia* and *sapientia*.[5] Where would a man like Alexander Pope, residing in Twickenham, surrounded by the most polite and learned wits of his age, encounter the lost world of magic? In texts. More precisely: during his work on the translation of Homer's *Iliad* and *Odyssey*.[6]

The modern editor of Pope's Homer-translations, Maynard Mack, describes the Homeric world as particularly remote from the intellectual climate of the early eighteenth century in England.[7] Homer spelled pastness in a world where "Cartesian, Hobbist, Spratian, Lockean forms of discourse insisted on the pastness of poetry as a serious vehicle of truth, and the obsolescence of poetic imagination [...]."[8] Alexander Pope deplored this devaluation of Homer, whom he regarded as a representative of universal poetic truth. For Pope, Homer was the chronicler of a world gone by as well as a maker of an enduring poetic world. But despite Pope's insistence on the universality of Homer, he had to admit that the world depicted in Homer's poetry was not easily to be reconciled with modern manners and values.

In France, the problem of Homer's lack of topicality had been at the centre of the second phase of the *querelle des anciens et des modernes*, which was fought, on one side, by the resolute Anne Dacier, an ardent ancient and author of a prose translation of Homer's epics, and, on the side of the moderns, by Charles Perrault and others.[9] Most of the charges brought against Homer referred to the rather rough manners depicted in his poetry, while the ancients – under the auspices of Madame Dacier – tried to defend Homer by reading him as a Christian moralist.[10] The *querelle* was not an exclusively French affair however; it was discussed throughout Europe, and thus also made its way into England. Here, the debate was conducted less severely and less viciously than in France.[11] Naturally, Pope was acquainted with the arguments on both sides, yet was relieved to find that the English were more relaxed about it all and generally remained loyal to Homer: "[...] as much as we abound in other miserable misguided Sects, we have at least none of the blasphemers of Homer. We steadfastly and unanimously believe, both his Poem, and our Constitution, to be the best that ever human wit invented."[12] Pope ultimately chose to follow his own *via media*.[13] He was undoubtedly an ancient, had strong sympathies for Madame Dacier and her cause, agreed with her in "preferring the simplicity of the ancient world to the luxury of ours,"[14] and held with the universal values expressed in Homer's poetry. Yet Pope was an ancient with a twist. Towards the end of his Homeric project, and disaffected by some misunderstanding between him and the learned woman, Pope made clear where he departed from Dacier's position: "She had said, the Manners of those times were so much the better the less they were like ours: I thought this required a little qualification, I confest that in my own opinion the world was mended on some points, such as the custom of putting whole nations to the sword, condemning Kings and their families to perpetual slavery, and a few others."[15] Pope argued – where necessary – for some notion of historical progress and thus for a sense of historical distance, rather than for blanket praise of Homer's world and culture. This distance had to be bridged by historical work and research into ancient or contemporary oriental cultures, all of which would then serve to account for the differences between contemporary manners and values and those encountered in Homer's poetry.

We must not forget, however, that Pope was a poet, not a philologist. Despite paying tribute to philology as an emerging academic discipline (a project initiated by the moderns) by compiling vast notes and surrounding his translation with scholarly paraphernalia and a decidedly learned air, Pope's project remained a backward- rather than forward-looking affair.[16] Up until the eighteenth century, Homeric scholarship had been dominated by allegorism, which was replaced by modern, historical, and philological Homeric scholarship only towards the end of the century.[17] Pope was situated between these two approaches and his translation marked, in the words of Maynard Mack, "the end of an old Homer and F. A. Wolf's *Prolegomena ad Homerum* (1795) the beginning of a new."[18] Wolf's *Prolegomena* heralded the emerging "desacramentalization of Homer," i.e. the academicisation of Homer at the expense of the amateurs (including poet-translators).[19] But even though the historicising of Homer has often been seen as a particular modern achievement, the ancients, as we can now see, also contributed to this development. The need to defend Homer prompted them to look for historical explanations and stress his historical alterity.[20] The ancients were thus preparing the ground for historicist work on Homer, even though their rather arbitrary and undifferentiated approach to source material and their insistence on universal values disqualifies them as historians in a modern sense.

How does the discourse of magic relate to the debate sketched above? Did the literary representation of magic in Homer's *Odyssey* pose a problem for Pope? If so, how did he overcome it? Could it be reconciled with his and his readers' Augustan values? How did he appropriate Homer? I shall argue that Pope was well aware of the improbable and – for most of his contemporaries – incongruous nature of the magic episodes in the *Odyssey.* For Pope, however, the main challenge did not so much result from the clash of cultures – the pre-modern and the modern age – and their competing notions of probability, but from the generic dignity of the epic poem, which he saw threatened by magic and marvellous episodes. Enchantment, magic, and marvels were not only features of epic poetry, but also characteristic of romance. Pope wanted to keep these genres separate at all cost, since he considered romance as irreconcilable with epic values.[21]

A description of Homer as pre-modern did not rock the foundations of Pope's beliefs. But the suspicion that Homer had been writing romance was an intolerable notion strongly to be refuted. In order to establish Homer's work as distinct from romance, Pope had to make clear that the marvellous was part of the epic tradition and functioned as a didactic tool in the epic project of specifying men's place in time and space. The following pages will be dedicated to the exploration of Pope's hermeneutic strategies. He chose a number of different approaches to reconcile Homer's marvellous episodes with modern notions of epic dignity and probability: he allegorised them, analysed their narrative strategies, and looked for historical evidence, thus saving these episodes from being 'mere' fiction or romance.

Defending the Marvellous in Epic Poetry

One way of accounting for the presence of magic and the marvellous in Homer's poetry was to call upon ancient authorities and their acknowledged insights into the matter. Since the early days of Homer criticism, critics had been concerned about the irrationality of much of Homer's poetry. Especially the treatment and representation of the Gods posed serious problems, since their activities were often more marvellous than plausible. The most influential discussion of the issue of probability in Homer's poetry was Aristotle's, whose arguments were taken up, repeated and expanded upon by later critics.

Throughout the *Poetics*, Aristotle offers insights into the nature of epic poetry, even though his main focus is on tragedy and he does not provide a fully developed theory and definition of the epic.[22] Epic poetry, according to Aristotle, works much along the same lines as tragedy: its aim is mimesis, the imitation of human action, and it is intended to delight ("surprise") the reader or listener and thus give pleasure.[23] One of the most important differences between epic and tragedy, however, lies in the fact that epic poetry is better suited for the narrative representation of marvels and the irrational, since "one is not looking at the person doing the action."[24] Hence, the epic poet – unlike the historian, who has to remain faithful to historical facts, and also unlike the dramatist, who depends on performance conditions – enjoys a certain creative freedom as a "maker of plots."[25] The only rule he has to follow in his imitative work is: Thou shalt be convincing. In order to achieve this, Aristotle recommends not to build a plot on irrational episodes and to "prefer likely impossibilities to unconvincing possibilities."[26] Ideally, poetry should represent 1) what was or is, or 2) what is commonly said and thought to be the case, or 3) what should be the case.[27] Thus, in the case of Homer – who was frequently accused of being irrational – one could argue that a) Homer depicts reality as it should be, or b) that he simply records what is commonly said, or c) that he is not irrational at all, since "things happen contrary to likelihood."[28] Aristotle, who was an ardent defender of Homer's art, thus managed to justify the marvellous in Homer's works and declare it an integral – even constitutive – element of epic poetry.

Later critics like Joseph Trapp and René le Bossu followed Aristotle's line of argument with regard to the vexed issue of the marvellous in Homer's poetry, albeit with slight shifts in emphasis. In Pope's time, Joseph Trapp, professor of poetry at Oxford from 1708-1718, rehearsed in his lectures on poetry Aristotle's main arguments concerning the necessity of the "wonderful" in epic poetry (as well as in tragedy) for achieving surprise in the reader. He argues that the "wonderful" does not by necessity have to be improbable or impossible: "The Moderns seem to mistake that Part of Epic and Tragedy which contains [...] the wonderful, confounding the wonderful with the improbable, and using those two Words promiscuously. [...] The Divine Power, and the Agency of the Gods, make all this agreeable to Reason. [...] In such Cases as these, whatever is possible, is probable: If you determine otherwise, Poetry is depriv'd of one of its best Ornaments, its greatest Fund of Surprize. And the same Observation extends to some Parts of the wonderful, which are not accounted for by a divine Power [...]."[29] Trapp here makes an important distinction between the "wonderful" as caused by "Divine Power, and the Agency of the Gods," and the "wonderful" in the human realm. Indeed, these two types of the "wonderful" differ in terms of the hermeneutic demands they make on the reader: if one accepts the notion of acting Gods ("machinery"), their marvellous deeds narrated in epic poetry seem highly probable because of their supernatural capacities.[30] With regards to human characters, however, the case is more intricate. Their marvellous deeds should be read as "hyperbolical," necessitated by the "boldness" of the "Nature of the Thing describ'd."[31]

René le Bossu, in his influential *Traité du poëme épique* (1675), also discusses the function and different types of divine intervention in Homer's poetry.[32] Generally speaking, Bossu follows Aristotle by claiming that the divine actions represented in the poem must be probable. Going one step further, however, he also states that these incidents "should be so disengag'd from the Action, that one might substract them from it without destroying the Action."[33] Thus, the Gods are reduced to an incidental status within the narrative, do not perform an essential function, and remain on the surface-level of the narrative. Over the following pages, it will become apparent that Alexander Pope – in his considerations of the matter – knew his Aristotle well, applied the same distinctions as Trapp, and followed Bossu to the letter.[34]

Magic in Book X of the *Odyssey*

A related problem can be encountered in a further sub-category of the marvellous (which is often – but not necessarily – related to the divine realm): magic and enchantment. Whereas the *Iliad* is not without its implausibilities (one need only think of Achilles' speaking horse[35]), the *Odyssey* abounds in marvellous tales and characters, many of them directly associated with magic (e.g. Proteus, the ships of Alcinous, Circe, the Cyclopes, Scylla and Charybdis, acts of necromancy,[36] etc.). Despite the fact that magic could easily be subsumed under the rules applying to the marvellous in epic poetry, it did not enjoy a particularly high status in epic theory, and secondary epics in particular were often criticised for exaggerating. In this vein, the afore-mentioned Joseph Trapp complained about Tarquato Tasso: "[...] he is too full of Magic, Enchantment, Machinery, and aerial Personages. Of the same Fault our Countryman Spenser is still more remarkably guilty, who treads almost perpetually upon enchanted Ground, and the greatest Part of whose Characters are Fairies, Ghosts, Magicians and Giants."[37] Nevertheless, Trapp admits that the latter is a wonderful poet and eminently worth reading.

The following sections will focus on the Circe-episode in Book X of the *Odyssey*, an episode which was translated and commented on by Pope himself. This episode, featuring the sorceress Circe, became one of the most popular stories interpreted by allegorists. Howard Clarke, in *Homer's Readers*, accounts for the popularity of the episode as follows: "The allegories derived from Odysseus's stay with Circe are much more extensive, as readers might expect from an incident that has all the ingredients: a god and a hero, a temptress and her victims, sex, magic, and metamorphosis"[38] – and, one might add, drugs.

After having landed on the island "Where *Circe* dwelt" (*TE* 9, 347), Odysseus' companion Eurylochus is appointed to explore the island accompanied by twenty-two men, while Odysseus and the rest of the men remain waiting on the shore.

The explorers come to the palace of Circe where they are greeted by ominously friendly wolves and lions, wagging their tails and licking the sailors' feet. Circe's "voice celestial" (*TE* 9, 355) resounds from within the palace and the men ask to be allowed in. Circe welcomes them and they – with the exception of Eurylochus, who remains sceptical – follow her into the castle. The men are entertained with delicious food, but their hostess has added some little extra: "[...] venom'd was the bread, and mix'd the bowl, / With drugs of force to darken all the soul" (*TE* 9, 355-356). A rather wild feast ensues, when: "Instant her circling wand the Goddess waves, / To hogs transforms 'em, and the Sty receives" (*TE* 9, 356). Eurylochus returns to Odysseus and informs him of what has happened, upon which Odysseus takes charge of the situation. With the help of Mercury, Odysseus overcomes Circe, the pigs/men are transformed back into their original shape, and Odysseus and Circe spend a year of love. At the end of the year, Odysseus leaves the island to journey on to the underworld.

Allegorising Homer

This episode lends itself well to an allegorising reading, which tends to stress the moral depravity of the heroine and neglects to examine her more helpful and supportive side. Circe is often presented as a whore, whose temptations Odysseus at first succumbs to, but finally and triumphantly overcomes.[39] Alexander Pope – despite repeatedly claiming to wish to distance himself from allegorism[40] – follows this line of interpretation (though he is less severe on Circe than other critics) and concludes: "There is a beautiful moral couch'd under this fable or allegory: *Homer* intended to teach, as *Eustathius* remarks, that pleasure and sensuality debase men into beasts." However, despite this clear moral of the episode, Pope is deeply concerned about the intensity of Odysseus' relations with Circe: "It is evident, that *Ulysses* had a very intimate commerce with *Circe*, for *Hesiod* writes [...] that he had two sons by her [...]." Even though an allegorical reading need not pay much attention to these intimacies (the degree of intimacy would not affect the "beautiful moral" as identified by Pope), Pope wants to avoid modern readers being alarmed by the immorality of the liaison and thus sets out to provide some historical background on ancient notions of "conjugal fidelity." The results of his historical research justify Odysseus' action as acceptable: "[...] the notions of Morality are now very different from what they were in former ages: Adultery alone was esteem'd criminal, and punish'd with death by ancient Heathens: Concubinage was not only permitted, but thought to be honourable [...]."[41] Thus, Odysseus' concubinage with Circe is historically legitimised and Odysseus still remains a hero, "a pattern of Wisdom, and imitation" (*TE* 9, 354). Whereas Odysseus' companions are "slaves" of pleasure, Odysseus himself enjoys pleasure (which is personified by Circe) "upon his own terms" (*TE* 9, 363). Pope concludes: "The moral of the fable is, that all pleasure is not unlawful, but the excess of it [...]" (ibid.).

Whereas Pope here uses both allegorical and historical interpretative strategies in order to maintain the hero's moral integrity, he restricts himself to a solely allegorising reading of Mercury's encounter with Odysseus.[42] Before Odysseus sets out to challenge Circe, Mercury passes a flower – Moly – to Odysseus, which is meant to protect him. Rather than launching into an erudite excursus on ancient botany, Pope limits his interpretation to allegory and explains that Mercury stands for "Reason," and the plant – which does "grow only in Poetical ground" (*TE* 9, 361-362) – is "instruction."

Allegorical readings of Homer had a long tradition and provided Homer's admirers with the possibility to defend their author and to reconcile his texts with Christian ethics or modern manners. With regard to the problem of probability in epic, allegorical readings could gloss over the more improbable passages in the text, but they could not justify them. The fantastic and marvellous remains visible and – for those unconvinced by allegory – ridiculous and anachronistic. Pope was concerned about this point, but not because of the improbable nature of the stories and their epistemological basis, which contradicted enlightenment values of rationality. He rather worried about the literary and aesthetic implications of representations of magic and the marvellous: the more marvellous stories are, the more they seem to belong to the realm of romance, a genre which, for Pope, stood in clear contrast to the values epitomised by epic poetry. Whereas romance is fabulous, fantastic, escapist, and not grounded in reality (and is thus non-mimetic), epic poetry aspires to true mimesis and follows didactic aims of the highest order.

Apart from allegorising the Circe episode, Pope made various efforts and did employ different hermeneutic strategies to save the episode from ridicule and to invest it with epic dignity. Firstly, he attempted to defend Homer's qualities as a poet against the criticisms voiced by Zoilus, 'Longinus,' and their respective followers. Secondly, he drew attention to the narrative framing of the marvellous episode; finally, he historicised the story by insisting on its *sensus litteralis* and by providing as much historical information as possible so as to ground the narrative in the realm of the plausible and make it more convincing.

Taking Homer Seriously

Transforming men into pigs is – even by the standards of divine and semi-divine "machinery" – a rather unusual accomplishment. Among the more rationally inclined critics, this passage had enjoyed a notorious reputation since the early days of literary criticism. Zoilus[43] was the first to comment on it, and his criticism was taken up by later commentators such as Pseudo-Longinus in his treatise *On Sublimity*.[44] Here, 'Longinus' declared the *Odyssey* to be the work of an old man: "[W]hen a great mind begins to decline, a love of story-telling characterises its old age."[45] According to 'Longinus,' the *Odyssey* is characterized by an addiction to the fable-lore, its author being "lost in the realm of the fabulous and incredible."[46]

However, the stories told are not without their sublime moments, but even in these episodes "the mythical element [...] predominates over the realistic." 'Longinus' concludes: "I digressed into this topic, as I said, to illustrate how easy it is for great genius to be perverted in decline into nonsense. I mean things like the story of the wineskin, the tale of the men kept as pigs in Circe's palace ('howling piglets,' Zoilus called them), the feeding of Zeus by the doves (as though he were a chick in the nest), the ten days on the raft without food, and the improbabilitie of the murder of the suitors. What can we say of all this but that it really is 'the dreaming of a Zeus'?"[47] For 'Longinus,' the true strength of the *Odyssey* resides in its concise depiction of manners. Despite the "decline of emotional power," its qualities as a "comedy of manners" are undisputable.[48]

Pope was a defender of Homer as well as an admirer of 'Longinus;' in this case, however, his affinity to Homer proved stronger. In his "Postscript" to the *Odyssey,* he takes 'Longinus' to task for his condescending treatment of Homer. Pope admits that the *Iliad* is more sublime than the *Odyssey*, but he argues that the *Odyssey* is, nevertheless, effective as poetry, since its fictions are well invented. In order to support his position, he evokes Aristotle, who did not make any distinction between the poems with regard to their respective literary merit. Instead, he "draws the rules and examples of Epic writing equally from both [...]" (*TE* 10, 384). Pope maintains that there is no difference in kind between the two epics, only one in degree, and that they are of equal literary quality: "And it is certain we shall find in each, the same vivacity and fecundity of invention, the same life and strength of imaging and colouring, the particular descriptions as highly painted, the figures as bold, the metaphors as animated, and the numbers as harmonious and as various." (TE 10, 385) Having thus defended the *Odyssey* and postulated its literary quality, Pope can continue to take the work seriously rather than dismiss it – as 'Longinus' did – as "nonsense."[49] Where 'Longinus' could only shake his head about the "*squeaking pigs of Homer*" (*TE* 9, 356), Pope treats them with respect and considers them worth one of his learned notes. But this note is fairly brief and less verbose than expected. Instead of commenting on the magic act, Pope discourses on the fate of Homer's first critic Zoilus, who, despite being a learned man who should have known better, criticised Homer in order "to make his Readers laugh." He had a talent for the "Burlesque," and since he was unable to refrain from writing in this mode, "he drew upon himself the indignation of all the learned world," which finally culminated in his execution (according to some authorities by stoning, according to others by crucifixion). Pope reads this "tragical instance" as an example "of the great value which was set upon his [Homer's] Poetry by antiquity, and of the danger of attacking a celebrated Author with malice and envy" (ibid.). Zoilus' mockery of Homer thus provides a *Lehrstück* for proper literary criticism as well as for the immense dignity and authority of Homer's works. The narration of Zoilus' grim fate should prevent the modern reader from being too amused by men being transformed into pigs.

Preparing the Augustan reader for a number of problematic passages yet to come, Pope explains the appropriate response to the more embarrassing episodes in Homer's *Odyssey* in his introduction to Book X. The lengthy introductory note to Book X amounts to a regular poetics of the "marvellous" and its defence (*TE* 9, 335-337). Pope sets out to argue Homer's case by stating that "Poetry is a mixture of History and Fable" (*TE* 9, 335). This allows Pope to reconcile conflicting tendencies in Homer's poems, such as the representation of historical and cultural facts alongside more fantastic plot-elements. But even though history and fable are – to a certain extent – reconciled in poetry, their relationship is constructed as hierarchical, because "the foundation is historical" (*TE* 9, 335).[50]

However, poetry cannot rely on facts alone, since – and here Pope echoes Aristotle – these "would not be sufficiently surprising; for the Marvellous ought to take place, especially in Epic Poetry" (*TE* 9, 335). Indeed, if the marvellous is a feature and requirement of epic poetry, the *Odyssey* easily qualifies as epic. But the question as to whether Homer is exaggerating remains: "But it may be ask'd, does not Homer offend against all degrees of probability in these Episodes of the Sirens, Scylla and Charybdis, Cyclops and Antiphates? How are these incredible stories to be reduc'd into the bounds of probability? 'tis true, the Marvellous ought to be used in Epic Poetry; but ought it to transgress all power of belief?" (*TE* 9, 335). In order to defend Homer, Pope draws on Aristotle's above-quoted line of argument, which Pope paraphrases as follows: "*A Poet*, says that Author, *ought to prefer things impossible, provided they are probable, before things possible, that are nevertheless incredible*" (*TE* 9, 335). This rather cryptic dictum occasioned numerous interpretations, and Pope decided to subscribe to the reading advanced by André Dacier – Anne Dacier's husband –, who summarised Aristotle's rule as follows: "A thing may be impossible, and yet probable" (*TE* 9, 335). In the depiction of the Gods, for example, the reader witnesses humanly impossible actions. But since they are performed by Gods, they are probable and thus convincing.[51]

But what about marvellous actions involving human protagonists as in the episode of Circe? Here, the case is more intricate: "[...] the Episodes of *Circe, Polyphemus, the Sirens, &c.* are necessary to the action of the *Odyssey*: But will any man say they are within the bounds of human probability? How then shall we save this difficulty?" (*TE* 9, 336). In order to tackle this problem, Pope draws on Bossu, who ingeniously suggested that Homer rendered these episodes plausible by using a particular narrative framing device: they are narrated as stories within the story, and thus their implied audience and the function of the narrative itself have to be taken into account as well. On these improbable actions, Bossu writes: "Yet *Homer* has artificially reduc'd them to human Probability, by the Simplicity and Ignorance of the *Phæacians*, before whom he causes those recitals to be made" (*TE* 9, 24). Pope takes up this point with reference to the marvellous episodes narrated in Book X and argues that the modern reader has to bear in mind that Homer "makes Ulysses relate them [the marvellous stories] before a credulous and ignorant assembly; he lets us into the character of the Phæacians, by saying they were a very dull nation, in the sixth book, Where never Science rear'd her laurel'd head. It is thus the Poet gives probability to his fables, by reciting them to a people who believ'd them, and who thro' laziness of life were fond of romantic stories. He adapts himself to his audience, and yet even here is not unmindful of his more intelligent Readers; he gives them (observes Bossu) in these fables all the pleasure that can be reap'd from physical or moral truths, disguis'd under miraculous Allegories, and by this method reconciles them to poetical probability" (*TE* 9, 336). Indeed, the adventures of Odysseus are not narrated by a third-person-narrator, but by Odysseus himself, who is stranded on the island of the Phæacians and encouraged by them to relate his adventures (Book IX ff.). The Phæacians are, according to Pope, "a most effeminate nation" (*TE* 9, 173), and he expends some considerable time on an analysis of their mentality. They are "simple" and "credulous" (*TE* 9, 203-4), "wholly given up to luxury and pleasures" (*TE* 9, 221) and – here, again, Pope follows Bossu – "they had all the qualities of a lazy people, who admire nothing so much as romantic adventures" (*TE* 9, 204).[52] The reader is supposed to be aware of this particular narrative frame and will analyse the stories accordingly. Only careful contextualising, however, allows the reader to realise the true intention of Odysseus' narration of his adventures within Homer's text: "[T]he truth intended to be taught by way of moral is, that a soft and effeminate life breaks the spirit, and renders it incapable of manly sentiments or actions" (ibid.). Read in this light, even the most marvellous and improbable story contains some edifying truth and moral lesson.

The Phæacians are the opposite to Pope's ideal, educated 18th-century readers who know how to contextualise and historicise. Whereas the Phæacians were fond of "romantic stories" (*TE* 9, 336) and "romantic adventures" (*TE* 9, 204) and thus not mature enough for true epic poetry, later readers have overcome this state of unrefinement and know how to differentiate between history and fiction.

Historicising Homer

While Pope's defence of the marvels invented for the Phæacians' benefit foregrounds and accepts the implausibility of these stories, he also tries – at the same time – to 'factualise' some of these stories and give them some basis in a (constructed) historical reality. These efforts are motivated by his assumption that poetry is fundamentally rooted in history and does not exist outside of a historical frame of reference. In his attempts to verify the historical truth and adequacy of some passages, Pope goes further than one might expect. When discussing Circe's use of "magical drugs" (*TE* 9, 355), for example, he raises the question as to whether enchanters existed in the first place. Strictly speaking, there would have been no need to bring up this question, since he has already explained that the Phæacians demanded such romantic stories and Odysseus, aiming to please, entertained them accordingly. But Pope does not seem to have been satisfied with this explanation alone; instead, he chooses rigorously to follow the logic of his argument (poetry is historical as well as fictional) and raises the issue in the accompanying note. There he argues: "I am not in the number of those who believe that there were never any Magicians who perform'd things of an uncommon nature: The story of Jannes and Jambres, of the Witch of Endor, and Simon Magus, are undeniable instances of the contrary. Magic is suppos'd to have been first practis'd in Ægypt, and to have spread afterwards among the Chaldaeans: It is very evident that Homer had been in Ægypt, where he might hear an account of the wonders perform'd by it. Dacier is of opinion, that these deluders, or Magicians, were mimics of the real miracles of Moses, and that they are describ'd with a wand, in imitation of that great Prophet" (*TE* 9, 355). This is a rather surprising statement from a man declaring and fashioning himself as a man of sense and reason. There are two ways of reading this passage: we can either take it at face value and acknowledge 18th-century indebtedness to earlier world views, or we can read it as a strategic attempt on Pope's part to strengthen the factual side of Homer's narrative (even if this might imply denying his own convictions). By emphasising the "undeniable instances" of magical acts, the evidence ("very evident") of them in Homer's time in Egypt (a commonplace in Homer studies since antiquity), and his own familiarity with magical practises, Pope foregrounds Homer's achievements as a historian and imitator of reality.

Pope also deploys this method of historicising, or 'factualising,' with regard to other passages in the *Odyssey* where it would often have sufficed to remind the reader of the Phæacian's appetite for "romantic stories" (*TE* 9, 336). Consider, for example, his note on the passage detailing the notorious brutality of Polyphemus. In this note, Pope remarks that "possibly the relation of the barbarity of Polyphemus, and Antiphates, with respect to their eating the flesh of men, may not be entirely fabulous: Modern history assures us, that savages have been found in parts of the world lately discover'd, who eat the bodies of their enemies: It is therefore no wonder that the more polite and civiliz'd nations of Antiquity, look'd upon such men as monsters, and that their Poets painted them as such, or perhaps aggravated the fierté, or fierceness of their features, struck with horrour at their brutal inhumanity" (*TE* 9, 346). Here, the concept of Odysseus as narrator is abandoned and Homer returns as the poet of one of the "more polite and civiliz'd nations of Antiquity," in the business of depicting and rendering plausible historical facts (facts, moreover, that have been confirmed as such by "modern history").

The Cyclopes, whom Pope could have sufficiently explained and rendered plausible by referring to Odysseus' catering to Phæacian taste, similarly receive the historicising treatment. Traditionally, the Cyclopes had been allegorized (for example by Eustathius), but Pope preferred to read them as literally as possible. He reckoned that the Cyclopes were wearing a special kind of helmet that made them appear one-eyed, which would have naturally terrified "a mariner who pass'd these coasts at a distance, observing the resemblance of a broad eye in the forehead of one of these *Cyclops*, might relate it accordingly, and impose it as a truth upon the credulity of the ignorant" (*TE* 9, 309).[53] Here, as well as in other passages, even the most fantastic and marvellous stories thus appear to have some basis in historical truth.[54] Such insistence on historical truth as the basis of Homer's poetry is primarily motivated by Pope's desire to draw a clear and unequivocal distinction between the genres of epic and romance. At the beginning of Book VII, he writes: "It is to be wish'd, that a faithful History of the Trojan war, and the voyages of Ulysses, had been transmitted to posterity; it would have been the best comment upon the Iliad and Odyssey. We are not to look upon the Poems of Homer as meer romances, but as true stories, heighten'd and beautify'd by Poetry: [...] I doubt not but we lose many beauties in Homer for want of a real history, and think him extravagant, when he only complies with the opinions of former ages. I thought it necessary to make this observation, as a general vindication of Homer; especially in this place, immediately before he enters upon the relation of those stories which have been thought most to outrage credibility: if then we look upon the Odyssey as all fiction, we consider it unworthily; it ought to be read as a story founded upon truth, but adorn'd with the embellishments of Poetry, to convey instruction with pleasure the more effectually" (*TE* 9, 232-233). Pope constructs a binary opposition between "meer romance" on the one hand and "true stories" on the other. But Homer properly belongs to neither. He certainly is not an author of romances, because of his indebtedness to historical truth. But neither is he a historian proper, since his stories are "heighten'd and beautify'd by Poetry" and – as Pope suggests – often more informed by what was thought to be true, than by what actually was true ("want of real history"). A true historical account of the Trojan war and of the voyages of Odysseus would thus make it easier for contemporary readers to discern where Homer acts as historian and where he acts as poet in the business of embellishing history. While these embellishments often border on the marvellous, the marvellous is nevertheless not dismissed as mere romance; instead, it is – with clear reference to Aristotle – recognised as an integral element of epic poetry. Whereas the marvellous in romance induces wonder for wonder's sake, its function in epic poetry is didactic, aimed at the Horatian *aut prodesse aut delectare*.

Despite his commitment to the factual and historical side of Homer's poetry, Pope did not forget that he was dealing first and foremost with a poet. His poetics, as derived from Homer, Horace, and Longinus, point to yet another defence of the marvellous in Homer's poetry: The "*Marvellous Fictions* [...] surpassing all the Bounds of Probability" are, after all, the work of an inspired poet, somebody under the spell of the *furor poeticus*, "*snatch*[ing] a *Grace* beyond the Reach of Art."[55]

"Rapture," "fire," and related images play a prominent role in the description of the creative process, and, to a certain extent at least, they seem to justify the transgression of "due Proportion"[56]: "Perhaps it may be with great and superior Souls as with gigantick Bodies, which exerting themselves with unusual Strength, exceed what is commonly thought the due Proportion of Parts, to become Miracles in the whole; and like the old Heroes of that Make, commit something near Extravagance amidst a Series of glorious and inimitable Performances. Thus Homer has his speaking Horses [...]"[57] But not only the poet himself appears as if spellbound; the poetically perceptive reader is equally overwhelmed by the achievements of the poetic imagination: "It is to the Strength of this amazing Invention we are to attribute that unequal'd Fire and Rapture, which is so forcible in *Homer*, that no Man of a true Poetical Spirit is Master of himself while he reads him."[58] Caught by the "poetical fire," the reader surrenders and "that Warmth of Fancy [...] holds the Heart of a Reader under the strongest Enchantment."[59]

Notes

[1] The final volume of the comprehensive *History of Magic and Experimental Science*, ed. Lynn Thorndike, is dedicated to the seventeenth century.

[2] See James Sambrook, *The Eighteenth Century. The Intellectual and Cultural Context of English Literature, 1700-1789* and *The Cambridge History of Science*, Vol. 4, *Eighteenth-Century Science*, ed. Roy Porter.

[3] See for example Philip *Ayres, Classical Culture and the Idea of Rome in Eighteenth-Century England* and Howard Erskine-Hill, *The Augustan Idea in English Literature.*

[4] See David B. Morris, *Alexander Pope: The Genius of Sense.*

[5] See Marjorie Nicolson, George S. Rousseau, *"This Long Disease, My Life": Alexander Pope and the Sciences.* On Pope's notion of *sapientia see Essay on Man*, especially Epistle I, section IV (*TE* 3, 113-30) and Epistle II, section I (*TE* 3, 53-62). (References to Pope's poems are from *The Twickenham Edition of the Poems of Alexander Pope*, abbreviated as *TE*. All references are to page numbers in this edition.)

[6] Both translations were products of collaborative efforts. William Broome (1689-1745) and Elijah Fenton (1683-1730) contributed to the translation of the *Odyssey*. Broome was responsible for books II, VI, VIII, XI, XII, XVI, XVIII, XXIII, Fenton for books I, IV, XIX, XX, Pope translated books III, V, VII, IX, X, XIII, XIV, XV, XVII, XXI, XXII, XXIV. On the "shabby business" accompanying the collaboration, see Maynard Mack, *Alexander Pope: A Life* 412-417.

[7] Maynard Mack, "Introduction," *TE* 7, xlix.

[8] Ibid.

[9] Nicolas Boileau-Despréaux, Jean de La Fontaine and Jean de La Bruyère were on Dacier's side, whereas Bernard Le Bovier de Fontanelle and Jean Desmarets de Saint-Sorlin supported Perrault. On the *querelle* and its English correlative, the "battle of the books", see Joseph Levine, *The Battle of the Books: History and Literature in the Augustan Age* and Donald M. Foerster, *Homer in English Criticism: The Historical Approach in the Eighteenth Century* 24.

[10] See Howard Clarke, *Homer's Readers: A Historical Introduction to the "Iliad" and the "Odyssey".*

[11] See Mack, "Introduction," *TE* 7, lxxviii.

[12] Alexander Pope, "Postscript. By Mr. POPE," *TE* 10, 397.

[13] The debate touched Pope only indirectly (see Foerster, *Homer in English Criticism*, chapters I and II).

[14] Pope, "Postscript," *TE* 11, 394.

[15] Pope, ibid. Pope also discussed Dacier's statement in his "Preface" to the *Iliad* (*TE* 7, 14).

[16] Pope's collaborators read and prepared the relevant sources for him. They had the university background which he – as a Catholic – lacked. On the sources see: Hans-Joachim Zimmermann, *Alexander Popes Noten zu Homer: Eine Manuskript- und Quellenstudie.*

[17] See Foerster, *Homer in English Criticism: The Historical Approach in the Eighteenth Century* 24.

[18] Mack, "Introduction," *TE* 7, lxxi. See also Clarke, *Homer's Readers: A Historical Introduction to the "Iliad" and the "Odyssey"* 97.

[19] Clarke, *Homer's Readers: A Historical Introduction to the "Iliad" and the "Odyssey"* 105. In England the rise of 'new' Homer-studies was heralded by: Richard Bentley, *Remarks Upon a Late Discourse of Free-Thinking: In a Letter to F. H. D. D.*, Thomas Blackwell, *An Enquiry into the Life and Writings of Homer*, and Robert Wood, *An Essay on the Original Genius and Writings of Homer: With a Comparative View of the Ancient and Present State of the Troade. Illustrated with a Map of Troy.*

[20] See the seminal article of Hans Robert Jauß on the issue of the ancients' contribution to the rise of historicism: "Ästhetische Normen und Geschichtliche Reflexion in der *Querelle des Anciens et des Modernes.*" On the issue of history see also Joseph M. Levine, "Ancients and Moderns Reconsidered" 72-89.

[21] 'Romance' was a rather diffuse category in the eighteenth century. In this essay it is presumed that Pope shared William Congreve's assumptions about this genre, which the latter had expressed in the preface to his novel *Incognita* (1692): "Romances are generally composed of the constant loves and invincible courages of heroes, heroines, kings and queens, mortals of the first rank, and so forth; where lofty language, miraculous contingencies and impossible performances elevate and surprise the reader into a giddy delight, which leaves him flat upon the ground whenever he gives off, and vexes him to think how he has suffered himself to be pleased and transported, concerned and afflicted, at the several passages he has read, viz., these knights' success to their damsels' misfortunes and such like, when he is forced to be very well convinced that 'tis all a lie" (Paul Salzman, *An Anthology of Seventeenth-Century Fiction* 474). Pope admired Congreve and dedicated his translation of the *Iliad* to him.

[22] Later Renaissance and neoclassical critics had to construct such a theory of the epic from the works of Aristotle, Horace, Plato, Strabo, Plutarch, Cicero, Quintilian as well as the poems of Homer and Vergil. On this process see Clarke, *Homer's Readers: A Historical Introduction to the "Iliad" and the "Odyssey"* 107.

[23] Aristotle, *Poetics* 51-90. Aristotle does not formulate moral or didactic aims for literature. Here, Horace had to fill the gap (see Clarke, *Homer's Readers: A Historical Introduction to the "Iliad" and the "Odyssey"* 110).

[24] Aristotle, *Poetics* 83. On further differences see ibid. 82-83.

[25] Ibid. 63.

[26] Ibid. 84.

[27] Ibid. Should – despite all precautions – absurdities occur in the narrative, they can be downplayed by brilliance of style and language, see ibid.

[28] Ibid. 85; 88.

[29] The *Prælectiones Poeticæ* were first published in Latin: vol. I in 1711, vol. II in 1715. The English translation appeared in 1742: Joseph Trapp, *Lectures on Poetry* here: 338-339.

[30] Trapp claims that epic poetry depends on the machinery, since "The Dignity of an Heroic Poem would scarce be kept up without it, especially since the [...] Marvellous depends upon it" (ibid. 341).

[31] Ibid. 340. See also Aristotle, *Poetics* 85.

[32] Translated into English in 1695. Extracts from Bossu's text were included in Pope's *Odyssey*: "A GENERAL VIEW of the / EPIC POEM / AND of the / ILIAD and ODYSSEY. / Extracted from BOSSU," in *TE* 9, 3-24.

[33] Ibid. 23.

[34] See Pope, *TE* 9, 336; 111.

[35] *Iliad*, Book XIX.

[36] On the implied theatricality of necromancy in Odysseus' descent into hell see Joseph Spence, *An Essay on Pope's Odyssey* 44-45.

[37] Trapp, *Lectures on Poetry* 351.

[38] Clarke, *Homer's Readers: A Historical Introduction to the "Iliad" and the "Odyssey"* 88.

[39] Ibid. 88-92.

[40] Pope complained about those interpreters of Homer "who can find mysteries in the plainest description, assure us, that this is not to be understood historically, but allegorically" (*TE* 9, 339). He characterised their approach as "an instance of trifling industry and strong imagination" (*TE* 10, 7), while he himself preferred "plain History to a dark Allegory [...]" (*TE* 9, 339).

[41] All quotations from *TE* 9, 353.

[42] "The whole passage is to be understood allegorically" (*TE* 9, 361).

[43] Zoilus of Amphipolis, fl. 350 B.C., the so-called "scourge of Homer" (see Clarke, *Homer's Readers: A Historical Introduction to the "Iliad" and the "Odyssey"* 107).

[44] 'Longinus,' *On Sublimity* 143-187.

[45] Ibid. 152.

[46] Ibid. 153.

[47] Ibid.

[48] Ibid.

[49] Ibid.

[50] Joseph Trapp also claims that a long literary work, such as an epic, should be based on some truth (see Trapp, *Lectures on Poetry* 340).

[51] However, these marvellous elements should be used by the poet with discretion and economically.

[52] The 12th-century Byzantine commentator of Homer – Eustathius – described the Phæacians "as a proud and luxurious people, delighted with show and ostentation" (*TE* 9, 240). The problem of the Phæacians was that they were (according to Madame Dacier) too happy: "[...] the *Phæacians* were so puff'd up with their constant felicity and the protection of the Gods, that they thought nothing impossible; upon this opinion all these Hyperboles are founded: And this agrees too well with human nature, the more happy men are, the more high and extravagant they talk, and are apt to entertain themselves with wild Chimæra's which have no existence but in the Imagination" (*TE* 9, 257-258). Further information on the Phæacians: *TE* 9, 261; 262; 267-268; 278; 294-295; 301; 325.

[53] Pope takes this idea from Sir Walter Ralegh (see *TE* 9, 232-233).

[54] See also *TE* 9, 113.

[55] Pope, "Preface," *TE* 7, 11; Pope, "Essay on Criticism," *TE* 1, 258.

[56] On Pope's theory of inspiration and invention see Kirsti Simonsuuri, *Homer's Original Genius: Eigh-*

teenth-Century Notions of Early Greek Epic (1688-1798) and Mack, "Introduction," *TE* 7, xlvii.

[57] Pope, "Preface," *TE* 7, 11.

[58] Ibid. 4.

[59] Ibid. 16.

Works Cited

Aristotle. *Poetics*, in *Classical Literary Criticism*. Ed. D. A. Russell and M. Winterbottom, Oxford: Oxford UP, 1989. 51-90.

Ayres, Philip. *Classical Culture and the Idea of Rome in Eighteenth-Century England*. Cambridge: Cambridge UP, 1997.

Bentley, Richard. *Remarks Upon a Late Discourse of Free-Thinking: In A Letter to F. H. D. D. By Phileleutherus Lipsiensis*. London, 1713.

Blackwell, Thomas. *An Enquiry into the Life and Writings of Homer*. London, 1735.

Bossu, René le. "A GENERAL VIEW of the / EPIC POEM / AND of the / ILIAD and ODYSSEY. / Extracted from BOSSU", in *Twickenham Edition of the Poems of Alexander Pope*. Vol. IX. General Ed. John Butt. London: Methuen; New Haven: Yale UP, 1967. 3-24.

Clarke, Howard. *Homer's Readers: A Historical Introduction to the "Iliad" and the "Odyssey."* Newark: Delaware UP, 1981.

Congreve, William. "Incognita" (1692), in *An Anthology of Seventeenth-Century Fiction*. Ed. Paul Salzman. Oxford: Oxford UP, 1991. 471-525.

Erskine-Hill, Howard. *The Augustan Idea in English Literature*. London: Edward Arnold, 1983.

Foerster, Donald M. *Homer in English Criticism. The Historical Approach in the Eighteenth Century*. New Haven: Yale UP, 1947. Repr. Archon, 1969.

History of Magic and Experimental Science (8 vols). Ed. Lynn Thorndike. New York: Columbia UP, 1923-1970.

Jauß, Hans Robert. "Ästhetische Normen und geschichtliche Reflexion in der *Querelle des Anciens et des Modernes*" in Charles Perrault. *Parallèle des Anciens et des Modernes en ce qui regarde les arts et les sciences*. Ed. Hans Robert Jauß. Munich: Eidos, 1964.

Levine, Joseph M. "Ancients and Moderns Reconsidered." *Eighteenth-Century Studies* 15 (1981): 72-89.

---. *The Battle of the Books: History and Literature in the Augustan Age*. Ithaca: Cornell UP, 1991.

'Longinus.' *On Sublimity*, in *Classical Literary Criticism*. Ed. D. A. Russell and M. Winterbottom, Oxford: Oxford UP, 1989. 143-187.

Mack, Maynard. "Introduction", in *The Twickenham Edition of the Poems of Alexander Pope*. Vol. 7. Ed. Maynard Mack. London: Methuen, 1967.

---. *Alexander Pope: A Life*. New Haven: Yale UP, 1985.

Morris, David B. *Alexander Pope: The Genius of Sense*. Cambridge, MA: Harvard UP, 1984.

Nicolson, Marjorie and George S. Rousseau. *"This Long Disease, My Life": Alexander Pope and the Sciences*. Princeton: Princeton UP, 1968.

Pope, Alexander. *The Twickenham Edition of the Poems of Alexander Pope*. General Ed. John Butt. 11 vols. London: Methuen, 1939-1969.

Porter, Roy, ed. *The Cambridge History of Science* (8 vols). Vol. 4. *Eighteenth-Century Science*. Cambridge: Cambridge UP, 2003.

Sambrook, James. *The Eighteenth Century: The Intellectual and Cultural Context of English Literature, 1700-1789*. London: Longman, 1986.

Simonsuuri, Kirsti. *Homer's Original Genius: Eighteenth-Century Notions of Early Greek Epic (1688-1798)*. Cambridge: Cambridge UP, 1979.

Spence, Joseph. *An Essay on Pope's Odyssey*. London, 1726. Repr. Hildesheim: Olms, 1968.

Trapp, Joseph. *Lectures on Poetry Read in the Schools of Natural Philosophy at Oxford*. London, 1742. Repr. London: Routledge, 1994.

Wood, Robert. *An Essay on the Original Genius and Writings of Homer: With a Comparative View of the Ancient and Present State of the Troade. Illustrated with a Map of Troy*. Dublin, 1776.

Zimmermann, Hans-Joachim. *Alexander Popes Noten zu Homer: Eine Manuskript- und Quellenstudie*. Heidelberg: Carl Winter, 1966.

The Dunciad and the City

Pope's Heterotopia

Brean S. Hammond

Writing on James Joyce, the critic Jeri Johnson points to the Irish writer's aspiration "to give a picture of Dublin so complete that if the city one day suddenly disappeared from the earth, it could be reconstructed out of my book."[1] This prompts for Johnson the question how far literary cities are imaginative constructions and how far, on the contrary, they reflect the material realities of the city at particular stages of historical development. Johnson's view is that the latter possibility is routinely neglected in favour of the former, which she claims is the usual starting-point of literary criticism. Another routine starting-point, however, is that this issue first becomes acute with respect to literary modernism, given its unprecedented attention to urban consciousness. Franco Moretti is influential, for example, in arguing that "what distinguishes the city [...] is that its spatial structure [...] is functional to the intensification of mobility; spatial mobility naturally enough, but mainly social mobility. The dazzling rapidity of success and ruin is the great theme of the nineteenth-century novel from Balzac to Maupassant: with it the city enters modern literature and becomes [...] its obligatory context."[2] Spatial and social mobility afforded by the city is not exclusively, however, a post-Romantic preserve. I want to contend in this article that Alexander Pope's *Dunciad* (1728-43) has a fair claim to being the first literary work to raise the question of the *production* of space in recognisably modern form. *The Dunciad* compels its readers to ask *both* questions posed by Jeri Johnson: how closely is Pope's London based on the material actuality of the Stuart and Georgian city, and what is the nature of the poet's imaginative transformation of it? That these are not *new* questions – that contemporary readers of Pope's poem were also compelled to ask them – is attested in John Dennis' exasperated reaction to Book II in his *Remarks upon Mr. Pope's Dunciad* (1729), where he is incensed by the affront upon verisimilitude committed by Pope's setting the heroic games near the church of St. Mary-le-Strand. Only with a crudely reductive notion of probability can Dennis combat Pope's mythologised space: "What Probability in the Games which take up a third Part of the Piece? Is it not monstrous to imagine any Thing like that in the Master Street of a populous City; a Street eternally crowded with Carriages, Carts, Coaches, Chairs, and Men passing in the greatest Hurry about Private and Publick Affairs?"[3] What Dennis responds to here is the *Dunciad*'s provoking way of overlaying upon one another different kinds and conceptions of space – a peculiar *layering* effect. Pope utilises the reader's personal knowledge of a particular material space, but superimposes on that space events that could not possibly take place within it – events that the reader's very knowledge of the nature of the space renders impracticable and impossible. What we might term Pope's use of 'hyperspacing' perhaps suggests that Jeri Johnson's dichotomy is a false one. One novel aspect of Pope's use of space is its rendering inseparable material reality and imaginative transformation. This essay's exploration of the *Dunciad*'s cultural geography will focus upon the conditions that produce novel conceptions of imaginative spatialisation in the poem.

I

Much of the best recent work on *The Dunciad* has shown that, rather than Pope's London representing something *other* than itself, encoding a set of ethical, political or metaphysical meanings, it represents *at least* itself. Biographical work on the main personae and charting, for example, of the routes taken by and the rituals attendant on the Lord Mayor's Procession and the Coronation have exposed the material vertebrae of the poem, its rich inspissation in contemporary material culture.[4] Yet it is at once obvious to readers who work their way through the voluminous prefatory matter that Pope intends the urban geography of *The Dunciad* to signify in a densely allusory and symbolic way. Based around a set of adjudications on inclusion and exclusion, *The Dunciad*'s very structure, the principles that govern it as a physical object and a piece of bookmaking, is determined by the question of material space. How much space is there in the poem, and who should fill it? In the section of the prefatory material devoted to "Testimonies of Authors," Pope quotes a letter from James Moore Smythe that indirectly claims Pope's *Memoirs of a Parish Clerk* to be a satire on Gilbert Burnet. In response, Pope cites the Earl of Peterborough as a witness that Smythe himself was the real enemy of Burnet, but he at once faces the question of the juxtaposition of such individuals, Peterborough and Smythe, on the same physical page, and hence the same imaginative universe: "Here in truth should we crave pardon of all the foresaid right honourable and worthy personages, for having mentioned them in the same page with such weekly riff-raff railers and rhymers; but that we had their ever-honoured commands for the same; and that they are introduced not as witnesses in the controversy, but as witnesses that cannot be controverted; not to dispute, but to decide."[5] The main objective of the prefatory material is to justify Pope's filling up of the space, who's in and who's out – "it is not every Knave, nor (let me add) Fool, that is a fit subject for a Dunciad" (*Dunciad* 78) – and to deal with *prima facie* objections to that. Should hack writers' poverty be a reason for pillorying them in a satiric work? "Poverty itself becomes a just subject of satyr, when it is the consequence of vice, prodigality, or neglect of one's lawful calling; for then it increases the public burden, fills the streets and highways with Robbers, and the garrets with Clippers, Coiners, and Weekly Journalists" (*Dunciad* 34-35). What the prefatory material makes clear is the extent to which the poem's space is a *contested* space, its spatial tactics calculated to reimpose cultural control in the face of perceived threats: "We shall next declare the occasion and the cause which moved our poet to this particular work. He lived in those days, when (after Providence had permitted the invention of Printing as a scourge for the sins of the learned) Paper also became so cheap, and Printers so numerous, that a deluge of Authors covered the land; Whereby not only the peace of the honest unwitting subject was daily molested, but unmerciful demands were made of his applause, yea of his money, by such as would neither earn the one, nor deserve the other" (*Dunciad* 70-71). *The Dunciad* is in itself a contested space, this is to say; and every aspect of it as a material object makes this clear, from the jostling of footnote against text for room on the page, to the way in which Pope has modeled the physical appearance of the finished product upon prestigious bookmaking in an earlier era.[6] The inclusion of such devices as "Testimonies of Authors" and in the appendices, "A List of Books, Papers, and Verses, in which our Author was abused, before the Publication of the Dunciad," alongside the multiplicity of commentators and editors, suggests that no single author is in overall control of this project. Surely, the reader asks, no self-respecting author would have included tables in which parallel lists of "names bestow'd on Mr. DRYDEN" and on "Mr. POPE" are drawn up? Individuals and their cultural values have infiltrated the poem, its structure suggests, against its author's express desire.

As well as *being* a culturally contested site, the poem also *represents* or *stages* cultural contest. In my own earlier writing on the poem, I put the argument that Pope, Swift, Gay, Fielding and the early Samuel Johnson are a group of writers upon whom identity is conferred by their opposition to many aspects of the 'whig' narrative of historical progress; and by their devising of parodic and ironic literary forms that express this opposition powerfully.[7] Ancients not moderns, militantly impolite, satirists rather than cultivators of the easy style, sensing oppression rather than liberty, the so-called Scriblerians are profoundly counter-cultural figures. Consonant with this conception of them, I have discussed *The Dunciad* as dramatising the struggle between the purveyors of low-brow, popular and irrational culture, and those who wish to prevent their infiltration into respectable vicinities. The struggle is represented as being fought over territory, over *Lebensraum*. One part of the city of London is set against another, as the "Smithfield muses" migrate westward toward St. James' Palace and Westminster.

We can pick up the overtones of class struggle in Martinus Scriblerus's introduction to the 1743 edition, where Pope is resuscitating anxieties about the domination and control of space that first became apparent during the post-conflagration rebuilding of London in the later seventeenth century – an issue to which we will return. As Martinus Scriblerus puts it in his essay on the poem: "[no Action] cou'd be more [remarkable] than [...] the restoration of the reign of Chaos and Night, by the ministry of Dulness their daughter, in the removal of her imperial seat from the City to the polite World; as the action of the Aeneid is the restoration of the empire of Troy, by the removal of the race from thence to Latium."[8] The 1729 variorum edition speaks of the "introduction of the lowest diversions of the rabble in Smithfield to be the entertainment of the court and town," even more graphically dramatising the stepping westward of the uncivilised hordes from north and east. Pope's overlaying of Virgil's cultural geography in *The Aeneid* with his own in *The Dunciad* makes the point that his literary London is both an actual, material city and a transformation of it for particular semiotic purposes. It will appear in Book III, when Cibber is being shown "the past triumphs of the Empire of Dulness" by Settle's ghost, that those are cast typologically in terms of earlier migrations from north and east to west by the uncivilised hordes:

The North by myriads pours her mighty sons,
Great nurse of Goths, of Alans, and of Huns!
See Alaric's stern port! the martial frame
Of Genseric! and Attila's dread name!
See the bold Ostrogoths on Latium fall;
See the fierce Visigoths on Spain and Gaul!
(*Dunciad* III.89-94)

Invading the pleasant purlieux of Westminster and St. James' therefore, the scribbling barbarians are completing work done by those who destroyed the civilisation of Rome, and reversing the heroic deeds chronicled in *The Aeneid* by those who began it.

II

The Dunciad, then, plays host to a condition of social conflict, engaging actors whose control of resources and access to power is differentiated according to their status. It is to this extent a *contested* space. But this is only one aspect of a broader sense in which the poem is novel and unique in its manipulation of space. It is a landmark in the imaginative invention of spatial practice. Representation of space in *The Dunciad*, and the respects in which the poem itself becomes a new space of representation, is an underexplored dimension of Pope's art; and my own perception of this is assisted by the insights afforded in theoretical writing in 'postmodern' geography. Recent work within the discipline of geography has emphasised that whereas we have a relatively sophisticated understanding of the dimension of *time*, the axis of space is not as adequately conceptualised. Henri Lefebvre in *The Production of Space* (first French publication 1974) was the first writer to speak of space as "produced," and "reproduced," as representing "the site and outcome of social, political and economic struggles."[9] Lefebvre draws a tripartite distinction between *experienced material spatial practices* that occasion physical infrastructures in transport and the built environment; the *rep-*

resentation of space in perception resulting in various semiotic and mapping systems; and *spaces of representation in the imagination* where dwell utopian plans, imaginary landscapes, mythologies of space and place, and poetics of space. Distinctions here are not entirely hard and fast, but our main concern is with the imaginative appropriation in art of processes actually occurring in material spatial practice, the engendering of imagined spatial representations by changing social formations. Here I would point to two separate but related processes – events actually taking place in the built environment of London immediately prior to and during the period of Pope's creative life – that might provide a material basis for the understanding of the poet's spatial reconceptualisation: the rebuilding of the City of London after the Great Fire, and the development of the 'Town' or west end of 'suburban' London.

T. F. Reddaway, the most authoritative historian of the rebuilding of London after the fire, writes: "statisticians were to compute the destruction at 13,200 houses, the Royal Exchange, the Custom House and the halls of 44 of the city Companies, the Guildhall and nearly all the City buildings, St. Paul's itself and 87 of the parish churches besides furniture and commodities valued at over three and a half million pounds."[10] For at least a decade after the final embers were extinguished, the City of London was an area unusually transparent, its infrastructure and the relation between social forms of living and the architectural environment peculiarly open to view. Plans presented by Wren, Hooke and others to reconceive the ravaged area on utopian lines were overtaken by the City's pressing need to reactivate commerce and retain its population, so the issues up for negotiation were the very down-to-earth ones of prescribing the heights and building materials of houses erected on the existing sites and streetplans, dictating widths of streets so as to eliminate alleys impassable to vehicular traffic and ensuring that the pre-conflagration encroachments on the walkways by jutties, shop signs, embellishments to buildings and so forth did not reoccur. Although much of the secular rebuilding took place in the decade following the fire before Pope was born, the religious building and some of the public building went on for much longer. St. Paul's, for instance, was not completed until 1710; and the evidence of transformation and renewal was everywhere in the City for many years to come. Defoe, for instance, could write in the 1720s of "new squares and new streets rising up every day to such a prodigy of buildings that nothing in the world does, or ever did equal it, except old Rome in Trajan's time."[11] And indeed, John Dennis' difficulty in reading Pope's location for the heroic games in Book Two is a direct consequence of architectural change following Queen Anne's churchbuilding programme. St Mary-le-Strand is erected on the former site of the maypole:

Amid that area wide they took their stand,
Where the tall may-pole once o'er-look'd the Strand;
But now (so ANNE and Piety ordain)

A Church collects the saints of Drury-lane.
(*Dunciad* II.27-30)

It might once have been a credible site for games-playing, Dennis might think, but surely cannot be now, consecrated ground as it now is? Pope's lines make clear, however, that a church located here can only be a sink for the prostitutes ("saints") of theatre-land around Drury. The reader is asked to receive an image of a well-known, if recent, landmark building, to consider the history of its site and to register the appropriation of that history for satirical purposes. This is altogether typical of the complexity of Pope's spatial awareness.

An infancy spent just off Lombard Street, one of the key streets in the rebuilding because it housed the main City centres of finance, must have alerted the young poet to the degree of urban transformation, even if only in stories told him by his parents. Imagining at the opening of his *Dunciad* the "Cave of Poverty and Poetry" situated, the poem tells us, close to Bethlem Hospital in Moorfields northwest of the City (rebuilt in the 1670s to designs by Robert Hooke), it is surely possible that Pope has in mind, amongst several other possible sites, Gresham College – very close to Lombard Street and the seat of City administration in the immediate aftermath of the Fire.[12]

After the Fire, some of the City's burnt-out richer denizens did not come back. A major anxiety for the City authorities was to limit the migration of its wealthier citizens, those who would traditionally accept public office, westward into the main development areas. Between the periods of Restoration and Regency, London's architectural development served to make of it a city more socially segregated than any other in Europe: a city in which address was the clearest possible indicator of social status. Speaking of John Nash's development of Regent Street and Regent's Park at the end of the eighteenth century, Roy Porter comments that "by restricting eastern access to Regent Street, Nash's route succeeded, in Sir John Summerson's phrase, in 'damming up Soho,' and London's grandest thoroughfare thereby became its social barrier, with Portland Place and Regent Street screening the fashionable West End from déclassé quarters. 'No family of ton can breathe eastward of Berkeley Square," remarked the *World* magazine in 1787: Regent Street embodied that observation.'[13] One can offer, as a reasonably straightforward example of the way in which imaginative and material processes of spatialisation interact, a passage from Dryden's 1672 play *Marriage à la Mode*: "DORALICE: That's very true. Your little courtier's wife, who speaks to the King but once a month, need but go to a town lady, and there she may vapor and cry, 'The King and I' at every word. Your town lady, who is laughed at in the circle, takes her coach into the city, and there she's called 'your honor' and has a banquet from the merchant's wife, whom she laughs at for her kindness. And as for my finical cit, she removes but to her country house and there insults over the country gentlewoman that never comes up, who treats her with frumenty and custard and opens her dear bottle of mirabilis beside for a gill glass of it at parting."[14] Sense of place is here stratified and hierarchised into a pecking order – court, town, city and country – that appropriates for the imagination processes actually occurring in material spatial practice, and perceived in representations of space. Outside the theatre as we have noted, the rebuilding of the fire-ravaged City of London was accelerating the process of suburbanisation and the migration west of established tradesmen, professionals and artisans into Lincoln's Inn Fields, Covent Garden and the Strand, which in turn shifted the gentle and the titled into Soho, St. James' and Mayfair. The 1660s and 70s saw large-scale development by aristocrats of great palaces and estates close to St. James' into fashionable residential housing – the areas of Piccadilly, St. James' St., the Haymarket and Pall Mall that Doralice is referring to when she speaks of the 'town.' In the year of the play's performance, the Duke of Buckingham's York House was sold to Nicholas Barbon for conversion into new streets. Restoration theatre, through the localisations of setting it includes and the imaginary exclusion zones it establishes – its spaces of fear, ritual and repression – is a transformation of the flows of people, goods, money and power that are apparent at a time and in a city where these processes are more than usually laid open to view. First Covent Garden, then the development of the great squares – Hanover, Berkeley and Grosvenor, Cavendish, Portman and Manchester, Bloomsbury, Bedford and Russell: by 1800, the east/west divide in London was accomplished and patent.

III

Envisaging as it does the breaching of that divide through migration out of the City, albeit not on the part of those social groups who were most apparently involved in so doing, *The Dunciad* taps into deep cultural anxieties. The poem's emphasis on acts of cultural policing, on guarding the barriers between segregated vicinities, is to an extent an imaginative embodiment of invisible demarcation lines instantiated in brick and stone that the rebuilding and development of London brought about. That development, however, also brought imaginative transformation of space, and it is in the almost-phantasmagoric mutations of its places and spaces that *The Dunciad* is most striking. The omniscient voice and the omnipresent focalisation of *The Dunciad*'s narrator seems to require an Olympian perch. In his influential essay "Walking in the City," Michel de Certeau speaks of the observer looking down from the [tragically now demolished] World Trade Centre onto the New York traffic circulating below: "An Icarus flying above these waters, he can ignore the devices of Daedalus in mobile and endless labyrinths far below."[15] Such a miniaturising perspective is part of the poem's effect, but there is more to it. Let us pick up on what we earlier saw John Dennis responding to in *The Dunciad*'s provoking way of overlaying upon one another different kinds and conceptions of space – this strange *layering* effect. The opening paragraphs illustrate the many ways in which the poem interrogates and challenges the reader's sense of the spatial:

The Mighty Mother, and her Son who brings
The Smithfield Muses to the ear of Kings,
I sing.
(*Dunciad* 1.1-2)

At once there is a particular location invoked: Smithfield cattle market, located in the densely crowded northwest of the city. Bartholomew Fair, sited on Smithfield, associates the location with the concept of the Muses, and this will be further developed in Book III; otherwise, it is the incongruous and mock-heroic association of cattle drovers and literati that the reader registers. Does the reader consider how Smithfield muses might find their way to St. James', through the bottlenecks of Temple Bar and Fleet Street? If that is too specific an act of the imagination, the opening assuredly announces the point about contested space, as refugees from overcrowding threaten to find their way into courtly London. The solidity of Smithfield is to be quickly evaporated, however, as by line 11 we are projected into a chronotope without coordinates: "In eldest time [...] Dulness o'er all possess'd her ancient right." "All" combines all terrains, presumably, and all created intelligences. The vast incompassable plains of eternity quickly give place to Swift's Ireland, classicised as "Boeotia." Valerie Rumbold points out in her note that this hellenised equivalent to Ireland works because Boeotia was a legendarily stupid and rustic part of ancient Greece as modern Ireland was supposed to be of the British Isles; but Pope is asking his reader not to activate the usual symbolic accretions of Ireland. England is Ireland now, under the transforming power of dullness. Invoking Swift laughing "in Rab'lais' easy chair" provides a domestic picture of Swift at home in the Deanery, a safe vantage point from which to "behold" (but not literally) the coming of the Age of Lead in England: but the exoticism of Renaissance France is also invoked as the erudite reader consults memories of *Gargantua and Pantagruel*.

Well before the poem's first significant mise-en-scène is introduced, the reader has been through many kinds of spatial experience, from the literal-material, to the metaphysical, to the metaphorical. The paragraph commencing at line 29 introduces the Queen of Dulness's "Cave of Poverty and Poetry." This is almost given a specific address: it is "Close to those walls where Folly holds her throne," that is close to Bethlehem Hospital in Moorfields, just north of London Wall on the site of modern Finsbury Park. As I have suggested earlier, Pope may have Gresham College in mind here, situated between Broad Street and Bishopsgate Street, a stone's throw from Bedlam. Given this approximate location, how, the reader wonders, can it also engender the production of "Curl's chaste press, and Lintot's rubric post," since both publishers were firmly located in Fleet Street? Space in the poem is, and is not, specific: it is where it is, and pacman-like, assumptive of other spaces. If this suggestion of location near Bedlam might help to anchor our experience of the Queen's "Cell" and help us to believe in the print factory or print nursery or insectarium that Pope will put there, we soon find any firm sense of place complicated by the allegorical figures of the "four guardian Virtues" that support the Queen's throne. Whether these are wood carvings, painted decorations, statues – or what on earth they are, perhaps simply a nonce reversion to the Spenserian allegorical tradition – is far from easy to discern. Neither is the ontological status of what is on show in the Cave easy to establish. Valerie Rumbold's note in her wonderful recent edition tells us that "Dulness, a parody of a fertility goddess, rejoices in a scene of perverse potential reminiscent of Milton's Satan looking into chaos" (*Dunciad* 106n), but although the parturitions and copulations and transitions that occur in II.55-78 seem at first to be a spectacle performed for the Queen's edification, we seem to learn in I.82 – "With self-applause her wild creation views" – that none of it has an objective existence separate from her own. What we are seeing is a theatrical masque of anti-creation conjured by the Queen herself. References to real places through the geographical blagues committed by the dunces in their writings – producing rain showers for Egypt, fruits for the Arctic wastes of Novaya Zemblya and flowers for the Libyan desert of Barca – further vex the reader's grasp on ontology in the passage.

Assuredly, the next act of 'seeing' undertaken by the Queen is a seeing with the mind's eye, a revolving in the memory of those desperately limited "City Swans" who had celebrated Lord Mayor's Days throughout the ages, but when she perceives Bays / Cibber as the epitome of all this roll-call, we seem to have moved physically into his garret though there were no deictic markers to point up the transition. Cibber's "Gothic Library! of Greece and Rome/ Well purg'd" functions partly as do the books in Swift's *Battle*: through personification and metonymy, books become their authors. But the space here is altogether more phantasmagorical, even hallucinatory, than that, calling to mind the underworld scene in *Rape of the Lock*, as literary genres metamorphose into "running Lead," and the mind that conceives them changes into a ricketty building, while playtexts transform into partially-digested lunches. Space here is at the far end of imaginative transformation, almost uncontrolled by reference to space in ordinary perception. Sequences from recent films such as *Trainspotting* or *The Naked Lunch* would seem to be closer parallels than anything in writing. Dulness appears to transport herself to Cibber's garret at I.259 in order to extinguish his funeral pyre with a sheet from Ambrose Philips's *Thule*, and then takes Cibber back with her to her Cave. Book 1 ends, however, with as specific and material a reference to real London topography as we find anywhere in it. A mock Coronation anthem sounds in the Chapel-royal in St. James' Palace, the Mexican soundwave making its way thence to White's coffee-house, Drury Lane, Mother Needham's brothel in Park Place, St. James' that was especially associated with the Court, the Devil Tavern in Fleet

Street, and fetching up in Hockley-in-the-Hole in Smithfield.

We could continue such an account of the spatial co-ordinates and modes of perceiving space throughout the entire poem, and this would certainly be rewarding: a typology of the poem's spaces would be a useful analytical tool. I have said enough, perhaps, to indicate its extraordinary complexity and deliquescence. I find myself reaching for Foucault's charismatic conception of the "heterotopia" to describe the kind of space that the poem represents. In a lecture given in 1967, but first published in 1984 in French as "Des Espaces Autres" ["Of Other Spaces"], Foucault speaks of real, rather than fictive, spaces that have a mirror-like relationship to utopias and that he designates "heterotopias": "There are also, probably in every culture, in every civilization, real places [...] which are something like counter-sites, a kind of effectively enacted utopia in which real sites, all the other real sites that can be found within the culture, are simultaneously represented, contested, and inverted. [...] The heterotopia is capable of juxtaposing in a single real place several spaces, several sites that are in themselves incompatible."[16] Thus the museum and the library are considered by Foucault to be heterotopias of the nineteenth century, characteristic of the era's passion for accumulation. Given that the theatre stage and the cinema screen are two further examples supplied by Foucault, and given what has been said here about *The Dunciad*'s appropriation of theatre as a site of transformation and anticipation of the cinema as capable of the radical transformation to which the poem seems to aspire, one imagines that Foucault might have accepted *The Dunciad*'s claim to represent the heterotopic. Nowhere is this claim stronger than in the series of Ovidian transformations between real creation and imaginary in Book III, through which Cibber is confronted by an alter ego – his own pantomimic Dr. Faustus, a "sable Sorc'rer" who conjures up an alternative Creation culled from popular theatre that, in a moment of supreme narcissism, the hero is invited to confront as his own and its creator as himself:[17]

Joy fills his soul, joy innocent of thought;
What pow'r, he cries, what pow'r these wonders wrought?
Son; what thou seek'st is in thee! Look, and find
Each Monster meets his likeness in thy mind.
(*Dunciad* III.249-252)

Cibber achieves self-definition by gazing at a mirror-image of himself reflected in the anti-creation of his artistic cosmos. He has reached the mirror stage of development. Some years ago, Emrys Jones published an essay, "Pope and Dulness," which, in stressing the sheer vivacity, the pullulating energy, the reckless irresponsibility of the poem, drew attention to the poet's immersion in the low and the little, the dabbling in the dirt, that is such a great part of the poem's appeal.[18] I think the overlaying of different spatialisations is another aspect of that. In the interests of a resounding conclusion, I will call *The Dunciad* the imaginative representation of the Popean heterotopia.

Notes

[1] Jeri Johnson, "Literary Geography: Joyce, Woolf and the City" 60.

[2] Quoted in Gary Bridge and Sophie Watson, *The Blackwell City Reader* 7.

[3] John Dennis, *Remarks upon... the Dunciad* (1729), *The Critical Works of John Dennis*, ed. Edward Niles Hooker 2 362.

[4] See in particular Pat Rogers, "Ermine, Gold and Lawn: *The Dunciad* and the Coronation of George II" 120-150; and Douglas Brooks-Davies, *Pope's* Dunciad *and the Queen of Night: A Study in Emotional Jacobitism* 108-111.

[5] Alexander Pope, *Dunciad* 56. All quotations from *The Dunciad* are from Alexander Pope, *The Dunciad in Four Books*. ed. Valerie Rumbold, hereafter *Dunciad*.

[6] See James McLaverty, *Pope, Print and Meaning*, for the best available treatment of the relationship between the physical and the semiotic features of Pope's printed book, especially chapter 4.

[7] Brean S. Hammond, "'Guard the sure barrier': Pope and the Partitioning of Culture" 225-239. See also, for a development of the argument about 'whig' narratives of progress, my "Is there a Whig Canon? The Case of Susanna Centlivre" 373-390.

[8] "Martinus Scriblerus of the Poem," in *Dunciad* 72.

[9] Henri Lefebvre, *The Production of Space*. I have quoted Tim Woods, *Beginning Postmodernism* 119.

[10] T. F. Reddaway, *The Rebuilding of London after the Great Fire* 26.

[11] Daniel Defoe, *A Tour through the Whole Island of Great Britain* (1724-1726), Letter 5 "Containing a Description of the City of London" 101.

[12] Rumbold notes that in earlier versions the "Cell" had been in Rag Fair near the Tower of London. She points to other possible associations: the disused foundry in present-day Tabernacle Street, used for Methodist worship post-1739; and Sion College.

[13] Roy Porter, *London: A Social History* 127.

[14] John Dryden, *Marriage à la Mode* (1673) 3.1.181-193. Quoted from Brian Corman's edition in J. Douglas Canfield, *The Broadview Anthology of Restoration and Early Eighteenth-Century Drama* 350.

[15] Michel de Certeau, *The Practice of Everyday Life*, trans. Steven Rendall 92.

[16] Michel Foucault, "Of Other Spaces" 24.

[17] This passage actually has more appropriate reference to the hero of the earlier *Dunciad*, Lewis Theobald, who was responsible for some of the specific pantomimes and special effects represented. See *Dunciad* 250n.

[18] Emrys Jones, "Pope and Dulness" 231-263.

Works Cited

Brooks-Davies, Douglas. *Pope's* Dunciad *and the Queen of Night: A Study in Emotional Jacobitism*. Manchester: Manchester UP, 1985.

Bridge, Gary and Sophie Watson, eds. *The Blackwell City Reader*. Oxford: Blackwell, 2002.

Canfield, J. Douglas, ed. *The Broadview Anthology of Restoration and Early Eighteenth-Century Drama.* Ormskirk: Broadview Press, 2001.

de Certeau, Michel. *The Practice of Everyday Life.* Trans. Steven Rendall. Berkeley: U of California P, 1984.

Defoe, Daniel. *A Tour through the Whole Island of Great Britain* (1724-1726). Ed. Pat Rogers. Devon: Webb and Bower, 1989.

Dennis, John. *The Critical Works of John Dennis.* 2 vols. Ed. Edward Niles Hooker. Baltimore: Johns Hopkins UP, 1943.

Foucault, Michel. "Of Other Spaces." *Diacritics* 16.1 (1986): 22-27.

Hammond, Brean S. "'Guard the sure barrier': Pope and the Partitioning of Culture," in *Pope: New Contexts.* Ed. David Fairer. Hemel Hempstead: Harvester Press, 1990. 225-239.

---. "Is there a Whig Canon? The Case of Susanna Centlivre." *Women's Writing*, 7.3 (2000): 373-390.

Johnson, Jeri. "Literary Geography: Joyce, Woolf and the City," in *The Blackwell City Reader.* Ed. Gary Bridge and Sophie Watson. Oxford: Blackwell, 2002.

Jones, Emrys. "Pope and Dulness." *Proceedings of the British Academy* 54 (1968): 231-263.

Lefebvre, Henri. *The Production of Space.* Trans. D. Nicholson-Smith. Oxford: Blackwell, 1991.

McLaverty, James. *Pope, Print and Meaning.* Oxford: Oxford UP, 2001.

Pope, Alexander. *The Dunciad in Four Books.* Ed. Valerie Rumbold. Harlow: Longman, 1999.

Porter, Roy. *London: A Social History.* London: Hamish Hamilton, 1994.

Reddaway, T. F. *The Rebuilding of London after the Great Fire* (1940). Repr. London: Edward Arnold, 1951.

Rogers, Pat. "Ermine, Gold and Lawn: *The Dunciad* and the Coronation of George II," in *Literature and Popular Culture in Eighteenth-Century England.* Ed. Pat Rogers. Brighton: Harvester Press, 1985. 120-150.

Woods, Tim. *Beginning Postmodernism.* Manchester: Manchester UP, 1999.

The Mummy's Curse

A Genealogy

Roger Luckhurst

This paper will explore the contention that the mummy's curse is part of what I want to delineate as an 'Egyptian Gothic,' a cultural formation that emerged and permeated popular culture from the 1880s to the 1930s. It is important to emphasise that this Egyptian Gothic is not just a literature, but is something wider, a set of beliefs or knowledges in a loosely occult framework: it existed in literature, journalism, and Orientalist scholarship, but also in theatres, museums, and the practices of the late Victorian magical revival. To get a full sense of it, one also needs to theorise something about *unofficial* and *undisciplined* knowledge, the role of folklore and those stories which are circulated unofficially as gossip, as spooky anecdotes that were widely known, often told, but rarely written down. Hearsay, as Hans-Joachim Neubauer puts it in his book *The Rumour* is 'an invisible literature, one that is forever changing shape.'[1] When it comes to supernaturalism with an Egyptian tinge this reliance on rumour becomes even more significant. This is undoubtedly because there is a legitimate and institutionalised discourse – Egyptology – that is trying to exert a cohesive authority and wishes to stamp on the proliferation of superstitions around Ancient Egyptian sites and objects. Indeed, archaeology has been identified as an exemplary discourse of modernity in its rational ordering, taxonomical knowledge and enthusiastic commitment to the eradication of superstition.[2] This is in parallel with a public language of diplomacy and international finance that is trying to resolve 'The Egyptian Question' by imposing rationality and order on the allegedly chaotic state of the native economy of Egypt. These official discourses only fed rumours of a body of hidden or shadow knowledges, however. It was often difficult to police the boundaries of Egyptological legitimacy, particularly in the traffic of archaeological artefacts and of stories about those artefacts. The Egyptian Gothic leaks from the corners of political economy and the archaeological documentation of Ancient Egypt: it is no accident that the repeated iconic scene, in fiction at least, for uncanny encounters is the Museum, at night, when orders of knowledge are subverted, and mummies become the Lords of Misrule.

All of this is relevant to beginning to sketch out a genealogy of the mummy curse. I mean genealogy in the loose sense of attempting to trace some of its larger branches back to possible origins, but also in the stricter sense that Michel Foucault invoked, as an exploration of both the history of science and what he called 'subjugated knowledges,' 'that whole series of knowledges,' he said, 'that have been disqualified as nonconceptual [...] insufficiently elaborated [...] [or]

below the required level or erudition or scientificity.' What he called genealogy was, in his evocative phrase, a 'coupling together of scholarly erudition and local memories'[3] which could open up history to new critical analysis. The mummy curse was a product of 'local memories' for sure, inarticulate and unscientific, but it also proffers a new way of interrogating official thinking about the Egyptian Question in this era.

When you think of Mummy Curses, you are immediately transported to the opening of the tomb of the pharaoh Tutankhamen in the winter of 1922-23 by the excavator Howard Carter and his wealthy patron, the Fifth Earl of Carnarvon. The tomb had been discovered in Carter's last-chance season of digging in the Valley of the Kings, in November 1922. No new tomb had been discovered for over a decade, and all the surviving valley tombs of the New Kingdom pharaohs were considered to have been unearthed. Carnarvon travelled from England immediately upon receiving word from Carter that an undisturbed tomb had been discovered. They secretly breached the sealed doors, and soon determined that the tomb would offer up treasures with that famous exchange: 'Can you see anything?' 'Yes, wonderful things.' The antechamber was opened at the end of November and news reports of the discovery swept round the world. They did not open the burial chamber until 17 February 1923, by which time Carnarvon had signed an exclusive reporting deal with *The Times* for £5000 and was in advanced negotiations with Sam Goldwyn for a Hollywood film of the discovery (he had even written the script). The world press that gathered for the tomb opening was seething with suppressed rage at their exclusion, and reporters from *The Daily Express* and *The Daily Mail* had been sent with the express purpose of undermining the exclusive deal with *The Times*. Tightly controlled at the site, they soon got another story when Lord Carnarvon fell ill at the beginning of March. What was assumed to be a mosquito bite had become infected – over the next month, Carnarvon progressed from fever to blood poisoning and then pneumonia before dying on 5 April. As he hovered at death's door, the famous popular novelist Marie Corelli told the *New York World*: "I cannot but think some risks are run by breaking into the last rest of a king of Egypt [...] According to a rare book I possess, which is not in the British Museum [...], the most dire punishment follows any rash intruder into a sealed tomb."[4] The day after Carnarvon died, Sir Arthur Conan Doyle stepped off a steamer in New York at the start of another tour to promote Spiritualism. Doyle, the *Morning Post* reported, "was inclined to support to some extent the opinion that it was dangerous for Lord Carnarvon to enter Tutankhamen's tomb, owing to occult and other spiritual influences. He said 'An evil elemental may have caused Lord Carnarvon's fatal illness'."[5] Rider Haggard, a considerable amateur authority on Ancient Egypt, publicly dismissed these stories, whilst mourning the sensationalism that embroiled the august Egyptian dead in this display of vulgar modernity. The Gothic writer Algernon Blackwood, in a commissioned piece for the *Daily Express* called 'Superstition and the Magic Curse,' managed to uphold the reality of curses whilst denying the likelihood of them operating in this case. A curse, he explained, was "a violent and concentrated strain of thought of which ordinary minds are, fortunately, hardly capable" – but their energy, he was certain, could not survive thousands of years. Whilst these Gothic writers seemed to control this sensational story, a different, authoritative source came from the Egyptologist Arthur Weigall. Weigall had been ousted from his official post in the Egyptian Antiquities Service in 1914, since when he had survived as a novelist, journalist and song-writer. He was sent to cover the opening of the tomb by the *Daily Express*; like all the other reporters, he had to watch in anger and frustration from outside the tomb. In his best-selling book of 1924, *Tutankhamen and other essays*, Weigall confessed he had been appalled by the lack of respect shown the tomb by Carnarvon. Weigall reported that on the day of the opening he had "turned to the man next to me, and said: 'If [Carnarvon] goes down in that spirit, I give him six weeks to live'."[6] The possibility of a curse was given further fuel by Weigall's accompanying essay, "The Malevolence of Ancient Egyptian Spirits," where he offered creepy stories from his own experience, with a teasing suspension of scientific judgment: "I will leave it to the reader's taste," he said, "to find an explanation for the incidents which I will here relate."[7] Artefacts including grave-jewellery and a mummified cat seemed full of poisonous intent and ill-omen. Or perhaps not. Perhaps it was just coincidence or Western credulity, after all. Some histories give Weigall the dubious honour of initiating stories of the curse – but he was clearly not responsible for it, nor for the way Tutankhamen's curse proliferated uncontrollably. Rumour or hearsay, to quote Neubauer again, "refers to other, absent narrators, to people who are not there. In the background of conversations in which this voice speaks there is a chain of anonymous speakers begun somewhere indeterminate and leading nowhere in particular. This series [is] what gives rumours their strange authority."[8] So it is with the proliferation of stories around the events told about the opening of Tutankhamen's tomb. *It was said* that a clay tablet had been discovered in the antechamber, with the inscription "Death shall come on swift wings to whoever toucheth the tomb of the Pharaoh." This had been destroyed by Carter and Carnarvon allegedly to prevent the superstitious fellahin, the Egyptian peasant workers, from abandoning the excavation. *It was said* another curse was inscribed on a statue at the door of the Treasury. *It was said* that Carter's beloved pet canary had been devoured by a cobra on the day they entered the tomb, the cobra being the embodiment of pharaonic power. *It was said* that Carnarvon's dog, back in England, howled mournfully and dropped dead at the precise moment of his master's death – with the appropriate time-difference built in, of course. Twenty-nine deaths were ascribed to Tutankhamen in the next few years.[9] These ripple out in an impressively exponential way. So, when Howard Carter's secretary, Richard Bethall, died suddenly, one breathless account reports: "When his father, the 87 year-old Lord Westbury, heard of his son's death, he plunged from the seventh storey of his London house. Then, on the way to the cemetery, the hearse ran over a little boy."[10] This offers an impressive associational chain of misfortune, common to these stories. And soon, we come to one of the principal laws of this literature: Curses are Recursive. That is, there is a tendency for those involved in disseminating curse narratives to get caught in its machinations. When Arthur Weigall died in 1934 his own paper, the *Daily Express*, carried the headline "A Curse Killed Arthur Weigall."[11] And when Christopher Frayling researched the material for his television series about the Tutankhamen story, in part intending to debunk the curse narrative, he was nevertheless happy to trade on several near death experiences during his stay in Egypt, situating himself as an unwilling participant of the logic of recursion.

In fact, the Tutankhamen story proliferated so fast and extensively not just because of the attentions of the global mass media of the 1920s, but also because it repeated at least two previous curse stories that had been circulating in Edwardian London. These are much less well documented and, on the rare occasions that they are written down, explicitly refer to their status as hearsay. The first narrative concerns Mr. T. Douglas Murray, and this is complicated to reconstruct because it is a story that starts in the 1860s, but is only brought into wider public circulation after a journalist broke the story in 1904. This journalist, Fletcher Robinson, had been lead reporter of the *Daily Express* during the Boer War, and rapidly rose to the editorship. When he told the story of a mummy curse that had befallen the Murray family, he finished his chilling tale by announcing that the curse seemed to end when the family presented the artefact to the British Museum in 1899. When Robinson died suddenly in 1907 at the age of 35, however, he was subject to the law of the recursive curse, and was assumed to have been its victim. Rumours about the evil mummy of the British Museum became a staple of London folklore well into the 1930s (a mummy was said to travel between the Egyptian Rooms of the British Museum and the now closed British Museum underground station using a secret tunnel, for instance, although the source for this rumour was the British film *Bulldog Jack*).

The success of Robinson's story was undoubtedly aided by the fact that Robinson was a close friend of Arthur Conan Doyle, who had dedicated *The Hound of the Baskervilles* to Robinson. Doyle recalled the fate of Robinson in his comments on Lord Carnarvon twenty years later. "I warned Mr. Robinson against concerning himself with the mummy at the British Museum," Doyle told the American press. "He persisted, and his death occurred [...] The immediate death was typhoid fever, but that is the way in which elementals guarding the mummy might act."[12]

The Douglas Murray story runs something like this – I say *something* like this advisedly, because the sources are completely unreliable and often contradict each other on virtually every detail. In 1868, Douglas Murray, an active member of the Royal Geographical Society, was wintering in Egypt with four friends, staying in Thebes, possibly with Lady Duff Gordon. Murray was in the market for an ancient Egyptian artefact to take home; Arab treasure hunters presented a particularly fine mummy case, painted with an image of a priestess of Amen-Ra. Although Murray purchased it, the Europeans drew lots as to who was to have the case. In some versions, Murray wins and hangs the case in his own house in London. In other versions, the case was won by a lawyer, A. F. Wheeler. Most stories agree on the instant consequences: Douglas Murray lost an arm in a shooting accident in Egypt; on the journey home, two members of the party contracted illnesses and died. Wherever the case was housed, it produced misfortune, illness, accident and death. Wheeler, it seems, returned either to Cairo or London to find his entire fortune had been lost in a financial disaster. By 1869, Wheeler gave the case to his sister, whose family also suffered. An attempt to photograph the case in 1887 resulted in the death of both the porter and the photographer, but not before a malign spectral image was allegedly recorded on the negative, hovering over the case. This ghostly presence was female in some accounts, male in others. The malignity was such that on the death of Wheeler in 1899, the case was presented to the British Museum. Here, however, it continued to induce nervous effects. The most sustained narrative of this story, by Ada Goodrich Freer in the *Occult Review*, reported: "There has been much talk of various misfortunes said to have befallen the officials of various grades in the British Museum who have come within the influence of the priestess" – rumours the Museum denied.[13] Follow up letters in the *Occult Review* attested to the continuing venomous influence of the case, including the death of a nervous student, or a fall, headlong down the Museum steps of a defiant society debutante. Another report also records: "It was rumoured that some unaccountable thing happened to anyone who attempted to make any drawing or sketch of this remarkable looking mummy case."[14]

The story of Douglas Murray and the priestess of Amen-Ra slowly intertwined with many personalities in late-Victorian occult circles. Madame Blavatsky, founder of The Theosophical Society, based in London in the late 1880s, was said to have "caught sight of it" at the Wheeler household "and at once begged the sister to send it out of the house, as the influence, she asserted, was malign."[15] Blavatsky's Theosophical movement was part of the hermetic tradition: her early tome, *Isis Unveiled* (1877), placed the prehistoric origins of a secret knowledge in the priesthood of Ancient Egypt. As an initiate of this tradition, Blavatsky claimed to have access to authoritative knowledge on such occult influences. Another key source for the Murray narrative comes from the well-known society palmist, Louis Harmon, who practiced under the name 'Cheiro.' Cheiro read the palms of many of the leading people of the late-Victorian and Edwardian age – he claimed to have read Oscar Wilde's hand in 1893 and predicted his downfall, for instance. He also improbably claimed to have read Douglas Murray's hand before his ill-fated trip to Egypt, and "an unaccountable feeling of dread and horror seemed to creep from it," he said, no doubt because this was the hand Murray would shortly lose in his shooting accident.[16] He records Murray saying that "as I looked at the carved face of the Priestess on the outside of the mummy case, her eyes seemed to come to life, and I saw such a look of hate in them that my blood seemed to turn to ice."[17]

Cheiro also later claimed to have been forewarned of Lord Carnarvon's death through the agency of a magical mummy hand he used in his divinations. Finally, we need to bring in another friend of Douglas Murray, the leading New Journalist of his day, W. T. Stead. Stead, after his hugely influential role in the politics of the 1880s as editor of the *Pall Mall Gazette*, had founded the *Review of Reviews* in 1890 but also turned to occult interests and Spiritualist belief, publishing his occult review *Borderland* between 1893 and 1897. His passion for anything weird or mysterious was to give the priestess her greatest single exercise of the curse.

When the *Titanic* sank on her maiden voyage in 1912, Stead was undoubtedly the most famous Englishman on the boat. Spiritualists claimed, of course, to have received the news of the sinking first – the dead communicate faster than the telegraph. But there was also intense interest in the last hours of the boat. One of Stead's dining companions that night recorded that "Mr. Stead talked much of spiritualism, thought transference and the occult [...] He told a story of a mummy case in the British Museum which, he said, had had amazing adventures, but which punished with great calamities any person who wrote its story. He told of one person after another who, he said, had come to grief after writing the story."[18] Stead, the journalist, probably had the fate of Fletcher Robinson in mind, and his emphasis on the dangers of *writing* the story of the mummy again foregrounds the role of rumour. Lower down the hierarchy of knowledge, merely gossiping about the curse did not appear to cause Stead worry about invoking the wrath of the priestess. Nevertheless, Stead's last conversation was to prompt the rumour that the mummy case was itself on board the *Titanic*. The British Museum, allegedly, had decided to be rid of the accursed object, and had sold it to a New York museum or collector. As one enthusiastic believer in this tale recounts: "The mummy was encased in a wooden crate, and, because of its great value, had not been placed in the hold of the Titanic but behind the command bridge. Did Captain Swain, too, look into those fatal radiant eyes?"[19] John Stephens offers a rather wonderful addendum to this story, which has the mummy case causing such havoc in America that it was decided to return it to England in 1914. It was stowed on the *Empress of Ireland*, which sank on the St. Lawrence River, with the loss of 1029 lives. When the English authorities decided to propitiate the dead priestess by returning the coffin lid to Egypt, the first stage of its journey was aboard the *Lusitania*, sunk by German torpedos in 1915.

The second curse story from this period concerns the Ingram family. Herbert Ingram founded the *Illustrated London News* in 1842, and his grandson, Sir Bruce Ingram, was to edit the paper continuously between 1900 and 1963, and carried his lifelong fascination with Egyptian archaeology into the journal. Details of the Ingram curse are often mixed up with the Murray narrative, and the two bleed into each other, but the story allegedly concerns the Ingram family role in helping finance the Gordon Relief expedition in 1885 – that iconic moment of the doomed attempt to rescue General Gordon from the dastardly dervishes of the Muslim uprising in Sudan in 1885. Some sources suggest that Herbert Ingram financed and travelled with the relief expedition on his own boat (unlikely, since he in fact died in 1860). As a souvenir of the journey, Herbert Ingram apparently decided to acquire a mummy, which he purchased for £50. Goodrich-Freer tells that "One of the fellahin, employed in its removal, said to him: 'Do you know what is going to happen to you? You will be torn to pieces and scattered like grass!'"[20] The case was also alleged to have an inscription, a curse on violaters at once "blood-curdling and exciting."[21] Henry Ingram's youngest son, William, was certainly killed on safari in Somalia, his body "scattered" by a rampaging elephant in 1888 (which was ascribed by some, the *Dictionary of National Biography* reports, to an Egyptian curse). Yet in other versions, it is Sir Bruce Ingram who suffers the consequences of the curse that this time promises "fire, water and pestilence." "Several months later," one memoir reads, "the editor's beautiful country house burned to the ground. It was rebuilt and a flood promptly swept through its ground floors. Sir Bruce did not wait for the pestilence. He sent the mummy [...] back to the Valley of the Kings."[22]

These accounts move the curse story backwards from 1923 to 1904, a key year for the start of the circulation of stories that themselves extend back into the 1860s. One urgent question still remains: *where* do these stories come from? I mean not just the fantasia of confusing, contradictory detail, but the very *idea* of the curse – because, as any academic Egyptologist will often wearily explain, Ancient Egyptian culture had no real concept of the curse. One or two feeble imprecations against grave-robbers have been gathered, but that is about it.[23] The curse, in other words, is a later cultural imposition. Usually the Egyptologist will go no further than issuing a blanket dismissal of such superstitions. Egyptology – at least until recently – had little interest in the mechanism of these stories, nor in asking about what kinds of cultural work they might do. Instead, they ask rather despairingly, as one recent essay title has it, "Is Egyptology Cursed by the Mummy's Curse?"[24] As this title suggests, impatient exasperation can't break open the persistence of such superstitions, and so therefore merely repeats it – presenting us with another form of recursion. A cultural history or genealogy, on the other hand, aims to explore the *relation* between official and unofficial knowledges, teasing out how occult discourse can flower by exploiting the resources of disciplinary science.

Surely one immediate resource feeding this outpouring of curse stories in the early twentieth century is the late Victorian Gothic revival: indeed, it is impossible to tell the stories of Douglas Murray or Lord Carnarvon without noting the presence of Conan Doyle, Rider Haggard, Algernon Blackwood, nor in noting just how often some of the principal actors were themselves partial to writing Gothic novels. The journalist Fletcher Robinson, for instance, issued his own Gothic farrago, *The Trail of the Dead*, in the fateful year of 1904. Whilst *The Illustrated Lon-*

don News carried details of the latest excavations, it also serialised Haggard's novel, *Cleopatra* in 1889. Arthur Weigall, too, had worked as a novelist once he was blocked from posts in the Antiquities Service. Is it that the Gothic provided the structure in which the curse could emerge? When Chris Baldick describes the Gothic as combining "a fearful sense of inheritance in time with a claustrophobic sense of enclosure in space," he could be describing an ancient Egyptian tomb – the ideal confining locale that evokes doomed inheritance and an inescapable curse on successive generations.[25]

There are some splendid examples of the Egyptian Gothic that come to our aid. Sax Rohmer, usually associated with Yellow Perilism, clearly included Egypt in his portrait of the monstrous Orient, and began his career with Egyptian tales in 1903. His remorseless iteration of the same plots, tropes, and racial typologies conveys a strong sense of a certain Edwardian pathology. By the time of *The Brood of the Witch Queen* (1918), there is no requirement for any motivation in the magical attacks launched on respectable London by the mysterious Anthony Ferrara, other than inherent malevolence. This "unnatural fiend," this "evilly effeminate thing," is, it transpires, "a wizard of Ancient Egypt, armed with the lost lore of that magical land."[26] This form of colonisation in reverse, finding Egyptian monstrosity at the imperial centre, had become typicalised in the 1890s. In 1894, Conan Doyle's "Lot Number 249" provides the first revived mummy that enacts the malevolent will of its reanimator – in this case, a rather suspicious Oxford student who conducts his fiendish experiments in college rooms stuffed with Egyptian grave artefacts and mummy parts. In Richard Marsh's *The Beetle* (1897), the horrific creature that oscillates between man and woman, human and scarab beetle, stalks London seeking revenge by exerting overwhelming mesmeric powers. In one adventure of the psychic detective Flaxman Low in 1899, an ancient ghost of an aristocratic mansion is rendered murderous through the vehicle of an ancient Egyptian mummy that has recently been added to the house's private museum. This is an elegant way of tracing how the fey ghosts of the English haunted house tradition become malevolent as they are Egyptianised.[27] Perhaps the most impressively sustained vision of Ancient Egyptian vengeance on London is depicted in Guy Boothby's *Pharos, the Egyptian* (1899), in which the English protagonist is made to suffer inordinately for the sin of his father, an Egyptologist who discovered and transported back to London the mummy of Ptahmes, Chief of Rameses the Second's Magicians. "'Thy father was it [...] who stole this body?'" the evil modern-day Egyptian incarnation of Ptahmes asks. "'If that be so, then may the punishment decreed against those guilty of the sin of sacrilege be visited upon thee and thine for evermore.'"[28] "'A time of punishment is coming'," Pharos promises, and the latter stages of the book detail an impressively punitive plague launched on Europe, carried by Forrester out of Egypt in a kind of reverse of the Biblical plague. London becomes a necropolis, the carrier Forrester agonisingly immune from the death he carries.

There *is* ample evidence of Egyptian curses, or malevolence at least, in the fiction of the late Victorian Gothic revival. But actually, the role of Egypt here is far more complicated than this, and any putative typology of an Egyptian Gothic would need to acknowledge that the curse tale is probably outnumbered by the mummy *romance*, in which the exotic, ancient Egyptian woman is revived either textually, through narratives framed as rediscovered and translated papyrus, or is literally reanimated through the scholarly re-enaction of ancient rites. Théophile Gautier's *Romance of a Mummy* and Rider Haggard's *Cleopatra* are key instances of textual resuscitation; probably the most famous attempt to reanimate a female mummy is Bram Stoker's *Jewel of the Seven Stars*, which followed in the wake of many similar narratives, such as H. D. Everett's *Iras: A Mystery* (1896). In Stoker, Ancient Egypt provides the metaphorical ground on which to imagine the archaeological strata of modern subjectivity, since the daughter of the Egyptologist is linked in fatal psychological and possibly biological ways to the mummy queen he is trying to revive. Egypt as a figuration of psychical sub-strata is something used masterfully by Algernon Blackwood's Egyptian-set tales. Blackwood's "A Descent into Egypt" enacts this subjective regression very precisely. Isley is drawn to the allure of the Egyptian past and literally begins to fade from modern-day existence as he burrows into his own Egyptianised unconscious. This is seen as an ascetic and disciplined process, and it is emphasised that Isley "remained unmoved by the commonplace mysterious; he told no mummy stories [and] ... he never in my hearing attached any possible revengeful character to the energy of an outraged past." Such stories, we are told, are for "superstitious minds or children."[29] The same regressions are explored at novel-length in Blackwood's *The Wave: An Egyptian Aftermath* in 1916. The Egypt evoked here is clearly one from the hermetic tradition and the scholarly revival of Egyptian magic at the end of the nineteenth century: it is a sober discipline, not a sensation fiction.[30]

It is not so straightforward, then, to trace the curse story back to the late Victorian Gothic, which uses Egypt – as you might expect from the genre – in a polyvalent and unstable way. Yet something *does* happen to the meanings ascribed to Egypt in the course of the nineteenth century. In the first wave of so-called 'Egyptomania,' following Napoleon's 1798 invasion, notions of a curse seems to be entirely absent. The Egyptian Hall, built in neopharaonic style in 1812 on Piccadilly housed Belzoni's recreation of Seti's tomb to spectacular effect in 1821. Like the famous Egyptian Court designed for the Crystal Palace in south London in 1854, these places were built as spectacles of *immersion*, encountered with a wonder seemingly unaccompanied by dread. Public demonstrations of mummy unwrapping were conveyed in numerous accounts from the 1820s and 1830s without any sense of transgression or doom.[31] Indeed, these shows were accompanied by a rash of comic farces on the subject. In fiction, the revived mummy was a device for satirical comment on contemporary society, as in Edgar Allen Poe's "Some Words with a Mummy" (1845). Jane Loudon's *The Mummy!* (1827) revived Cheops, no less, builder of the Great Pyramid, but rather than some vicious prehistoric heathen he proves to be a Christian moralist who lambasts contemporary society for its laxity. The only cloud here is Thomas de Quincey's disordered orientalised opium fantasies which incorporate Egyptian iconography. In *The Confessions of an English Opium Eater* de Quincey records: "Thousands of years I was buried in stone coffins, with mummies and sphinxes, in narrow chambers at the heart of pyramids [...] and was laid, confounded with all unutterable abortions, amongst reeds and Nilotic mud."[32] I'm not yet convinced, though, by John Barrell's view that this is a generalisable pathology – at least in relation to Egypt. This nightmare vision would be a weird thing to see transposed into a wide-scale craze for Egyptianised architecture, design and interiors.

To identify the factors that render the curse narrative culturally significant in the later nineteenth century, we need to be able to read a conjuncture of factors that strikingly few critics have yet put together. Leaving aside internal disciplinary histories of Egyptology, cultural criticism tends to allegorise the mummy as a locus of disturbance to the late Victorian sexual order (Smith), or as a liminal, abject being that concerns death and memorialisation, and thus as a locus for reflections on contemporary technologies like cinema that might preserve the dead (Lant), or as the vehicle for what Nicholas Daly has called "a descriptive commodity theory."[33] These are suggestive whilst curiously abstracted modes of allegorisation, which could often be transposed from the mummy to any other form of Gothic monster – the vampire, most obviously.[34] Meanwhile, the remarkable overdetermination of factors in Egypt tends to be downplayed, reduced to sketchy background, just when Egyptian *specificity* needs to be emphasised.

Here, then, in very compressed form are some of the factors that need to be put into play. The profession of Egyptology – a term first used in English only as late as the 1860s – became front page news in 1881, with the discovery of the cache of Royal mummies, thirteen pharaohs whose remains had been removed from tombs and reconcealed by a priesthood anxious to prevent graverobbers destroying the integrity of the bodies. The romance of the discovery was an involved spy plot against corrupt Arab tomb-raiders and illegal dealers, foiled by British daring-do. The cache famously included Rameses the Second, considered then as the pharaoh at the time of the Jewish Exodus. This prompted a renewed focus on the archaeology of Biblical history, and the setting up of the Egyptian Exploration Fund, by Amelia Edwards and others, explicitly to gather evidence for the Bible as literal history. It would initiate intense anthropological debate about the origins of Western civilisation; not least because an ancient civilisation in Africa remained a problematic anomaly in developmental theories. These events informed Rider Haggard's first visit to Egypt, and the beginning of his mournful reading of doomed Egyptian history. In *Cleopatra*, the betrayal of an august pharaonic tradition by the foreign queen announces the arrival of the Christian era with a peculiar sense of melancholic loss. Haggard's repeated return to lost empires and reincarnation plots has suggested to Richard Pearson that Haggard "was engaged on a project to restructure the Gothic around the new discipline of archaeology."[35] From the 1880s on, the Royal cache provoked the golden age of tomb-hunting in the Valley of the Kings, with a sequence of wealthy prospectors using new industrialised techniques to dig for glory. Carnarvon was the last representative of this entrepreneurial era. The attitudes of this generation were evoked at the opening of Amelia Edwards' book *Pharaohs, Fellahs and Explorers* in 1891: "It has been said that all Egypt is but the façade of an immense sepulchre," she says. "This is literally true; for the terraced cliffs that hem in the Nile [...] are everywhere honey-combed with tombs. [...] The mummied generations are everywhere."[36] This representation of Egypt as an uncanny land of the dead, a wasteland laced with hidden treasure, works exactly to obscure the other blindingly obvious element of this conjuncture: the revolt of *living* Egyptians.

The Egyptian debt crisis, worsening through the 1870s, had resulted in a decade of viciously enforced tax collection from the poor population, and dual control of Egyptian finances by France and England. Some fifty per cent of Egyptian revenue was paid to the eleven European powers that had financed the collapsed industrial and infrastructural development of the country under the Khedive, Ismail. With Ismail deposed and the country being run by international banks, the spark for rebellion was the union of the disaffected Egyptian army officer core with emergent nationalist and Islamic political movements, resulting in a rebellion that began in Alexandria in June 1882, with fifty Europeans killed. The British Expeditionary Force, led by Garnet Wolsley invaded and occupied Cairo by September that year. Evelyn Baring, soon to be Lord Cromer, was installed as British Consul-General, and immediately added four million pounds to the debt to repay the damage done to European interests during the rebellion. This moment has frequently, if controversially, been taken to be the turning point in British imperial history, where a harsher ideological forward policy emerged, the tipping point toward military occupation that escalated into the Scramble for Africa.[37] Although Egypt did not become a British protectorate until 1914, the British dominated Egyptian policy through Lord Cromer, and the security of the southern border of Egypt generated twenty years of religious war between Islamic insurgents in Sudan and the British occupiers. This war incorporated the botched evacuation of Sudan in 1884 and the subsequent death of Gordon in Khartoum. The war was terminated by Kitchener at the Battle of Omdurman in 1898, after which the victorious troops conducted their own excavations, digging up the body of the dead Mahdist leader, Mohammed Ahmed, and displaying his humiliated remains.

Wherever there is imperial occupation, there is a reserve of supernaturalism, an occult supplement to allegedly enlightened rule that becomes one of the popular currencies for acknowledging and perhaps even beginning to negotiate the consequences of colonial violence and guilt. This occult currency is what makes Rudyard Kipling's early gossip tales about India so compelling, and once telepathy is coined in 1882, that mechanism instantly provides the apparatus for a parallel communication system that works on an imperial measure.[38] The mummy curse emerges from complex superimpositions of the colonial context. The way that Douglas Murray describes being turned to ice by the malignant stare of the priestess of Amen-Ra of course tempts one to theorise in terms of the structure of paranoia. The mechanism of paranoia, Freud contended, is a suppressed internal thought that, "after undergoing a certain kind of distortion," returns as an external and persecutory threat.[39] No wonder, then, that the British Museum becomes the locus of the Murray story, because the museum is what Bruno Latour terms a "centre of calculation," the hub of a Victorian network of routes and pathways that passes artefacts back to the imperial centre to transform them into knowledge.[40] Inert stones or wrapped bodies become animated by the Western museological context only at the expense of turning into persecutory objects. This is a repeated scenario in the Egyptian Gothic tale. But this is not just a two-term relationship, an anxiety about the kind of plunder by Western powers of ancient Egypt – a plunder that was already being criticised as early as 1841, when Gliddon wrote an appeal against the "Destruction of the Monuments of Egypt."[41] The discoveries in the 1880s and 1890s of the New Kingdom pharaohs in the Valley of the Kings certainly did open up an uncanny history of an ancient empire extended and then collapsed, just as England was beginning its major territorial expansion into Africa. But we also need to include a third term in this relationship, the living Egyptians, the nationalists and Islamic revolutionaries that haunted the English imagination with their fanaticism, murderous cruelty and, of course, their huge reserves of native superstitions. The 'evil eye' of the mummy case is surely also a transposition of those elaborate Arabic beliefs in the malevolence of the bewitching gaze.[42] The more Orientalist discourse condensed a stereotype of the credulous and child-like Arab cowering in superstitious awe, the more the mummy curse narrative tells us of a Western investment in very similar beliefs undergoing complex cross-cultural translation below the level of official imperialist discourse.

If I have objected to critics turning too quickly to abstract allegorisation of the mummy, has this move to imperial history left genealogy behind and begun to indulge in my own allegorical reading of the curse? I'm not sure it has, given how *literally* the specific mummy curse tales end up being concerned with the mechanics of Empire. Howard Carter and Lord Carnarvon's management of the opening of the tomb of Tutankhamen was disastrous, partly because their imperious contempt for the native press and Egyptian Antiquities Service was outdated in the very year when Britain granted Egypt formal, if limited, independence in 1922, and a nationalist government was in power under the prime ministership of a man the British had only recently imprisoned as a political subversive. The tomb was opened against the turmoil of negotiations to secure Egyptian independence, and in a charged atmosphere where a number of English people had been assassinated in Cairo and elsewhere. Carnarvon's death is never related to this context in curse narratives, despite being perhaps suitable material for a generational curse, given that his father, Henry Herbert, the 4th Earl of Carnarvon, had been Colonial Secretary in governments of the 1860s and again in the 1870s. Douglas Murray, similarly, was bound up with colonial expansion. Although all the narratives of the family curse rarely give any details of his life, Murray did co-author a memoir about Sir Samuel Baker, and was named as Baker's sole executor when the latter died in 1893. Baker had been knighted for his exploration of Upper Egypt and the tracing of the White Nile back to its source in the lake he christened Albert in 1863. He then went on to carve out and rule a kingdom called Equatoria in what is now Sudan, in the name of the suppression of the slave trade. "Although I am no 'nigger-worshipper'," Baker wrote in 1869, "I have great hopes of effecting a vast improvement over all those countries belonging to the Nile Basin."[43] It was this intervention into the slave routes across Sudan that helped stir the Mahdist revolution of the 1880s, and the lone stand against the dervish hordes by Baker's successor in Equatoria, General Charles Gordon. The Expedition to rescue Gordon was of course the reason for the third curse story that I told, about the Ingram family, cursed for buying up a mummy at the time of the Gordon Relief campaign.

This, then, is the thoroughly materialist, maybe even rather leaden, explanation of the origins of the curse story, which we can trace back into those legendary white spaces of the map of central Africa in the 1860s. Mummies are liminal, abject, sexually ambiguous, alluringly veiled, disciplinarily subversive, magical and monstrous objects, and these categories allow them to do a lot of their cultural work. Gothic criticism has delighted in exploring the "technology of monstrosity" that condenses in the figure of the mummy.[44] But they also carry the distinct residues of a colonial context that needs to be understood in all its complex ramifications in a critical passage of British imperial history.

Notes

[1] Hans-Joachim Neubauer, *The Rumour: A Cultural History* 2.

[2] See Julian Thomas, "Archaeology's Place in Modernity."

[3] Michel Foucault, *Society Must Be Defended: Lectures at the Collège de France 1975-6* 7 and 8.

[4] Marie Corelli, cited Christopher Frayling, *The Face of Tutankhamun* 43. Note the emphasis on knowledge which falls beyond the bounds of that 'total archive,' the British Library.

[5] Arthur Conan Doyle, cited Frayling, *The Face of Tutankhamun* 46.

[6] Arthur Weigall, *Tutankhamen, and other essays* 59.

[7] Ibid. 138.

[8] Neubauer, *The Rumour: A Cultural History* 2.

[9] These are listed in loving, credulous detail in Philipp Vandenberg, *The Curse of the Pharaohs.*

[10] Philipp Vandenberg, *The Curse of the Pharaohs* 28.

[11] Headline cited Julie Hankey, *A Passion for Egypt: Arthur Weigall, Tutankhamun and the "Curse of the Pharaohs"* 340.

[12] Cited Frayling, *The Face of Tutankhamun* 47.

[13] Goodrich-Freer, "The Priestess of Amen-Ra" 16.

[14] "Cheiro," *True Ghost Stories* 54.

[15] Goodrich-Freer, "The Priestess of Amen-Ra" 15.

[16] "Cheiro," *True Ghost Stories* 48.

[17] Ibid. 52.

[18] Fred Seward, cited John Richard Stephens, *Into the Mummy's Tomb* 11.

[19] Philipp Vandenberg, *The Curse of the Pharaohs* 197.

[20] Goodrich-Freer, "The Priestess of Amen-Ra" 19.

[21] G. St. Russell, "The Mysterious Mummy" 171.

[22] Cited Frayling, *The Face of Tutankhamun* 56. Frayling details the factual inaccuracies of this account.

[23] See the brief discussion of these in Salima Ikram and Aidan Dodson, *The Mummy in Ancient Egypt: Equipping the Dead for Eternity.*

[24] Carter Lupton, "'Mummymania' For the Masses: Is Egyptology Cursed by the Mummy's Curse?".

[25] Chris Baldick, "Introduction" to *Oxford Book of Gothic Tales* xix.

[26] Sax Rohmer, *The Brood of the Witch Queen* 23 and 196.

[27] Kate O'Brien Prichard and Hesketh V. Prichard, *Ghosts* (1899).

[28] Guy Boothby, *Pharos the Egyptian* 56.

[29] Algernon Blackwood, "A Descent into Egypt" 278.

[30] The hermetic tradition of "Egyptosophy" has been traced by Erik Hornung and the late Victorian, early Edwardian occult revival by Alex Owen.

[31] See, for instance, William Osburn, *An Account of an Egyptian Mummy.*

[32] Thomas De Quincey cited Barrell xii.

[33] Nicholas Daly, *Modernism, Romance, and the Fin de Siècle* 114. See also particularly Antonia Lant's translation of Egyptian contexts into general metaphors for early cinema.

[34] See David Glover, *Vampires, Mummies and Liberals: Bram Stoker and the Politics of Popular Fiction.*

[35] Richard Pearson, "Archaeology and Gothic Desire: Vitality Beyond the Grave in H. Rider Haggard's Ancient Egypt" 218-219.

[36] Amelia Edwards, *Pharaohs, Fellahs and Explorers* 3 and 12.

[37] The focus of this discussion has long been Ronald Robinson and John Gallagher with Alice Denny, *Africa and the Victorians: The Official Mind of Imperialism.*

[38] See Roger Luckhurst, *The Invention of Telepathy, 1870-1901.*

[39] Sigmund Freud, "Psychoanalytic Notes upon an Autobiographical Account of Paranoia" 204.

[40] Bruno Latour, *Science in Action* 214ff.

[41] Discussed in Donald Malcolm Reid, "Rediscovering Ancient Egypt."

[42] For more detailed comparative discussion, see Alan Dundes, *Evil Eye: A Folklore Casebook* and Rupert Sheldrake, *The Sense of Being Stared At, and other aspects of the extended mind.*

[43] Cited in T. Douglas Murray and A. Silvia White, *Sir Samuel Baker: A Memoir* 147.

[44] See Judith Halberstam, *Skin Shows: Gothic Horror and the Technology of Monsters* for a general theory of overdetermined meanings of late Victorian Gothic monsters.

Works Cited

Baldick, Chris. "Introduction," in *Oxford Book of Gothic Tales*. Oxford: Oxford UP, 1992.

Barrell, John. *The Infection of Thomas de Quincey: A Psychopathology of Imperialism*. New Haven: Yale UP, 1991.

Blackwood, Algernon. "A Descent into Egypt," in *Incredible Adventures*. London: Macmillan, 1914.

---. *The Wave: An Egyptian Aftermath*. London: Macmillan, 1916.

Boothby, Guy. *Pharos the Egyptian*. London: Ward, Lock and Co., 1899.

"Cheiro." *True Ghost Stories*. London: London Publishing Co., 1928.

Conan Doyle, Arthur. "Lot 249," in *Round the Red Lamp: Being Facts and Fancies of Medical Life*. London: Methuen, 1894.

Daly, Nicholas. *Modernism, Romance and the Fin de Siècle*. Cambridge: Cambridge UP, 1999.

Douglas, Theo. *Iras: A Mystery*. London: Blackwood, 1896.

Dundes, Alan. *The Evil Eye: A Folklore Casebook*. New York: Garland, 1982.

Edwards, Amelia B. *Pharaohs, Fellahs and Explorers*. London: James R. Osgood, 1891.

Foucault, Michel. *"Society Must Be Defended": Lectures at the Collège de France 1975-6*. Trans. David Macey. London: Allen Lane, 2003.

Frayling, Christopher. *The Face of Tutankhamun*. London: Faber, 1992.

Freud, Siqmund. "Psychoanalytic Notes upon an Autobiographical Acount of Paranoia" (1911), in *Penguin Freud Library*, vol. 9. Harmondsworth: Penguin, 1979.

Gautier, Théophile. *The Romance of a Mummy*. Trans. M. Young. London: John and Robert Maxwell, 1886.

Glover, David. *Vampires, Mummies and Liberals: Bram Stoker and the Politics of Popular Fiction*. Durham: Duke UP, 1996.

Goodrich-Freer, Ada. "The Priestess of Amen-Ra: A Study in Coincidences." *Occult Review* XVII (1913): 11-19.

Halberstam, Judith. *Skin Shows: Gothic Horror and the Technology of Monsters*. Durham: Duke UP, 1995.

Hankey, Julie. *A Passion for Egypt: Arthur Weigall, Tutankhamun and the "Curse of the Pharaohs."* London: I. B. Taurus, 2001.

Hornung, Erik. *The Secret Lore of Egypt: Its Impact on the West*. Trans. David Lorton. Ithaca: Cornell UP, 2001.

Ikram, Salima and Aidan Dodson. *The Mummy in Ancient Egypt: Equipping the Dead for Eternity*. London: Thames and Hudson, 1998.

Lant, Antonia. "The Curse of the Pharaoh, or How Cinema Contracted Egyptomania." *October* 59 (1992): 87-112.

Latour, Bruno. *Science in Action*. Cambridge, MA: Harvard UP, 1987.

Loudon, Jane Webb. *The Mummy! A Tale of the Twenty-Second Century* (1827). Ann Arbor: U Michigan P, 1994.

Luckhurst, Roger. *The Invention of Telepathy, 1870-1901*. Oxford: Oxford UP, 2002.

Lupton, Carter. "'Mummymania' for the Masses – Is Egyptology Cursed by the Mummy's Curse?" in *Consuming Ancient Egypt*. Ed. Sally MacDonald and Michael Rice. London: UCL Press, 2003.

Marsh, Richard. *The Beetle* (1897). Ed. Julian Wolfreys. Peterborough, Canada: Broadview, 2004.

Murray, T. Douglas and A. Silva White. *Sir Samuel Baker: A Memoir*. London: Macmillan, 1895.

Neubauer, Hans-Joachim. *The Rumour: A Cultural History.* Trans. Christian Braun. London: Free Association, 1999.

Osburn, William. *An Account of an Egyptian Mummy, Presented to the Museum of the Leeds Philosophical and Literary Society by the late John Blayds.* Leeds: Robinson and Hernaman, 1828.

Owen, Alex. *The Place of Enchantment: British Occultism and the Culture of the Modern.* Chicago: Chicago UP, 2004.

---. *Lord Cromer: Victorian Imperialist, Edwardian Proconsul.* Oxford: Oxford UP, 2004.

Pearson, Richard. "Archaeology and Gothic Desire: Vitality Beyond the Grave in H. Rider Haggard's Ancient Egypt," in *Victorian Gothic: Literary and Cultural Manifestations in the Nineteenth Century.* Ed. Ruth Robbins and Julian Wolfreys. London: Palgrave, 2000.

Prichard, Kate O'Brien Ryall and Hesketh V. Prichard. *Ghosts: Being the Experience of Flaxman Low.* London: C. Arthur Pearson, 1899.

Reid, Donald Malcolm. "The Urabi Revolution and the British Conquest, 1879-82," in *The Cambridge History of Egypt*, vol. 2. Ed. Martin W. Daly. Cambridge: Cambridge UP, 1998.

---. *Whose Pharaohs? Archaeology, Museums, and Egyptian National Identity from Napoleon to World War I.* Berkeley: U California P, 2002.

Robinson, B. Fletcher and J. Malcolm Fraser. *The Trail of the Dead: The Strange Experiences of Robert Harland.* London: Ward, Lock and Co., 1904.

Robinson, Ronald and John Gallagher with Alice Denny. *Africa and the Victorians: The Official Mind of Imperialism.* 2nd ed. London: Macmillan, 1981.

Rohmer, Sax. *The Brood of the Witch-Queen.* London: C. Arthur Pearson, 1918.

---. *Tales of Secret Egypt* (1920). Amsterdam: Fredonia, 2002.

Romber, John. *Valley of the Kings.* London: Michael Joseph, 1981.

Russell, G. St. "The Mysterious Mummy." *Pearson's Magazine* 28 (1909): 162-172.

Sheldrake, Rupert. *The Sense of Being Stared At, and Other Aspects of the Extended Mind.* New York: Crown, 2003.

Smith, Andrew. "Love, Freud, and the Female Gothic: Bram Stoker's *The Jewel of the Seven Stars.*" *Gothic Studies* 6.1 (2004): 80-89.

Stephens, John Richard, ed. *Into the Mummy's Tomb.* New York: Berkeley, 2001.

Stoker, Bram. *The Jewel of the Seven Stars* (1903). Stroud: Alan Sutton, 1996.

Thomas, Julian. "Archaeology's Place in Modernity." *Modernism/Modernity* 11.1 (2004): 17-34.

Vandenberg, Philipp. *The Curse of the Pharaohs.* Trans. Thomas Weyr. London: Hodder and Stoughton, 1975.

Weigall, Arthur. *Tutankhamen, and other essays.* New York: George Doran, 1924.

Magic Pastoral

Tina McElroy Ansa

Walter Göbel

The history of the novel in England mimics the history of secularisation and enlightenment individualisation, culminating on the one hand in the subjection of human nature to mundane and ungodly natural laws in naturalism and on the other hand in the solipsistic self-reflections of modernist *stream-of-consciousness* novels and of postmodern meta-fiction. It is within the naturalist mode that the late-born African American novel first became known to a wider audience. The protest novels of the 1940s present their protagonists as dehumanised victims of a stifling ghetto existence who live on the darker side of a predominantly white definition of the economy and of modernisation, and who suffer from emasculation and the blotting out of the African American experience, hence the pervasive use of the theme of invisibility. Horkheimer and Adorno's dialectics of the enlightenment must have appeared familiar for the mass of African Americans living in ghettos and leading lives of quiet desperation.[1] Quite quickly, however, the African American novel moved beyond its naturalist beginnings to modernist experiments, to postmodernism and magic realism.

Given this trajectory of the African American novel, it is little wonder that, while partly following the trends of mainstream literature, the genre also subverts and partly inverts the main line of development in Western fiction, that from romance to novel. Undoubtedly this line of development has been challenged more than once from within the mainstream, whether in the gothic mode or in more recent explorations of science fiction, utopian and fantasy literature. In the African American novel, however, the discovery of romance and of the supernatural has been at the centre of the formal evolution of the genre, last but not least because of the cultural links the adoption of magic elements has allowed and allows with imagined Africas.[2] The subversion of the novel and its realistic tradition can be interpreted as part of an antagonistic self-fashioning which is inspired by Latin American explorations of magic realism.[3] Three examples may suffice to illustrate this aesthetic agenda: The inversion of Western enlightenment discourses by a culture of magic, music and dance in Ishmael Reed's novel *Mumbo, Jumbo* (1972), the rejection of the industrialised and commodified North for a mythic and enchanting vision of the South in Toni Morrison's *Song of Solomon* (1977) and the contrast between alienated city life in New York and the magic powers that are ever active among the blacks of pastoral Georgia in Gloria Naylor's *Mama Day* (1988). Taking a closer look at Tina McElroy Ansa's *Baby of the Family*, I want to point out how the subversion of enlightenment discourse has finally also allowed for a non-antagonistic version of what I would like to call the 'magic pastoral.' In the course of this development the figures commanding magical powers are promoted from the traditional status of outsiders to a central position, and magic is presented not as dangerous but as beneficial.

If we first take a short look at the development of African American literature outside the novel genre, the rediscovery of romance, magic and the supernatural gains in definition. In the short story and in poetry we can indeed find some instances of the use of the supernatural and of magic before the heyday of magic realism in the 1970s and 1980s, for example in Charles Wadell Chesnutt's story "The Goophered Grapevine," published as early as 1887 and often interpreted as a *local color* story. In the manner of Washington Irving and of the authors of the so-called *plantation school*, the story is presented from the point of view of an elderly storyteller, who has access to the folklore of North Carolina. Like Joel Chandler Harris' famous storyteller Uncle Remus, Uncle Julius McAdoo spins a tale for the gullible listener, here for a Northerner who wants to move to the South because of his wife's poor health. He is considering buying an abandoned vineyard which Uncle Julius, as finally becomes apparent, had been squatting on for a number of years. Uncle Julius enlightens this Northerner, the narrator of the frame tale, in his broad dialect that the vineyard has been bewitched or 'goophered' by its former owner to keep thieving blacks at bay and that the vines were additionally spoilt by the disastrous advice of a presumptuous Northerner who posed as an expert concerning methods of cultivation. At the end of the story, however, it transpires that the vineyard is bought in spite of Uncle Julius' warnings, and that it proves highly profitable for the new owner. Uncle Julius' tale of magic is finally viewed as an attempt to scare the prospective new owner off and to keep the vineyard for himself. The romance elements and the magic are thus shown to be no more than a background of *local color*, supporting a somewhat *tall tale* in an otherwise realistic setting, and the Northerner functions as a commonsensical demystifier in a backward world in which the black trickster figure wants to outwit him.[4] Magic belongs to a bygone, more unenlightened age.

It is only in the Harlem Renaissance of the twenties and thirties that black magic is taken more seriously, that is, in the course of the discovery and idealisation of Africa and African art, a discovery that was, by the way, to quite some extent ushered in by European expressionism.[5] Visions of imagined Africas inspired art and poetry in the Harlem Renaissance, for example in Claude McKay's poem "Outcast," where African culture and the Western gods (the holy trinity? economic gain?) are set apart:

My soul would sing forgotten jungle songs
I would go back to darkness and to peace,
But the great western world holds me in fee,
And I may never hope for full release

While to its alien gods I bend my knee [...][6]

While many authors of this era were more interested in European art forms than in Africa, the turn towards African and Caribbean Voodoo is central for Zora Neale Hurston, who became initiated to Voodoo cults in New Orleans at the age of 26 and learnt incantations of the famous Marie Leveau.[7] So, before the birth of the novel, the exploration of African culture and magic had definitely set in. In the naturalist phase, however, the supernatural and magic were anathema. In Wright's naturalist novel *Native Son* the supernatural is only present in a few allusions to gothic effects. The white world appears ghostlike and threatening, like a white wall, and Mrs. Dalton, the blind wife of Bigger's white employer, appears to be a ghost when Bigger first meets her: "Bigger paused, bewildered; then he saw coming slowly toward him a tall, thin, white woman, walking silently, her hands lifted delicately in the air and touching the walls to either side of her. Bigger stepped back to let her pass. Her face and hair were completely white; she seemed to him like a ghost."[8] And in the key scene of the novel, when Bigger is alone in the bedroom with Mrs. Dalton's daughter, this ghost suddenly appears again: "Bigger turned and a hysterical terror seized him, as though he were falling from a great height in a dream. A white blur was standing by the door, silent, ghostlike. It filled his eyes and gripped his body."[9] The Dalton's white cat adds to the diabolical atmosphere that emasculates and terrifies Bigger and provides an adequate setting for his involuntary killing of Mary Dalton and her subsequent dismemberment. Supernatural elements are here introduced as traditional gothic features and are associated with a white society which endangers the main protagonist. At the same time they characterise Bigger's distorted perception, the fear bordering on madness that does not allow him to acknowledge the white other as an individual. The gothic phenomena are projections of Bigger's overheated and obsessed imagination. In more recent novels, however, the supernatural becomes a part of black society worth exploring and preserving; in the modern and postmodern African American novel black magic becomes more and more an ideological issue with positive overtones.[10]

I would now like to turn to Tina McElroy Ansa's novel *Baby of the Family* as a striking example of a positive use of black magic – so entirely opposed to the presentation of ghosts and magic as threatening elements in "The Goophered Grapevine" and "Native Son." The heroine of the novel, Lena McPherson, is from the very outset introduced as a special person, born with a caul (a veil) on her face and thus gifted with supernatural powers, e.g. the ability to see ghosts or to "put the magic" on things that are out of order, such as a car, a television set or a stuck door.[11] The supernatural is thus not part of a threatening and intimidating power that acts upon the main protagonists, as in "The Goophered Grapevine," in the gothic tradition and in *Native Son* – it activates and inspires the main figure and is interpreted as a "gift from God" from the outset. Lena is able to see ghosts that are invisible to others, and she can even talk to her dead grandmother. In a key scene young Lena meets the ghost of a slave girl who had many years ago committed suicide rather than enduring the hardships of slavery any longer, and from her she learns a lot about the plight of the former slaves. By her sympathy for the ghost Rachel and by listening to her history, which is a tale of woe, Lena is able to free this ghost once more and through her to find access to communal cultural memories. In a first act of frenzied liberation, the slave-girl Rachel had chosen to commit suicide by drowning herself in the sea off the Georgia coast, where Lena is now on vacation. Lena learns self-respect from the ghostly Rachel and the fact that the seaside and the coast belong to everyone, also to colored people. As a sign of liberation and of the budding of an intimate understanding between the two, the ghost laughs loudly at the end of the episode when Lena romps around her in the water.

Ghosts and the theme of the supernatural are not central to *Baby of the Family*, however. They merely add to the special value of Lena, who is treated like a princess by her parents in any case, and they emphasise the theme of female empowerment which inspires all of Ansa's novels. Most of *Baby of the Family*, however, is a realistic presentation of a family idyll with very little tension. As a member of a well-to-do middle class family, Lena enjoys a sheltered childhood, though because of her precocity and extrasensory powers she does attain somewhat of an outsider status for a while and has few friends of her own age. The generally idyllic atmosphere and the lack of tension is emphasised by static chapter headings like "Veil," "Tea," "Picture," "Night," "Table," "Hair," "Beach," "School," etc. The chapters are generally sketches with a marked ecphrastic element: "The summer vacation ... to the Georgia coast the year she was seven could have been called *Colored Folks Going to the Beach*, like a painting" (*Baby of the Family* 128); this is also a phrase which Lena's grandmother repeats with aversion, as she old-fashionedly maintains that colored folks shouldn't go to the beach.[12] The grandmother's dislike for the beach holiday may serve as an example for the kind of everyday and easily solved tensions that serve to move the episodic chapters forward and to link the scenes in what appears otherwise to be a collection of autobiographic tableaus:

'Grandma acts like she takes it personally that we're going to the beach,' Lena told her mother, whom she found at the big white metal kitchen table shucking long ears of sweet white Silver Queen corn. Her father could never get enough sweet fried corn in the summer when it was plentiful, cooked just the way Nellie did it: First, holding the corn upright on the pointed end, she went over the outside of each shucked ear with a very sharp butcher knife, barely shearing off the very tips of each kernel. Each of the firm pieces of corn fell into the bottom of the big black iron skillet with a little plink. Then she went back over each ear with the flat

> side of the knife, scraping the cobs dry of all their milky juices. 'This is the only way to fix fried corn,' her mother told her as she spit a long strand of silk from her mouth over her shoulder. The smooth reddish-brown skin on her face, chest, and bare arms would be dotted with spatters of corn juice. It seemed to Lena that her mother was polka-dotted with bits of corn milk all summer long. 'Now you watch how I do this,' her mother said as she added a handful of flour to the corn, washed the remaining flour off her hand into the mixture with about a cup of water, stirred the whole thing around in the big black skillet with her fingers, and put it over a low flame on the stove. 'Heaven help you, you may get a man who loves fried corn the way your daddy does.' Finally she poured some hot fatback bacon grease into the corn, where it sizzled, and covered it with a big lid. 'I'm watching,' Lena said as she sat with her chin on the table watching a fat green worm inch across the surface. But she persisted, 'Mama, why is Grandmama so mad about us going to the beach?'
> (*Baby of the Family* 130f.)

This scene is a good example of ecphrasis or of the use of the tableau, of the loving presentation of visual detail from a child's point of view and of slowly moving time in idyllic fiction. A long descriptive paragraph that lingers on the preparation of a well-known dish – smells and tastes seem to be savoured in anticipation – is inserted before Lena's question is repeated, and it will take another half page till it is tentatively answered. Such a presentation of slowly moving time in an idyllic childhood is noteworthy in two respects. Firstly, it implicitly denies and subverts the classic tale of deprived childhood and dramas of violence in an all-black community, as prototypically presented in Richard Wright's fictional autobiography *Black Boy*. It turns against the classical stereotypes and insists on the happiness of the new black middle class. Secondly, the idyllic form transforms the function of magic and the supernatural in fiction. To the extent that it can be assimilated to the laws of the idyll, magic is not dangerous, it can even contribute to felicity and happiness, most obviously so when the grandmother's ghost appears to give Lena some good advice, or when Rachel enlightens Lena about the more atrocious aspects of slavery. Naturally, these scenes also strike subversive notes, but they are given an idyllic setting and do not lead to any kind of confrontation with the absent – and thus in a sense invisible – white society.

In *Baby of the Family* a knowledge of magic is conducive to a good education and to the broadening of a child's mind. To talk with the dead can lead to enlightenment about the living; like Lena's veil, magic appears to be a "gift of God." Thus the marginalisation of magic – e.g. because it belongs to the tradition of the *tall tale* ("The Goophered Grapevine") or because it is a gothic element that must be exorcised ("Native Son") – is questioned by Tina McElroy Ansa. In *Baby of the Family* magic is a central part of lived experience and conducive to the understanding even of everyday things.

The re-interpretation of the classic story of black childhood as a battle for survival becomes palpable if we take a short glance at Richard Wright's *Black Boy*. The martyrdom of the first person narrator leads to reflections on the striking absence of the softer and kinder emotions in African American families who have to fight for a living: "... I used to mull over the strange absence of real kindness in Negroes, how unstable was our tenderness, how lacking in genuine passion we were, how void of great hope, how timid our joy, how bare our traditions, how hollow our memories, how lacking we were in those intangible sentiments that bind man to man, and how shallow was even our despair. After I had learned other ways of life I used to brood upon the unconscious irony of those who felt that Negroes led so passional an existence! ... I knew that Negroes had never been allowed to catch the full spirit of Western civilization, that they lived somehow in it but not of it. And when I brooded upon the cultural barrenness of black life, I wondered if clean, positive tenderness, love, honour, loyalty, and the capacity to remember were native with man."[13] Tina McElroy Ansa's *Baby of the Family* can be interpreted as a re-writing of the classical story of African American infant martyrdom, as a counter-narrative which acknowledges the existence of tenderness and love in black families, albeit within the middle class and in a somewhat fictionalised black community (Mulberry, Ga.). Scenes of disorder, violence and injustice are in *Baby of the Family* erased and displaced by household order, neatness and an atmosphere of loving care and tenderness. Re-writing traditional black experiences of childhood on the one hand, *Baby of the Family* also turns against the Western Christian and enlightenment discourses which have traditionally paired magic with blasphemy and heresy.

Tina Mc Elroy Ansa's most recent novel, *You Know Better*, continues the centering of magic and the supernatural against the marginalisation so characteristic of the Western enlightenment and of colonialist ideology. Three generations of women from the Pines family are, by a series of inner monologues and dialogues, linked with ghosts who appear as saving angels to each of them, in a night in which the youngest member of the family, LaShawndra, has committed a serious blunder and finally undergoes a crucial experience of reform. The angels, all deceased elderly women of the community, act as psychotherapists and confessors for the Pines women who spill out their life stories and are thus able to cast off the burdens of care and woe that are weighing them down. The 'ghostly angels' manage to help the story move towards a harmonious ending in which the careless, uninhibited and often shameless LaShawndra learns to take her mother's and grandmother's feelings into consideration and to assume responsibility for her own life. Again magic and the supernatural have a positive function, here: of sorting out the mess of African American lives, of helping the blundering Pines family to cope with woe, disunity and guilt that have built up over the years. It appears that only with supernatural help can the tensions of African American existence be endured and a catastrophic turn of nineteen-year-old LaShawndra's life be averted. As the lives of the Pines women are patched up, an ideal of resilience and of female bonding, which encompasses the ghostly angels of the three Pines women, is established. The community of powerful women reaches out beyond the grave and is able to save LaShawndra from the dangers of drug abuse, unprotected sex with numerous partners and other forms of careless or even criminal behaviour.[14] The charmed lives the Pines women lead have utopian overtones which are emphasised by the fact that the white world is so obviously absent from Ansa's fictive Mulberry, the setting for all her novels. The question why the black couples fail so often, why the males disappear and leave the women with the burden of parenthood, or the question how important the influence of a white-dominated labour market is, lies beyond the scope of Ansa's world. She focuses on states of disorder, but not on causes. Her female figures fight for survival, but they don't actually intend to change the world. However, by offering an ideal of female bonding and decency which reaches beyond this world, Ansa does sketch an ideal female ethics of care and responsibility as a conceptual basis for black survival. The three black angelic figures guiding the Pines women could also be interpreted as midwives for a female ethics or for the better black life.

At the centre of the novel is the opposition between concepts of order and of disorder. Disorder reigns in the outside world, where young people of both sexes want to take advantage of LaShawndra or simply don't take proper care of their bodies, their health and their often misplaced affections. Such exterior disorder is balanced by a concept of female bonding and protection and of family harmony and love, the latter an extension of the idyllic atmosphere suffusing *Baby of the Family.* In *You Know Better* the Pines household is tidy, orderly and painstakingly clean, a fact that to LaShawndra appears like magic. She takes her grandmother Mama Mama for a good spirit which can magically provide a haven for love and affection: "'You ought to see Mama Mama when she have one of her big sit-down dinner parties with all her good china and glasses and stuff. She go to all that trouble cooking and seeing that the house is clean and everything, then she spend all evening making sure nobody feel funny if *they* don't know the right fork to use. So everybody come out feeling good. It's like magic. But Mama Mama say it ain't no magic. She say it's just hard work. I mean, she so upright and righteous and real.' I surprised my own self with those words, 'cause they sounded so real good together, so I repeated 'em to myself three times."[15] With the emphatically repeated terms 'upright' and 'righteous' and with the concepts of order and cleanliness, traditional Puritan values are called upon and claimed for a black Mulberry family. What appears like magic to young LaShawndra is no more than the creation of an aura of care and protection, though coloured by a bit of silver fork etiquette. This may appear magical to a child, especially if compared to classic descriptions of havoc, need and dereliction in traditional stories of black childhood in the South or in northern ghettos. The term 'magic' here emphasises the impact of the idyll when it transforms black life and black art, but it does not, strictly speaking, approach the supernatural. This sphere is restricted to the ghostly advisers of the Pines women and anticipates a black female utopia beyond the sphere of the everyday. At the same time, magic and the supernatural transcend the world of righteousness and uprightness which they are protecting and supporting. It remains open to debate whether the idyll of normal caring love and homeliness can, in its ephemeral nature, only be evoked with the help of black magic, or whether black magic is added to the traditional commonplaces of the idyllic and provides elements of a specific African American identity.

In the fictional all-black community of Mulberry the ghosts are usually protective and can even be beautiful: "When Miss Moses turned to me and smiled, I was struck with how lovely she looked. Miss Moses had a beautiful face. I could see that even in the darkened interior of the car. Not beautiful like a lovely elderly woman, but beautiful like a woman in her prime. And although her face was relatively unwrinkled for a woman her age, it was not her youthful skin that made her beautiful. It was the glow of spirit underneath seeping through the very pores of her skin. Just beautiful. The women of my family have good skin, too. But Miss Moses looked like something in an old painting. Luminous. The glow of her skin lit up the dark inside the car like an aureole" (*You Know Better* 30). The benevolent ghost with divine attributes (the aureole, the luminosity) is a new figure in Western literature and is also largely unknown in African mythology. The ghosts of ancestors in Igbo or in Voodoo mythology, for example, have usually to be appeased and offered gifts, in order to forestall possible malevolent acts. In *You Know Better*, however, they assume divine powers and become guarding angels for the Pines women, thus compensating for the suffering and oppression, which is often presented in African American novels, in a somewhat utopian sphere.

What saves *You Know Better* from sentimentality are the utopian dimension and the intricate presentation of personal feelings in inner monologues, as well as the fact that the idyllic fiction is poised against a tradition of protest fiction in Black and White and thus has an intertextually subversive quality: any piece of all-black African American idyllic fiction must have utopian overtones and – at least implicitly – partake of subversion. Formally speaking the most remarkable feature is the coupling of the laws of idyllic fiction with magic and the supernatural. Within the Western tradition, pastoral may have itself generally had a utopian dimension, but it was traditionally thisworldly, idealising a simpler form of life. In Tina McElroy Ansa's novels simplicity and the supernatural are conjoined, thus indicating that the idyll is not easily accessible for African Americans, but has an elusive quality and must be conjured up. In *You Know Better* the happiness of the Pines family and the harmony of the idyllic female community can only be maintained because the guardian angels intervene and save LaShawndra from a life of drug abuse, exploitation and probably eventual prostitution. The fusing of the idyllic and the supernatural – perhaps Ansa's novels could be called magic pastorals, as suggested above – has two additional dimensions: firstly it emphasises the spiritual dimension of the African American experience, the importance of transcendence, magic and superhuman powers, to be found, for example, in myths of flying in African American and African folklore and art. In magic pastoral links are formed with the African tradition, while at the same time a more positive exploration of the black American experience is suggested. Centrifugal and centripetal tendencies, a distancing afrocentric gesture and a desire for new forms of self-fashioning in the US, are interwoven. Secondly, the magic pastoral subverts enlightenment discourses, turning against the dichotomy of reason and imagination, common sense and fantasy and indicating possibilities of a fuller, more complex use of human potential. Female susceptibility, female ethics and a tolerant admittance of possible worlds, of the imaginative and the supernatural, implicitly question the marginalisation of magic in Western cultures. In this respect Ansa's novels neatly dovetail with Gauri Viswanathan's investigations into the function of Christian religion in Western discourse, as an agent for the suppression of the occult and of more heterogloss forms of spiritual behaviour.[16]

The magic pastoral can be regarded as a new articulation of magic realism, and as a new genre in its own right. The fusion of black magic with an idyllic projection of black everyday life liberates magic from the Christian indictment of the blasphemous and the heretical and demands a new awareness of and tolerance for dissident black spiritualities. Recent investigations into the construction of blasphemy as foreign have exposed the mechanisms by which Western epistemology has for centuries affirmed the opposition between the religious and the secular and has condoned the deification of scientific knowledge. The epistemological displacements initiated by Thomas Kuhn's *The Structure of Scientific Revolutions* have questioned our certainties about such oppositional concepts and have allowed for a new tolerance between scientific, religious and supposedly blasphemous discourses. Or, to quote Gauri Viswanathan, "The presumed archaism of blasphemy is invoked to separate cultures of modernity from those of premodernity, as if to suggest that blasphemy is operative today only in the Third World."[17] Gauri Viswanathan explores the paradoxes between the marginalising of the blasphemous on the one hand and its embracing on the other, as a sign by which we measure modern ideas of tolerance. Tina McEnroy Ansa's novels explore the interstitial space between a modern everyday world and the magical powers of the spiritual, implicitly questioning the lack of tolerance and of latitude granted to non-Christian interpretations of minority religions. Such a latitude would implicitly decenter Western demands for epistemological prerogatives and also the basis of a Western politics of identity. By the evocation of the magical and the occult the scope of postcolonial and diaspora subversions is articulated, as it aims at the rediscovery of a world beyond white supremacy. Such invocations map a global sphere of cultural antecedence which definitely expands Paul Gilroy's somewhat afrocentric concept of a Black Atlantic.[18]

The mixture of pastoral and benevolent magic shapes a new mode of expression for the African American novel, in which white magic achieves a fairy-tale quality and thus allows for a fusion with idyllic elements. In Tina McElroy Ansa's novels idealised female versions of Caliban regain lost powers because they have Ariel's help and Prospero has become insignificant.

Notes

[1] Max Horkheimer und Theodor W. Adorno, *Dialektik der Aufklärung*.

[2] On imagined Africas or, as Salman Rushdie has it, "Imaginary Homelands," see, instead of many, Marion Berghahn, *Images of Africa in Black American Literature*.

[3] Lois P. Zamora and Wendy B. Faris, *Magical Realism: Theory, History, Community*. For a definition see Franz Roh's essay in this collection.

[4] Uncle Julius is an early and somewhat ineffective example of the black trickster as defined by H. L. Gates in *The Signifying Monkey: A Theory of Afro-American Literary Criticism*.

[5] Cf. Walter Göbel, "Constructive Dialogues: European Traditions and African American Fiction".

[6] Claude McKay, "Outcast" 987.

[7] Cf. Robert Hemenway, *Zora Neale Hurston: A Literary Biography* 120ff. Marie Leveau is explicitly mentioned in Tina McElroy Ansa's novel *The Hand I Fan With* 131. Ansa calls this novel "a tribute not only to *Their Eyes Were Watching God* but to everything Hurston embodied," cf. "An Interview with Tina McElroy Ansa," ibid. 473. Hurston is mentioned in this novel several times and her black magic is a pervasive subtext (82; 211; 344; 473).

[8] Richard Wright, *Native Son* (1940) 47f.

[9] Wright, *Native Son* 84.

[10] Especially in Toni Morrison's and Gloria Naylor's novels, cf. G. Michelle Collins, "There Where We Are Not: The Magical Real in *Beloved* and *Mama Day*" 680-685.

[11] Tina McElroy Ansa, *Baby of the Family* 53. Henceforth quoted as *Baby of the Family* in the text.

[12] McElroy Ansa, *Baby of the Family* 128.

[13] Richard Wright, *Black Boy* 31.

[14] The theme of the powerful and resilient black woman is common also to Toni Morrison's and Alice Walker's novels, cf. Anne Koenen, *Zeitgenössische Afro-Amerikanische Frauenliteratur* and the interpretations of *Sula* collected in Harold Bloom, *Modern Critical Interpretations of Toni Morrison's* Sula, esp. Christian and Demetrakopoulos.

[15] Tina McElroy Ansa, *You Know Better* 230. Henceforth quoted as *You Know Better* in the text.

[16] Gauri Viswanathan, *Outside the Fold: Conversion, Modernity, and Belief*.

[17] Ibid. 240.

[18] Paul Gilroy, *The Black Atlantic: Modernity and Double Consciousness*.

Works Cited

Ansa, Tina McElroy. *Baby of the Family*. New York: Harcourt Brace, 1991.

---. *The Hand I Fan With*. New York: Doubleday, 1996.

---. *You Know Better*. New York: Harper Collins, 2003.

Berghahn, Marion. *Images of Africa in Black American Literature*. London: Macmillan, 1977.

Bloom, Harold, ed. *Modern Critical Interpretations of Toni Morrison's* Sula. Philadelphia: Chelsea House, 1999.

Collins, Michelle G. "There Where We Are Not: The Magical Real in *Beloved* and *Mama Day*." *Southern Review* 24 (1988): 680-685.

Gates, Henry Louis. *The Signifying Monkey: A Theory of African American Literary Criticism*. Oxford: Oxford UP, 1988.

Gates, Henry Louis and Nellie Y. McKay, eds. *The Norton Anthology of African American Literature*. London: Norton, 1997.

Gilroy, Paul. *The Black Atlantic: Modernity and Double Consciousness*. Harvard: Harvard UP, 1995.

Göbel, Walter. "Constructive Dialogues: European Traditions and African American Fiction," in *Pioneering North America: Mediations of European Literature and Culture*. Ed. Klaus Martens. Würzburg: Königshausen and Neumann, 2000. 193-202.

---. *Der Afroamerikanische Roman im 20. Jahrhundert: Eine Einführung*. Berlin: Erich Schmidt, 2001.

Hemenway, Robert. *Zora Neale Hurston: A Literary Biography*. Urbana: U of Illinois P, 1977.

Horkheimer, Max and Theodor W. Adorno. *Dialektik der Aufklärung*. Frankfurt am Main: Suhrkamp, 1944.

Koenen, Anne. *Zeitgenössische Afro-Amerikanische Frauenliteratur*. Frankfurt am Main: Campus, 1985.

Martens, Klaus, ed. *Pioneering North America: Mediations of European Literature and Culture*. Würzburg: Königshausen and Neumann, 2000.

McKay, Claude. "Outcast," in *The Norton Anthology of African American Literature*. Ed. Henry Louis Gates and Nellie Y. McKay. London: Norton, 1997.

Viswanathan, Gauri. *Outside the Fold: Conversion, Modernity, and Belief*. Princeton: Princeton UP, 1988.

Wright, Richard. *Native Son*. New York: Harper and Collins, 2001.

---. *Black Boy*. London: Longman, 1978.

Zamora, Louis P. and Wendy B. Faris, eds. *Magical Realism: Theory, History, Community*. Durham: Duke UP, 1995.

Scientific Perspectives

Physics and Magic

Disenchanting Nature[1]

Gregor Schiemann

Introduction

A widespread view of the natural sciences holds that their historical development was accompanied by a constantly widening gap between them and magic. Originally closely bound up with magic, the sciences are supposed to have distanced themselves from it in a long-drawn-out process, until they attained their present magic-free form.[2] I would like, in this essay, to discuss some arguments in support of this plausible view. To this end, I shall begin with a definition of magical and scientific concepts of nature – a definition appropriate to the considerable length of time from the beginnings of science (which can plausibly be placed in Greek antiquity) to the present day.

One can define as 'magical' a concept of nature which asserts the possibility of gaining knowledge of secret natural forces, and the possibility of man's influencing some of them. These forces are 'secret' in several senses. Their presumed efficacy springs from a hidden, meaningful nexus that comprehends the whole of nature, and is often dependent upon the knowledge of this nexus. Also, one can know of these forces, and in some cases influence them, only in the context of actions not accessible to everybody.[3] This definition by no means covers the whole spectrum of concepts included in the idea of magic, but can be put forward to gain an initial orientation to the relationship between many of the varieties of magic and the sciences.[4] Natural science, in contrast to magic, denies the existence of secret forces. From the scientific standpoint, a force is 'secret' only as long as it remains unknown. Scientific statements about nature may not be founded on assumptions or practices that are restricted to only a small circle of the initiated. Scientific knowledge should be testable under conditions that can be repeated, and it claims unlimited inter-subjective validity.

However, the last century's historiography of science has taught us that the application of such a systematic distinction between magic and science can be very problematic. The magical and the scientific understanding of nature influenced one another so closely that it seems questionable whether a terminological differentiation between them can be sustained. This is, for example, the case with a good deal of physical and chemical research in the late Middle Ages and the Early Modern period. Thus alchemical notions, for example, classifiable as belonging to the sphere of magic, played an important role in the formation of modern scientific theories in these subjects.[5] However, the historically important areas of overlap between science and magic go only some way towards qualifying the thesis of a steadily widening gap between the two modes of understanding of nature. The links between magic and natural science in medieval and modern alchemy – to stick with the example already mentioned – involved only some aspects of these two lines of research, and were already clearly different from the very close links that had subsisted between these two areas of knowledge in ancient Greece. Besides, they were of short duration.

To exemplify the gap which, over several epochs, widened between the magical and the scientific understanding of nature, I would like to examine two concepts in natural science, both assumed by physics, in historical sequence. My inquiry thereby restricts itself to concepts fundamental to the formation of theory in a scientific discipline, and selects only two themes illustrative of the separation of this science from magic. Physics is suitable for this purpose, because it has functioned as a leading discipline from the beginnings of natural science to the present day, dealing with matter in its various conditions of state and motion and its reciprocations.

The first concept I select is Aristotle's concept of *physis* (φυσις). It was fundamental to the emergence of physics, and set its mark on thought in this field right up to the beginning of the modern period. By means of this central concept of his physics Aristotle distanced himself from earlier magical notions of nature, though he was unable to prevent his concept being invoked by the magic both of the Middle Ages and of the Modern Period. The reason for the continuing ambivalence of the concept *physis* towards magic derives, in my view, essentially from the ambiguity of the contrast thought to subsist between it and art, i.e. *techne* (τεχνη) (section 2). *Physis* and *techne* are, in Aristotle, contrasting terms, and denote two mutually exclusive principles and spheres of reality. They still correspond closely enough to present-day modes of thought in everyday life to be translatable by the words 'nature' and 'art.'[6] (However, as they refer to only two of the meanings contained in the semantic complexes of both terms, they should actually be called 'Aristotelian nature' and 'Aristotelian art.') To show the result, in history, of this opposition of *physis* to *techne*, I will show how, in the case of alchemy, both its defenders and its critics availed themselves of this opposition (section 3).

The distinctive characteristic of the second concept of nature is negative, consisting in the elimination of the Aristotelian distinction between *physis* and *techne*. The criticism was directed, though by no means exclusively, against arguments which have recourse to Aristotelian physics to support a magical view of nature. Champions of magic asserted that their *techne* of magic perfected *physis* for man's purposes. Against this, the champions of modern thought hold that all technical operations are subject to the laws of nature, and can therefore only modify nature within set limits. I consider Galileo Galilei to be a trail-blazer for this anti-magical position, as well as a co-founder of experimental science with his mechanical and astronomical works (section 4).

With Galileo's rejection of the distinction between *physis* and *techne* my inquiry is essentially at an end. As a conclusion I will say something of the relationship of magic to concepts of nature typical of the following period, both in physics and other natural sciences (section 5).

In as much as my investigation confines itself to questions concerning the understanding of nature, I exclude from consideration the contexts in which natural sciences and magic were actually practised. There are to date hardly any analyses dealing with a comparable period and focusing on the practical modification of nature.[7] However different practically oriented investigations may be in other respects from those undertaken from the perspective of the history of ideas, I believe the two kinds of inquiry would come to remarkably similar conclusions as regards the development of the relationship between natural science and magic. For, for one thing, the present-day marginalisation of magic as a means of investigating nature and its almost complete insignificance for natural science is obvious and a fact without historical precedent. For another, the disenchanting of nature that has led to this situation is to a large extent undisputed. It can be described by reference to various factors comprising conceptions and practices, three of which I wish to stress:

- The disenchanting of nature results from a historic process of rationalisation which affects the acquisition and processing of knowledge of nature, accumulates increasing social significance as it goes on, and assumes an understanding of rationality, which is conceived more and more instrumentally.
- It is also among the consequences of an increasing empiricisation of the bases of natural science, by which means the relevance of a universally accepted fund of empirical statements continues to grow.
- Finally, it has been furthered by a progressive mathematisation of the knowledge of nature, which changes the empirical basis into quantified and measurable data to render it calculable and therefore predictable in its future development.[8]

Interpretations of the term 'nature' constitute, to be sure, only one facet of these general tendencies. But their historical extent (over several epochs) allow such interpretations to offer the advantage of a criterion applicable over a long period. It is all the more surprising that in investigations of the relationship between magic and natural science in the history of ideas, a merely incidental importance was attributed to the concept of nature.[9]

Aristotle's Contrasting of Nature and Art[10]

Aristotle's physics provided the natural sciences with a classical rational foundation that remained dominant until well into the nineteenth century. Aristotelian natural science is characterised by a systematic structure that claims to achieve completeness, general validity, truth.[11] Its object, *physis*, is characterised by the principle of self-movement, claimed, in physics, to be obvious to everybody.[12]

Self-movement means that all natural things have in themselves "a beginning of change and durability, in part related to space, in part to growth and decay, in part to change of condition."[13]

In contrast to modern science, Aristotle has no concept of a law of nature. Natural processes on earth follow their structural principles only "with regularity."[14] If, in the sublunary world, what is normally expected does not happen, then some "hindrance" has prevented it.[15] In the context of physis as teleologically conceived, hindrances occur either as mistakes or by chance. Mistakes differ from usually expected results, Aristotle argues, adducing freak-births and miraculous appearances as examples.[16] Accidental events differ from what normally happens not in the result, but in the lack of necessary cause. The accidental result has no inner relationship to what happened before.[17] It comes about spontaneously, remains inexplicable and unpredictable. Although Aristotle, in contrast to his predecessors (whom he also calls "magicians"[18]) undertakes a complete rationalisation of knowledge of nature, nevertheless the teleology of his nature leaves blank slots to be filled by accidents and mistakes.

In terms of Aristotle's physics, human art also represents a divergence from the regularity of nature. In contrast to natural things, those that result from art are not moved by themselves but by something external to them: "On the contrary, a bed, or article of clothing, or whatever other classes of things there may be besides, (in so far as it meets this designation and is an object made by art) has no inner impulse towards alteration."[19]

The operations of art are located in between the sciences characterised by generalised knowledge – physics being one of them – and experience, which is typified by particular knowledge about individual things.[20]

The relationship between *techne* and *physis* is treated by Aristotle in his physics, where he describes it as imitation and completion. "In general terms, art sometimes completes what nature cannot bring to term, and sometimes emulates nature."[21] *Techne* as the imitation of *physis* does not imply a reproduction of *physis*; rather, it means that art shares structural principles of processes with nature, *physis* being the source of these.[22] Both these spheres of reality can be investigated by means of the same fundamental conceptual categories (matter-form-deprivation, potentiality, actuality) and explained by the main kinds of causality (impulse to change/movement, matter, form, goal). In this sense therefore the products of art are rationally comprehensible.[23] *Techne* succeeds in completing *physis* when it closes gaps in the teleology of nature for the accomplishment of human intentions.[24] That is, *techne* not only repairs the flaws and coincidences of nature in cases where they run counter to human intentions – it also brings phenomena to pass which cannot be produced by nature. Water flows downhill by nature, but it is sometimes desirable for man that it should flow uphill. Art brings this about by the construction of wells.[25] As the extrapolation of what is already potentially present in *physis*, the completion shows a relationship with imitation. Thus nature is primary, art is secondary and derived from the primary. But *techne* is also essentially alien to natural processes. This is already expounded in texts ascribed to Aristotle himself, albeit less clearly than in those that probably come from later authors. A good example of this is the *Mechanical Problems*, which shows that the artificial operations carried out against *physis* are unlimited in scope.[26] The completion of *physis* goes beyond *physis* to occupy a sphere of artifice to which *physis*, in its imperfection, is subordinate.

Although the concept of *techne* originating in Aristotle is ambiguous in its relation to *physis*, the post-Aristotelian interpretations of this relation generally take as their starting point a difference between *physis* and *techne* in a way that remained fundamental to medieval and early modern thinking.

Medieval Alchemy as an Example of the Ambivalence of Aristotelian Physics and Magic

In order to discuss the ambivalent relationship between Aristotelian physics to magic exemplifying it by means of medieval alchemy, I must first clarify whether the magical, alchemical and Aristotelian concepts of nature current at that time are sufficiently related to one another thematically.

I characterised as 'magical' an understanding of nature that asserted the possibility of recognising hidden forces immanent in a meaningful nature, and the possibility that some of these might be influenced by man. Many of the thought processes in medieval alchemy meet this definition. Alchemists generally considered themselves as members of secret societies whose thinking was guided by a comprehensive symbolism of nature, and whose goal was the manufacture of the 'philosopher's stone.' In so far as the procedures applied to this end were comparable to the crafts of the artisan, alchemy was considered, in the Middle Ages, to be an *"ars mechanicae."*[27] In addition, it was often associated with magic by outsiders, as the reason alchemists could influence the hidden forces was thought to be their participation in supernatural forces. The alchemists themselves by no means always considered themselves to be magicians. In particular many of them rejected the classification of their art as demonic, or so-called 'black magic.' On the contrary several alchemists considered themselves committed to the contrasting 'white' or 'natural' magic whose goal was the improvement of human life through the understanding and modification of nature.[28]

Along with Plato's philosophy of nature and Neo-Platonism, Aristotle's physics can also be reckoned among the theoretical bases of medieval alchemy. Admittedly no unified and self-consistent alchemical system can be documented for any period.[29] But in the various conceptions, explicit references to Aristotle's physics can be found nearly everywhere. They range from the individuation principles of matter and form to the doctrine of the elements and the teleological concept of the whole of nature right through to the opposition of *physis* to *techne*.[30]

But Aristotle's physics not only constituted an important point of reference for alchemical ideas. His work was also of decisive authority for those who criticised alchemy. This double function of Aristotle's physics – which played a role also in the divergent evaluations of the relation between *physis* and *techne* – is reflected in the ambivalence of Aristotelian physics to magic.

As a representative author, who, applying Aristotle's thought, rejected alchemy, I would like to name the Persian philosopher Avicenna (approx. 980-1037).[31] His argument, conducted in his work "On the congelation and coagulation of stones" ("*De congelatione et conglutinatione lapidum*"),[32] says that no human art can transform a naturally occurring baser metal into a more valuable and also naturally occurring metal, because, according to Aristotle, artificial and natural objects are essentially different. The reasons brought to bear against this view – and these also appeal to Aristotle – exalt the power of the art of alchemy above that of nature. Thus Albertus Magnus (about 1200-1280) explains the possibility of alchemical transformation of the properties of metals as being the exchange of forms which are compatible with a basic material common to all metals. The *techne* of alchemy is supposed to replace the impure by the purer form. Albertus compares the procedure of the alchemist with that of the art of medicine, which Aristotle, too, had classified as a *techne*. Just as doctors purified the body of a sick patient, so did alchemists purify material to put it into a better condition.[33] While Albertus's justification of alchemy restricts itself to its rather instrumental character, Roger Bacon (about 1214-1292), in his Aristotle-oriented philosophy of the *opus tertium* (1266), elevates that art to the status of a basic science because of its ability to lay bare and to alter nature, arguing that the whole science of medicine and of nature should spring from alchemy.[34]

Further examples of the argument over alchemy could be cited until well into the sixteenth century.[35] Arguments both for and against magical concepts avail themselves of the conceptual ambiguity of the relations between the contrasting notions of *physis* and *techne*. While the idea of *physis* permits the appearance of phenomena which contradict the general run of natural occurrences, the idea of *techne* conceives all technical operations together as a non-natural procedure recognising no natural limits to its ability to change reality. On the other hand, both *physis* and *techne* are subordinated to shared rational principles which explain phenomena – principles that can be adduced against magical concepts.

Galileo's Elimination of the Aristotelian Opposition of *Techne* to *Physis*

As the opposition of *physis* to *techne* was such a fundamental determinant in medieval and early modern thought, so the process of dealing with and criticising it in physics was correspondingly prolonged. This process can be traced back to the medieval theory of impetus,[36] it determines the foundation of modern mechanics, and does not near its end until the early nineteenth century with the formulation of the principle of the conservation of energy. The whole process is an argument about basic physical concepts and theories, an argument so important as to provide us with a criterion to distinguish one epoch from another in the history of science. The elimination of the opposition between *physis* and *techne* allowed nature to be investigated by means of technical constructions, without restriction. Technology as an object of natural science first opened the door to the development of the experimental method which led to where science is now. By this method the laws of nature were discovered and/or tested under artificial, repeatable laboratory conditions. Conversely, it became possible to think of nature not only as a model of human art, but to think of nature as actually being a technical construct. Nature became a mechanism, and mechanics was promoted to the status of the leading science. Compared to these innovations, which determined the further development of the natural sciences, the devastating consequences for magic of the elimination of the categorical difference between *physis* and *techne* seem merely an incidental matter. In a world in which man's ability to modify nature is subordinated to generally comprehensible laws, there is no longer any place for secret magical knowledge.

In Galileo, whose work I shall use to exemplify the critique of the Aristotelian opposition of *physis* to *techne* a lack of interest in magical inquiries is already to be seen – inquiries which still dominated the minds of his contemporaries. Alexandre Koyré, the science historian, aptly if with some exaggeration typified the culture of the time as one in which "gloomy superstition was dominant, magic and witchcraft [...] were far more widespread than in the Middle Ages," and "astrology [...] played a far greater role than astronomy."[37] Yet we find Galileo living in this time and free of all enthusiasm for magic. In his work he deals with themes belonging to the wider field of magical concepts only incidentally, and in the context of the mechanical and astronomical topics that interested him.[38] In some places he mentions astrology positively, sometimes using arguments which can be matched in the writings of Johannes Kepler.[39] But his astronomical discoveries and theories do not derive their claim to validity from the assumption of hidden influences emanating from the stars, but from phenomena in the sky that anybody with a telescope can observe. In other sections of his work he distances himself strictly from alchemical interpretations of nature. To discredit the Aristotelian critics of his discoveries and theories, he scorns (in, for example, the "Dialogue Concerning the two Chief World Systems") the search for the 'philosopher's stone' that enjoyed among them such high repute. How ridiculous is their belief that they can find in texts from bygone epochs the secret of making gold. "Nothing is funnier," he writes, than to hear the alchemists' commentaries "on ancient poets."[40]

Galileo wrote no elaborated critique of alchemy or any other related form of magic. His most important contribution to the dispelling of the concepts basic to these procedures was of a practical nature. He practised and propagated technical-experimental research into nature and mathematical models for the results thus gained. Accordingly, in the discussion of mechanical instruments and their potential for modifying nature, he came to terms with the concepts of *techne* that went back to Aristotle. In the introduction to his early treatise "Mechanics," 1593, he criticised the belief of the "Mechanici," that they "can move and raise the heaviest weights with little effort, intending thereby, with their machines, to cheat nature, to some degree."[41] They deceived themselves concerning the "immutable characteristics" of nature which is such "that no resistance can be overcome by a force weaker than itself is."[42]

Using simple mechanical devices he showed that the work expended on them did not depend on the procedures used each time, but only on the results attained. Looking back on these investigations, he wrote in a letter to Ciampoli (1625) that through many experiments he had convinced himself that "nature cannot be conquered or deceived by art."[43]

Galileo had recourse to experiments with technical apparatus and thought experiments relating to these, to help him formulate invariably valid laws of nature which all arts obey as well. His orderly nature knew of no hidden forces whose efficacy revealed itself to initiates alone. The comprehensibility of his physics, achievable for anybody, corresponds to the epistemological status of the technical constructions by means of which natural laws are discovered and/or tested – for the constructions are fully understandable. By becoming part of a generally available technology, natural research parts company with the notion of secret knowledge, which includes magic.

Galileo is only one of the founders of modern science. Others, like Johannes Kepler or Giordano Bruno before him or Isaac Newton after him, attach more importance to magical concepts. The initial restriction of the revolution in physics to research into astronomy and mechanics left room, on the one hand, for various concepts in the philosophy of nature among which the magical systems were to persist. On the other hand the narrower definition of the research areas and methods proper to physics resulted in a specialisation which, in its subsequent development, totally excluded every function of comprehensive world-pictures from the forming of scientific theories.

Conclusion

Galileo formulated neither an elaborated critique of magical interpretations of nature, nor any alternative to Aristotle's concept of nature. New and fundamental definitions of nature were not worked out until later, by philosophers like René Descartes, Baruch Spinoza and Gottfried Wilhelm Leibniz, to name only a few important thinkers. Among these, Descartes with his distinction between nature and mind achieved an importance that has remained relevant to discussions of natural philosophy to the present day. Descartes integrates into his concept of nature the whole field of Aristotle's *techne*, and of his *physis*, too, except for that part of the human mind which he sees as the mind of a thinking 'I,' and which he sets up as an opposite principle to nature. In contrast to mind, nature is, as simply extended substance, completely predictable, mathematically describable, and belongs entirely in the subject-area of mechanics.

A discussion of Descartes's dualism and the natural philosophies that came after him could show that these, not only in their original substance-theoretical form, but also in tempered-down but still dualistic variants, leave room for magical thought. Within the realm of mind – thought of as completely independent of nature – the belief in the possible existence of extra-sensory forces able also to influence physical things, remains irrefutable. Such magic-related conceptions play, however, no role in Descartes's scientific reflections, which are not able to include the human mind, opposed as it is to the realm of nature. In Cartesian thought, magic has already been fully excluded from the realm of natural science.

The footholds still left for magical thinking as a result of the opposition between nature and mind do not disappear until the advent of a concept of nature which interprets all manifestations of consciousness and action as natural phenomena. This naturalistic concept, characteristic of present-day research, has gained influence in physics as well. I would like merely to mention here that in the last hundred years physics has lost its role of leading discipline that it had enjoyed ever since antiquity. The completely novel ways of discussing magic scientifically which follow from the naturalistic concept are more important for the relationship between scientific and magical interpretations of nature. These two interpretations no longer oppose one another as two different modes of knowing which relate to one another reciprocally. Rather, the occurrence of magical beliefs becomes a phenomenon to be investigated by science. Thus the relationship between scientific and magical interpretations of nature is turned upside down. The question is no longer whether magic can possibly influence nature, but rather, what sort of natural phenomenon is the belief in magic. What scientific explanation can be found for the fact that people believe in the efficacy of forces – and in the possibility of influencing these – when these forces are, from a scientific point of view, non-existent? The legitimacy of such questions opens a further – and presumably not the last – chapter in the history of the relationship between magical and scientific interpretations of nature.

Notes

[1] Translated from the German original by John Fowler, Universität Stuttgart.

[2] Examples of this interpretation are, among others, James Frazer and Lynn Thorndike. Also, Jean-Jacques Rousseau: "Indeed, one may consult the annals of the world [...], but one will never find the origins of the sciences to be as one would wish them. Astronomy took its rise from superstition, [...] and natural sciences form idle curiosity" (*Discours sur les sciences et les arts* 45). Friedrich Wilhelm Joseph Schelling: "Mankind's first relationship with nature was in fact magical" (*Sämtliche Werke*, 2. Abteilung,

vol. 3, 363), and Norbert Elias: "The liberation from the double bind [Doppelbinder] that had held mankind so long on the magical-mythic level of life in the state of nature, could hardly have happened as a short-term event" (*Der Fischer im Mahlstrom* 116). All English versions here by John Fowler.

[3] The qualifications are intended to enable astrology to be included in the concept of magic. The influence of the stars is neither dependent upon the knowledge of their constellations, nor can it be influenced.

[4] For the concept of magic, cf. Kurt Goldammer, *Magie*, Bert Hansen, *Science and the Magic* 484ff., Bronislaw Malinowski, *Magie, Wissenschaft und Religion* 71, Kurt Goldammer, *Der Göttliche Magier und die Magierin: Natur, Religion, Naturmagie und die Anfänge der Naturwissenschaft vom Spätmittelalter bis zur Renaissance* 14ff., Claus Priesner, "Magie."

[5] This insight derives in large measure from Lynn Thorndike, *The History of Magic and Experimental Science* and Frances A. Yates, "The Hermetic Tradition in Renaissance Science." For a criticism, cf. Brian Vickers, "Introduction" 3ff.

[6] Gregor Schiemann, *Natur, Technik, Geist: Kontexte der Natur nach Aristoteles und Descartes in lebensweltlicher und subjektiver Erfahrung.*

[7] The secondary literature on the relationship of natural science to magic is predominantly oriented towards the history of ideas. Cf. Lynn Thorndike, *The History of Magic and Experimental Science*, Bert Hansen, "Science and the Magic," Bronislaw Malinowski, *Magie, Wissenschaft und Religion*, Brian Vickers, *Occult and Scientific Mentalities in the Renaissance*, Jean-François Bergier, *Zwischen Wahn, Glaube und Wissenschaft: Magie, Astrologie und Wissenschaftsgeschichte*, Wayne Shumaker, *Natural Magic and Modern Science*, Richard Kieckhefer, *Magic in the Middle Ages.*

[8] On the process of mathematisation, cf. Eduard Jan Dijksterhuis, *Die Mechanisierung des Weltbildes.* On the concept of quantification and measurement, cf. Gernot Böhme, "Quantifizierung – Metrisierung."

[9] Cf. footnote 7. The importance of the concept "nature" for medieval magic is pointed out by Hansen, *Science and the Magic* 484ff., Goldammer, *Der Göttliche Magier und die Magierin: Natur, Religion, Naturmagie und die Anfänge der Naturwissenschaft vom Spätmittelalter bis zur Renaissance* 8ff., Christa Habiger-Tuczay, *Magie und Magier im Mittelalter* 176. The last two do not deal with the conceptual relation between nature and art or skill, which Newman discusses for alchemy. In Vickers, *Occult and Scientific Mentalities in the Renaissance*, Wayne Shumaker, *Natural Magic and Modern Science* and Richard Kieckhefer, *Magic in the Middle Ages*, the concept is not part of the theme.

[10] 'Art' (*techne*) as in 'art of medicine,' 'art of house-building,' etc.

[11] On the difference between the classical and modern concept of science, cf., in general, Alwin Diemer, *Der Wissenschaftsbegriff in historischem und systematischem Zusammenhang*, and on physics, Gregor Schiemann, "Was heißt moderne Physik?".

[12] "ὡς δ' ἔστιν ἡ φύσις, πειρᾶσθαι δεικνύναι γελοῖον· φανερὸν γὰρ ὅτι τοιαῦτα τῶν ὄντων ἐστὶν πολλά": "It would be ridiculous to try to prove that there is such a thing as a natural condition. It is obvious that many of the things that exist are of this nature" (Aristotle, *Physics* II 1 193 a 3f.). Quotations from Aristotle are indicated by book, chapter and paragraph according to Bekker's edition.

[13] "τούτων μὲν γὰρ ἕκαστον ἐν ἑαυτῷ ἀρχὴν ἔχει κινήσεως καὶ στάσεως, τὰ μὲν κατὰ τόπον, τὰ δὲ κατ' αὔξμσιν καὶ φθίσιν, τὰ δὲ κατ' ἀλλοιωσιν" (Aristotle, *Physics* II 1 192 b 13ff.), cf. corresponding passage in III 1 200 b 12, also VIII 3f., 253 b 5, and 254 b 17. Cf. for what follows, Schiemann, *Natur, Technik, Geist: Kontexte der Natur nach Aristoteles und Descartes in lebensweltlicher und subjektiver Erfahrung*, chapter 1.1.1.

[14] Aristotle, *Physics* II 8 198 b 35f.

[15] Ibid. II 8 199 a 10f. and b 17f.; IV 8 215 a 23f.

[16] Ibid. II 7 197 b 32ff.

[17] Ibid. II 8 197 b 19f. and b 36f.

[18] Aristotle, *Metaphysics* XIV 4 1091 b 10.

[19] "κλίνη δὲ καὶ ἱμάτιον, καὶ εἴ τι τοιοῦτον ἄλλο γένος ἐστίν, ᾗ μὲν τετύχηκε τῆς κατηγορίας ἑκάστης καὶ καθ' ὅσον ἐστὶν ἀπὸ τέχνης, οὐδεμίαν ὁρμὴν ἔχει μεταβολῆς ἔμφυτον" (Aristotle, *Physics* II 2 192 b 16ff.). On Aristotle's opposition of Nature and Art cf. in addition to Schiemann, *Natur, Technik, Geist: Kontexte der Natur nach Aristoteles und Descartes in lebensweltlicher und subjektiver Erfahrung*, also Hans Blumenberg, "Das Verhältnis von Natur und Technik als philosophisches Problem," and Joachim Schummer, "Aristotle on Technology and Nature."

[20] "γίγνεται δὲ τέχνη, ὅταν ἐκ πολλῶν τῆς ἐμπειρίας ἐννοημάτων μία καθόλου γένηται περὶ τῶν ὁμοίων ὑπόληψις," "Art arises when, on the basis of many observations made from experience, a general concept of similar instances develops" (Aristotle, *Metaphysics* I 1 981 a 5ff.).

[21] "ὅλως δὲ ἡ τέχνη τὰ μὲν ἐπιτελεῖ ἃ ἡ φύσις ἀδυνατεῖ ἀπεργάσασθαι, τὰ δὲ μιμεῖται" (Aristotle, *Physics* II 8 199 a 15ff.).

[22] Ibid. 2 194 a 21f.

[23] On this view of the standard interpretation of Aristotle, e.g. Michael J. Moravcsik, *What Makes Reality Intelligible?*

[24] Robin Smith, "Filling in Nature's Deficiencies."

[25] Aristotle, *Meteorology* II 1 353 b 27ff.

[26] Cf. Fritz Krafft, *Dynamische und statische Betrachtungsweise in der antiken Mechanik.*

[27] Bernhard Dietrich Haage, *Alchemie im Mittelalter: Ideen und Bilder: Von Zosimos bis Paracelsus* 44ff. Mircea Eliade deals with the medieval and later association of craftsmanship and alchemy in *Schmiede und Alchimisten.* Magic (as distinct from alchemy), along with prophesying the future, was included among the 'forbidden arts.' Cf. also Habiger-Tuczay, *Magie und Magier im Mittelalter* 177, and Goldammer, *Der Göttliche Magier und die Magierin: Natur, Religion, Naturmagie und die Anfänge der Naturwissenschaft vom Spätmittelalter bis zur Renaissance* 14. The closeness of medieval magic to technique has been stressed in Hansen, *Science and the Magic* 495ff. and William Eamon, *Technology as Magic in the Late Middle Ages and the Renaissance* 171-212, and is documented and exemplified by Joachim Schummer in "Aristotle on Technology and Nature" and Habiger-Tuczay in *Magie und Magier im Mittelalter* 184ff.

[28] Claus Priesner, *Magie* 227ff.

[29] Claus Priesner and Karin Figala, *Alchemie: Lexikon einer hermetischen Wissenschaft* 8.

[30] Haage, *Alchemie im Mittelalter: Ideen und Bilder: Von Zosimos bis Paracelsus* 18ff.

[31] In what follows I am indebted to William Newman, "Technology and Alchemical Debate in the late Middle Ages," note 59.

[32] Avicenna, "*De congelatione et conglutinatione lapidum*: Being Sections of the Kitâb al-Shifâ," in William Newman, "Technology and Alchemical Debate in the late Middle Ages" 427ff.

[33] Newman, "Technology and Alchemical Debate in the late Middle Ages" 431f., Martha Baldwin, "Albertus Magnus" 21f. and Karl-Heinz Göttert, *Magie: Zur Geschichte des Streits um die magischen Künste unter Philosophen, Theologen, Medizinern, Juristen und Naturwissenschaftlern von der Antike bis zur*

Aufklärung 114f.

[34] Newman, "Technology and Alchemical Debate in the late Middle Ages" 432f.

[35] Cf. the references provided ibid., note 59.

[36] Hans Blumenberg, *Die Genesis der kopernikanischen Welt* 174ff.

[37] Alexandre Koyré, *Galilei: Die Anfänge der neuzeitlichen Wissenschaft* 84 and 82.

[38] One of the texts that deal with a theme belonging to the broader field of magical concepts considers the possible astral influence of the moons of Jupiter (Letter to Biero Dini 1611, in Galileo Galilei, *Le Opere* XI 105-116). Galileo there counters the objection that the moons of Jupiter discovered by him cannot exist, as they don't appear anywhere in astrological practice, and this takes everything into account. Galileo does not contest the basic astrological claim that the stars exert a manifold influence upon earthly affairs. Instead, he adopts the position that all such influences are conveyed by means of light. But the light of Jupiter's moons is so weak that they need not be taken into account in astrological practices – yet could none the less exist (cf. Volker R. Remmert, *Ariadnefäden im Wisseschaftslabyrinth: Studien zu Galilei: Historiographie – Mathematik – Wirkung* 207-209; I follow the account there given.). Galileo's openness *vis à vis* astrology does not contradict his critique of the Aristotelian opposition (of *physis* to *techne*) as that only contested any effective instrumental influencing of nature, which is not claimed by astrology.

[39] Darrel Rutkin, "Celestial Offerings: Astrological Motifs in the Dedicatory Letters of Kepler's *Astronomia Nova* and Galileo's *Sidereus Nuncius*."

[40] Galileo Galilei, *Schriften, Briefe, Dokumente* I 209.

[41] Ibid. 68.

[42] Ibid.

[43] Letter to Ciamoli of 1625 in Galileo Galilei, *Le Opere* VIII 571 ff., cit. in Stillman Drake, *Galileo at Work* 297ff.

Works Cited

Aristoteles. *Metaphysik*. Trans. H. Seidl. Hamburg: Meiner, 1970.

---. *Physik*. Trans. H. G. Zekl. Hamburg: Meiner, 1987.

Avicenna. *De congelatione et conglutinatione lapidum: Being Sections of the Kitâb al-Shifâ*. Ed. and trans. E. J. Holmyard and D. C. Mandeville. Paris: Geuthner, 1927.

Baldwin, Martha. "Albertus Magnus," in *Alchemie: Lexikon einer hermetischen Wissenschaft*. Ed. Claus Priesner and Karin Figala. Munich: Beck, 1998.

Beckermann, Ansgar. *Analytische Einführung in die Philosophie des Geistes*. Berlin: de Gruyter, 1999.

Bergier, Jean-François, ed. *Zwischen Wahn, Glaube und Wissenschaft: Magie, Astrologie und Wissenschaftsgeschichte*. Zurich: Verlag der Fachvereine, 1988.

Blumenberg, Hans. "Das Verhältnis von Natur und Technik als philosophisches Problem." *Studium Generale* 4 (1951): 461-467.

---. *Die Genesis der kopernikanischen Welt*. Frankfurt am Main: Suhrkamp, 1975.

Böhme, Gernot. "Quantifizierung – Metrisierung," in *Am Ende des Baconschen Zeitalters*. Ed. Gernot Böhme. Frankfurt am Main: Suhrkamp, 1993.

Diemer, Alwin. "Der Wissenschaftsbegriff in historischem und systematischem Zusammenhang," in

Der Wissenschaftsbegriff: Historische und systematische Untersuchungen. Ed. Alwin Diemer. Meisenheim am Glan: Hain, 1970.

Dijksterhuis, Eduard Jan. *Die Mechanisierung des Weltbildes.* Berlin: Springer, 1956.

Drake, Stillman. *Galileo at Work: His Scientific Biography.* Chicago: U of Chicago P, 1978.

Eamon, William. "Technology as Magic in the Late Middle Ages and the Renaissance." *Janus* 70 (1983): 171-212.

Eliade, Mircea. *Schmiede und Alchimisten.* Stuttgart: Klett-Cotta, 1980.

Elias, Norbert. "Der Fischer im Mahlstrom," in *Engagement und Distanzierung.* Ed. Norbert Elias. Frankfurt am Main: Suhrkamp, 2003.

Galilei, Galileo. *Le Opere.* Edizione Nazionale. 22 vols. Florence: Tipografia di G. Barbera, 1890-1909.

---. *Schriften, Briefe, Dokumente.* 2 vols. Ed. Anna Mudry. Munich: Beck, 1987.

Göttert, Karl-Heinz. *Magie: Zur Geschichte des Streits um die magischen Künste unter Philosophen, Theologen, Medizinern, Juristen und Naturwissenschaftlern von der Antike bis zur Aufklärung.* Munich: Fink, 2001.

Goldammer, Kurt. "Magie," in *Historisches Wörterbuch der Philosophie.* Ed. Joachim Ritter and Karlfried Gründer. Darmstadt: Wissenschaftliche Buchgesellschaft, 1971ff.

---. *Der Göttliche Magier und die Magierin: Natur, Religion, Naturmagie und die Anfänge der Naturwissenschaft vom Spätmittelalter bis zur Renaissance.* Stuttgart: Steiner, 1991.

Haaqe, Bernhard Dietrich. *Alchemie im Mittelalter: Ideen und Bilder: Von Zosimos bis Paracelsus.* Darmstadt: Wissenschaftliche Buchgesellschaft, 1996.

Habiger-Tuczay, Christa. *Magie und Magier im Mittelalter.* Munich: Diederichs, 1992.

Hansen, B. "Science and the Magic," in *Science in the Middle Ages.* Ed. David Lindberg. Chicago: U of Chicago P, 1978.

Kieckhefer, Richard. *Magic in the Middle Ages.* Cambridge: Cambridge UP, 1990.

Koyré, Alexandre. *Galilei: Die Anfänge der neuzeitlichen Wissenschaft.* Berlin: Wagenbach, 1988.

Krafft, Fritz. *Dynamische und statische Betrachtungsweise in der antiken Mechanik.* Wiesbaden: Steiner, 1970.

Malinowski, Bronislaw. *Magie, Wissenschaft und Religion.* Frankfurt am Main: Fischer, 1983.

Moravcsik, J. Michael. "What Makes Reality Intelligible? Reflections on Aristotle's Theory of Aitia," in *Aristotle's Physics.* Ed. Lindsay Judson. Oxford: Clarendon Press, 1991.

Newman, William. "Technology and Alchemical Debate in the late Middle Ages." *Isis* 80 (1989): 423-445.

Newman, William and Anthony Grafton, eds. *Secrets of Nature: Astrology and Alchemy in Early Modern Europe.* Cambridge, MA: MIT Press, 2001.

Priesner, Claus. "Magie," in *Alchemie: Lexikon einer hermetischen Wissenschaft.* Ed. Claus Priesner and Karin Figala. Munich: Beck, 1998.

Priesner, Claus and Karin Figala. "Vorwort der Herausgeber," in *Alchemie: Lexikon einer hermetischen Wissenschaft.* Ed. Claus Priesner and Karin Figala. Munich: Beck, 1998.

Priesner, Claus, ed. *Alchemie: Lexikon einer hermetischen Wissenschaft.* Munich: Beck, 1998.

Remmert, Volker R. *Ariadnefäden im Wissenschaftslabyrinth: Studien zu Galilei: Historiographie – Mathematik – Wirkung.* Bern: Lang, 1997.

Rousseau, Jean-Jacques. "Abhandlung über die Wissenschaft und Künste (Discours sur les sciences et les arts)," in Jean-Jacques Rousseau. *Schriften.* Ed. H. Ritter. Vol. 1. Munich: Hanser, 1978.

Rutkin, H. Darrel. "Celestial Offerings: Astrological Motifs in the Dedicatory Letters of Kepler's Astronomia Nova and Galileo's Sidereus Nuncius," in *Secrets of Nature: Astrology and Alchemy in Early Modern Europe*. Ed. William Newman and Anthony Grafton. Cambridge, MA: The MIT Press, 2001.

Schelling, Friedrich Wilhelm Joseph. *Sämtliche Werke* [Part I: vol. 1-10; Part II: vol. 1-4. Ed. K. F. A. Schelling]. Stuttgart, 1856ff.

Schiemann, Gregor. "Was heißt moderne Physik?," in *Emergence of Modern Physics: Proceedings of a Conference Commemorating a Century of Physics*. Ed. Dieter Hoffmann, Fabio Bevilacqua and Roger H. Struewer. Berlin 23 March-24 March 1995. Pavia: La Goliardica Pavese, 1996.

---. *Natur, Technik, Geist: Kontexte der Natur nach Aristoteles und Descartes in lebensweltlicher und subjektiver Erfahrung*. Berlin: de Gruyter, 2005.

Schütt, Hans-Werner. "Georg Ernst Stahl," in *Alchemie: Lexikon einer hermetischen Wissenschaft*. Ed. Claus Priesner and Karin Figala. Munich: Beck, 1998.

Schummer, Joachim. "Aristotle on Technology and Nature." *Philosophia Naturalis* 38 (2001): 105-120.

Shumaker, Wayne. *Natural Magic and Modern Science: Four Treatises, 1590-1657*. Binghamton, N.Y.: Center for Medieval and Early Renaissance Studies, State University of New York, 1989.

Smith, Robin. "Filling in Nature's Deficiencies," in *Aristotle's Ontology*. Ed. Anthony Preus and John P. Anton. Albany. NY: State UP, 1992.

Thorndike, Lynn. *The History of Magic and Experimental Science. I-VIII*. New York: Columbia UP, 1923ff.

Vickers, Brian. "Introduction," in *Occult and Scientific Mentalities in the Renaissance*. Ed. Brian Vickers. Cambridge: Cambridge UP, 1984.

---, ed. *Occult and Scientific Mentalities in the Renaissance*. Cambridge: Cambridge UP, 1984.

Yates, Frances A. "The Hermetic Tradition in Renaissance Science," in *Art, Science, and History in the Renaissance*. Ed. Charles S. Singleton. Baltimore: Johns Hopkins UP, 1967.

The Techno-Magician

A Fascination Around 1900

Robert Stockhammer

I

"Magick," Aleister Crowley decrees, "includes all acts soever. Anything may serve as a Magical weapon; [...] a Magical Operation [...] may be defined as any event in Nature which is brought to pass by Will. We must not exclude potato-growing or banking from our definition."[1] At around 1900 *magic* becomes attractive to the degree that it was impossible to define it by way of a *definitio e contrario*, i.e. by indicating what it was not. Ill-defined or somewhat arbitrarily redefined again and again, the signifier *magic* is less than a *concept*. It is, however, more than simply a *metaphor*, since magical notions of language precisely state the impossibility of distinguishing between metaphorical and literal usages of language. It is more than simply a word since it appears in well-ordered syntactical co-texts and pragmatic con-texts whose rules can be analyzed, and in which it can even be substituted by other words: in many contexts, for instance, it works almost synonymously with 'energy.' Rather than searching for a common denominator of its conflicting definitions, I will therefore point out several conditions for its attractivity and its omnipresence around 1900, addressing *magic* (somewhat arbitrarily in my turn) as a *figure* in a specific *configuration*.

Even a superficial first look reveals that, from 1880 onwards, *magic* begins to play a crucial role in various disciplines. Master disciplines for the study of magic are, of course, the history of religion and anthropology, and one nucleus of its career is formed when these disciplines meet in an analysis of 'primitive religions' – as is the case with Sir Edward Burnett Tylor or Sir James George Frazer. One of the earliest (and certainly one of the most interesting) *Outlines of a General Theory of Magic*, the *Esquisse d'une Théorie générale de la Magie* (1902/03), is the product of a collaboration between a classicist, Henri Hubert, and an anthropologist, Marcel Mauss. Since these studies point out the involvement of our 'own' European tradition with what is called *magic*, they nourish a wide-spread interest in anthropology, as, for instance, in the works of Bronislaw Malinowski, whose Trobriand tetralogy is one life-long effort to come to terms with *magic*. Cultural philosophy and the philosophy of history, from Ernst Cassirer's *Philosophy of Symbolic Forms* to Max Horkheimer and Theodor W. Adorno's *Dialectic of Enlightenment*, will continue to rely on tripartite models of historical evolutions as conceptualised by Frazer, with *magic* as a first stage. The Melanesian word *mana* – introduced into European languages by Max Müller as a synonym for magical power and discussed in detail by Hubert/Mauss – plays a crucial role for all of the texts referred to as well as for many more.

Moreover, other disciplines point out the involvement of the *contemporary* European situation with magic. This is true not only of occultism, which is extremly fashionable in all its varieties: new varieties such as Theosophy (Mme. Blavatsky) or Spiritism, as well as reconfigured prestigious traditions, such as 'Kabbala' (e.g. Papus) or 'Alchemy.'[2] It is also true of disciplines and discourses usually regarded as belonging to the core of Western enlightenment, such as psychology, theories of language and the 'philosophy of technology' ('*Technikphilosophie*'). Even if *magic* is not the main subject of these discourses, and even if it enters them in a seemingly more casual way, the presence of *magic* in these contexts is a striking symptom of its attractivity around 1900 as well as an important reason for it. I will give a somewhat more detailed account of these aspects by sketching out the relation of *magic* to three other terms, two of which are enlisted in the title of our symposium: *psychology*, *technology* and *science*. At least two of these terms may also serve as *definitiones e non-contrario* (rather than as *definitiones per analogiam*), i.e. as indicators for what *magic* is *not* not [sic!]. In this modern configuration, *magic* has, so to speak, lost some of its traditional opposites:

1. *Magic* is no longer the contrary of *psychology*. An enlightenment tradition presupposes that seemingly supernatural occurrences are 'in fact' subject to psychological explanations. The new configuration, however, strives to include *magic* rather than to dissolve it. Phenomena formerly – but no longer – called 'supernatural' or 'supernormal' are integrated into psychology as a *subject* and as a *practice*. Psychical research (*Métapsychologie* in France, *Parapsychologie* in Germany) investigates phenomena such as telepathy or clairvoyance "in the same spirit of exact and unimpassioned inquiry which has enabled Science to solve so many problems, once not less obscure or less hotly debated"[3] – to quote from the programmatic introduction to the *Proceedings of the Society for Psychical Research*, founded in London, 1882. In the very same year, the *Académie des Sciences* in Paris concedes the existence of hypnotism to Jean-Martin Charcot. The same institution that, a century before, had declined entrance to Franz-Anton Mesmer now integrates a practice even while it remains unexplicable. The number of publications on hypnotism explodes in the following decades. The neologism *psychomagic* ("*Psychomagie*") – coined by the astrologist Alfred Fankhauser in a book simply entitled *Magic* – is helpful to cover a large scope of phenonema and practices. A boundary commonly accepted before the emergence of this configuration and re-established later on – a boundary distinguishing 'normal,' albeit strange psychic phenonema such as schizophrenia from 'paranormal' phenonema like telepathy – does *not* work as a distinctive boundary in this configuration. Telepathy, for instance, is most widely accepted to exist 'in fact' – so that it is not astonishing that Sigmund Freud, for instance, was among its apologists.

2. Fankhauser differentiates *psychomagic* from *technomagic* ('*Technomagie*'), thereby pointing out that *magic* is not the contrary of *technology*, but very com-

patible with it. Inventions like the phonograph, film, or wireless telegraphy, the discovery of X-rays or radioactivity were not so much perceived as 'disenchanting the world' than as enlarging the realm of 'marvellous' phenomena. Psychomagic and technomagic mutually support each other: since wireless telegraphy has become indisputable, why should we exclude the possibility of a human being sending and receiving signals over large distances even without an apparatus? And even if occultists and technicians disagree with regard to details, they share the same configuration that locates *magic* in the centre of culture as a whole (the "*Kulturganze*"), in the intersection of everything else, with technology as the field closest to it. The diagram depicted on the picture here below, with M., the magician, at the centre of everything, is taken (not from an occultist book, but) from a *Philosophy of Technology* (*Die Philosophie der Technik*), unsuspicious of occultist tendencies.[4]

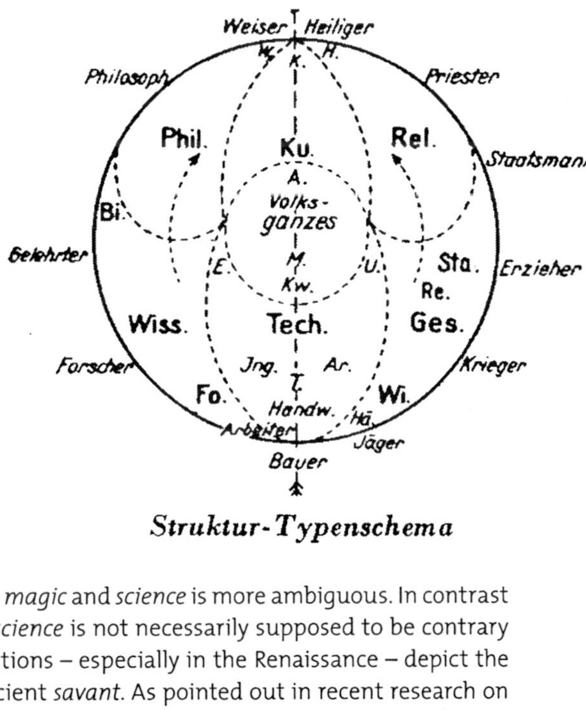

Struktur-Typenschema

3. The relationship between *magic* and *science* is more ambiguous. In contrast to *psychology* and *technology*, *science* is not necessarily supposed to be contrary to magic, since venerable traditions – especially in the Renaissance – depict the magician as an erudite, omniscient *savant*. As pointed out in recent research on Early Modern knowledge (from John Dee via Francis Bacon to Isaac Newton), even the kind of knowledge involved in magic is not necessarily opposed to scientific knowledge, since a stable distinction between these kinds of knowledge is established only in the 18th century. Modern occultists claiming to inherit these Western or other (Eastern) traditions, 'Alchemists,' 'Kabbalists,' Theosophists, and Yogi, stick to this Prospero-like concept of magic based on the study of books, of innumerable books (as stored in caves in the Himalayas), or of *the* one and only se-

cret book. – In other aspects of its modern configuration, however, *magic* inherits the more popular connotation as 'something inexplicable.' And it is precisely this connotation that is enforced by the liaison of *magic* and *technology*. Max Weber may well define 'disenchantment' as the confidence in the existence of somebody able to explain how a tram works.[5] In contrast to Weber's optimism, however, the amount of *non-knowledge* is increasing with respect to many things in the contemporary environment. Transforming the title of Dirk Vanderbeke's paper for this volume, one might say that *Technology is Magic that works – without a science to explain how*. One of the reasons for this development relates to the fact that, as early as in the late 19th century, engineering begins to separate even from sciences like physics, which are supposed to deal with the 'same thing.' The title of Carl du Prel's two-volume book *Magie als Naturwissenschaft* (*Magic as Science*, 1899) may serve as a *pars pro toto* of the configuration as a whole, but the title of Ludwig Staudenmaier's book of 1912 (republished in a revised edition, 1922) is still more revealing: *Magie als experimentelle Naturwissenschaft* (*Magic as Experimental Science*). In many cases, experimentation overflows into what science would be able to account for: techno-magicians such as Thomas Alva Edison, but also psycho-magicians such as Jean-Martin Charcot, are engineers proceeding by trial-and-error rather than scientists whose practice is based on received knowledge.

II

But what about the forth term in the title of our symposium, *literature*? I will try to give an answer to this question in two steps. First, I will present an early example of how a literary text participates in – and even contributes to – the attractivity of *magic* in its modern configuration, especially in its sub-configuration as *techno-magic*. In a second step, I will reflect on these very notions of 'literary example,' 'participating in,' 'contributing to,' in order to sketch out a more precise account of the systematic role of literature in this configuration.

My example is Philippe-Auguste Villiers de l'Isle-Adam's novel *L'Ève Future*, first published in 1886 (and translated into English as *Tomorrow's Eve*).[6] Its protagonist is Thomas Alva Edison, less than forty years old at that time, but already famous as the inventor of the phonograph, and the light bulb, as well as several other items improving already existing technologies such as electricity, telegraphy, or sound transmission. Moreover, he has already begun to occupy contemporary imagination, a fact explicitly recalled in Villiers's *Avis au lecteur*, which preceeds the novel. Villiers's preface already points to the "legend" that has formed itself in "the popular imagination" and lists Edison's common epithets "*Magicien du siècle, le Sorcier de Menlo Park, le Papa du Phonographe*," frequently repeated in the novel itself; at one point of the novel, Edison insists on calling himself, metonymically, "electricity" (*L'Ève Future* 205). The plot is easily summarised: Lord Ewald, visiting his friend Edison, declares his decision to commit suicide because of his beloved

Alicia. Since she is overwhelmingly beautiful, he is not able to split up with her, but he cannot bear her silly behavior, and her hackneyed way of speaking, which are so contradictory to her outward appearance. Her 'soul,' in other words, is not adapted to her 'body,' and when Lord Ewald exclaims "Who could remove her soul from her body?," Edison's answer is that he will do precisely this (*L'Ève Future* 169): He constructs an artificial copy of Alicia, identical in every outward aspect, but decisively improved in her 'spirit' or 'soul,' especially with regard to what she says.

Hadaly, as the android is called, is a multimedia system in every respect: 'media' in our post-Second-World-War sense of technical devices, as well as in the sense of personal medium as in psychical research and Spiritism around 1900. Instead of lungs, two golden phonographs are implanted into Hadaly, on which the "words [...] of the greatest poets, the most subtle metaphysicians and the most profound novelists" are recorded (*L'Ève Future* 282), yet also retaining "the accent, timbre and intonations" of Alicia's voice (*L'Ève Future* 213); thus, Edison 'removes soul from body' in the realm of language, separates the signified from the signifier. In the course of Hadaly's construction, optical media are introduced as well, for example the *"photographie successive"* (*L'Ève Future* 265), anticipating Edison's invention of cinematography by almost a decade. Obviously, Villiers relies not only on Edison's work, but also on the experiments of his close friend, the poet and inventor Charles Cros who constructed machines for converting pictures into writing and vice versa. Villiers's/Edison's complex multimedia system is based on the analysis of its elements and their mutual convertibility. The replacement of 'natural' with 'artificial' life appears as one particular case in this general investigation of convertibility. A 'philosophy of technology' such as Ernst Kapp's theory of 'organ projection' (1877) would cover the production of artificial flesh as well as of artificial voices.

The construction of "Tomorrow's Eve," however, also involves human media. The techno-magician Edison shares some features of the psycho-magician, in his ability to control human beings by hypnotising them or by pushing them from one of their split personalities into another one (the 'dissociation of personality' being a contemporary obsession, in experimental psychology as well as in stories such as *Dr. Jekyll and Mr. Hyde*, published one year after *L'Ève future*). In Edison's case, even the distinction between techno- and psycho-magician becomes blurred, since his theory of psychic forces has a distinctly physical penchant, as may be expected from 'Mr. Electricity.' In a chapter somewhat ironically entitled "*Explications rapides*" and headed by Hamlet's often-quoted sentence "There are more things in heaven and earth, Horatio, than are dreamt of in your philosophy," Edison not only states the indisputable existence of a *"fluid nerveux"* on which magnetism as a "positive science" is based (*L'Ève Future* 396), but also claims the possibility of a fusion between *"fluid nerveux"* and *"fluid électrique"* (*L'Ève Future* 403).

Villiers's Edison explicitly refers to William Crookes, then famous as a chemist and physician as well as an adept of psychical research (*L'Ève Future* 323). As inventor of the 'Crookes radiometer,' he explained the phenomena observed in this tube by a theory of 'radiant matter' understood as a forth state of matter (beyond solid, liquid and gaseous states). As an investigator of 'paranormal' phenomena, he introduced the notion of 'psychic force' as early as 1871 – a peculiar force allowing his then famous medium Daniel Dunglas Home, for example, to play "a well-known sweet and plaintive melody" on the accordion while "not moving a muscle."[7] At the end of the 19th century Crookes, simultaneously President of the *British Association for the Advancement of Science* and the *Society for Psychical Research*, Crookes uses his twofold 'authority' to stress, in his addresses to the respective associations, the "connexion between my old-standing interest in pyschical [sic!] problems and such original work as I may have been able to do in other branches of science,"[8] and "something like continuity between those unexplained ['psychic'] forces and laws already known."[9]

One problem lurks in this formulation, because it contains the admission that phenomena like paranormal messages, telepathy, levitation etc., all seem to have been established beyond doubt according to the standards of positivist epistemology, but cannot *yet* be integrated into the framework of natural laws. At the same time, however, they *no longer* appertain to the older framework of supernatural laws, since the vast majority of psychical researchers dismiss the hypothesis of spirits as a religious ideology incompatible with unprejudiced scientific attitude. Villiers's Edison quarrels about similar problems. All of his "rapid explanations" end with the concession that the fusion between nervous and electric fluid has led to "a synthesis of unknown nature and power" (*L'Ève Future* 403). Edison's peculiar problem is that he has succeeded *too* well, to the degree that the android Hadaly appears as much more human and natural than her allegedly human model Alicia, so that Lord Ewald mixes them up – and "the artifact terrifies the artificer" (*L'Ève Future* 399) Edison, since he is no longer able to give scientific explanations for the results of his own operations.

One of the paranormal messages transmitted by the medium Home and recorded by the scientist Crookes runs: "This is unscientific science." Evidently this is a commentary on the very setting in which it is uttered. However, it allows at least two readings: either it means, as adepts of psychical research would prefer to interpret it, 'This is a science not yet sufficiently accepted as a serious science in the academia;' or it means, as I would prefer to read it, 'This is a kind of practice in accordance with the standards of positivist science and resulting in positive phenomena that, however, do not allow scientific explanations.' In another context, Crookes concedes: "I never said that it was possible, I only said it was true."[10] When Lord Ewald, in the chapter "Explications rapides" already quoted, asks if a creature like Hadaly is actually possible, Edison answers: "No: but it is" (*L'Ève Future* 401).

III

So much for the example – but how about its 'exemplarity'? Or, more generally: how does literature participate in, contribute to, or criticise the reconfiguration around *magic* as a whole?

On the one hand, Villiers's inventions are almost uncannily close to those of his contemporaries outside of the realm of fiction (Thomas Alva Edison, William Crookes or Charles Cros). On the other hand, scholars of literature are inclined to defend the autonomy of literature, at least in certain respects. Therefore, a more detailed analysis of the state of literature within the reconfiguration of *magic* should start with distinguishing several meanings of 'literature.'

1. If 'literature' is understood as anything that is written down, the reconfiguration of magic around 1900 is essentially 'literary,' since it is highly dependent on writings of all kinds. While this might be a trivial fact for historical analysis, it marks a difference from what *magic* is traditionally supposed to be: a ritual, based on oral performances, surviving especially in 'magic societies,' i.e. those defined precisely by their lack of writing (*'schriftlose Gesellschaften'*). The German word for *spell* is *'Zauberspruch,'* a composite noun implying its oral performance (a 'magic saying'). Its equivalent within the realm of writing, *'Zaubertexte,'* does not seem to exist, since it is not enlisted among the several hundreds of composites beginning with *'Zauber-'* contained in *Grimm's Wörterbuch*. When entitling my book *Zaubertexte*, however, I did so in accordance with one reference I found and whose context is revealing: Hugo Ball, when preparing the famous performance of his onomatopoetic poems ("jolifanto bambla...") at the Cabaret Voltaire in 1916, commented on the power of "*uralte Zauber*texte" where the 'correct' term would be *'Zaubersprüche.'*[11] Thus, perhaps unconsciously, Ball accounts for the fact that his knowledge of medieval or contemporary Pacific spells is due to their transmission through writing. The power of the magic word is mediated by *The Power of the Written Tradition* – to quote a title of a book by Jack Goody. (Peculiarly enough, the situation is somewhat different for the English language, since *spell* seems to be etymologically related to the alphabet, so that *'Zaubertexte'* might be translated as *spelt spells*).

2. When restricting 'literature' to its widely accepted contemporary sense, we enter the field of widespread notions of literature as an 'inheritor' or 'restoration' of magic. In a lineage dialectically intertwining secularisation and re-sacralisation, literature is on the one hand conceived as residue, on the other hand as a restitution of magic. Theories about this relationship multiply around 1900, in the realm of literature and cultural studies as well as in essays by poets. Among the various genres of literature, it is of course poetry that seems to be most directly linked to the spell, especially the kind of poetry allegedly 'meaningless' like the dadaistic "jolifanto bambla..." (Hugo Ball) or "fmsbw..." (Raoul Hausmann). Drama comes in on another level of this relationship since the theatre seems to promise a

re-enactment of magic rituals, in the prominent tradition of Shakespeare's *Tempest*.[12] Drama and poetry coincide in the presupposition that both are recited or performed, so that the written word is reconverted into an oral one, thereby reestablishing a sort of ritual, even if this ritual is displaced from the context of a society organised by the belief in magic.

3. Novels, on the contrary, are not likely to be read aloud, and they do not seem to have an equivalent in the context of magical societies. Hubert/Mauss, however, point to the importance of narratives for the image of the magician in popular imagination, and even use the expression "*romans magiques*" (magic novels) at one point of their *General Theory of Magic*.[13] While it is not sufficiently clear to which novels they refer, their expression fits novels like *L'Ève future* perfectly well, that participate in constructing 'legends' of modern magicians. A category like 'magical novels' is much more suitable for including narratives in the modern configuration of *magic* than is the widespread notion of 'fantastic literature.' Theories of fantastic literature conceive of it as a *deviation* from a zero degree of 'reality,' whether this deviation is conceptualised as a "rupture" (Roger Caillois) or as a "hesitation" (Tzvetan Todorov). But, to begin with, this zero degree is subject to historical transformation, not only in the long run – as is the case for the so-called transition from 'magical' to 'scientific world view' – but also with regard to relatively small periods – as is the case with telepathy or other 'paranormal' phenomena that were much more widely accepted in the period in question than they were before or are accepted nowadays.

Moreover, there is no such thing as a degree zero from which literature deviates, since literature is always already part of a discourse and an imagination forming a degree x of 'reality.' Literature itself influences common notions of 'reality;' it partakes in its construction, just like the "words [...] of the greatest poets, the most subtle metaphysicians and the most profound novelists," recorded on the golden phonographs, partake in the construction of artifical Hadaly; and "novelistic episodes [are part of] the domain of Science" ("*des épisodes... romanesques... dans le domaine de la Science,*" *L'Ève Future* 362). Villiers, it is true, will not convince most of his readers in 1886, of the possibility of constructing artificial women as perfect as Hadaly, and his intention as a novelist most probably does not consist in persuading his readers to accept this possibility; he even explicitly concedes in his *Avis au lecteur* that "the Edison of the present work, his character, his habitation, his language and his theory are – and need to be – at least fairly distinct from reality." ("*l'Edison du présent ouvrage, son caractère, son habitation, son langage et ses théories sont – et devaient être – au moins passablement distincts de la réalité,*" *L'Ève Future* 96). But it is even more unlikely that Villiers intended to criticise the hubris of the modern scientist constructing artificial beings he cannot control: Hadaly, locked in a coffin during her passage to Europe, sinks in a shipwreck (along with her model Alicia), doing harm only to Lord Ewald who suffers from losing her much more than from losing Alicia. *L'Ève future* is hardly open for a reading in the tenor of '*Technikfolgenabschätzung*' (estimation of technological risks) that might be suitable for a reading of Mary Shelley's *Frankenstein* or Heinar Kipphardt's *In der Sache J. Robert Oppenheimer.*

Rather, Villiers draws consequences from Edison's epithet *The Wizard of Menlo Park* that will inspire other consequences, or feedback effects: "When Edison went to the Paris exhibition in 1889 he became acquainted with Villiers's novel. He read it at a sitting and expressed himself thus: 'That man is greater than I. I can only invent. He creates.' He did not meet the author, who was mortally ill, though an attempt was made to bring the Frenchman and American together."[14] Evidence for this anecdote is, like for so many other details of Edison's life or famous sentences attributed to him, hard to establish. *Se non è vero, è ben trovato.* In any case, Edison never lost his confidence in the progress of techno-magic. In 1920, when he was seventy-three years old, rumour had it "that the great inventor was working on a device, or apparatus as he prefers to term it, to communicate with personalities which have passed on to another existence or sphere. Immediately the press of the United States and Europe announced that Thomas A. Edison had joined the ranks of the spiritists, which now number many a prominent scientist, author, inventor, physicist, engineer, clergyman, and so on."[15] Alarmed by this news, the prominent magazine *Scientific American*, usually unsuspicious of occultism, sent a correspondent to interview Edison. To the relief of the decreasing number of scientists who had *not* yet "joined the ranks of the spiritists," Edison depicts himself as an agnostic rather than as an adept of spiritism: "Now follow me carefully; I don't claim that our personalities pass on to another existence or sphere. I don't claim anything because I don't know anything about the subject. For that matter, no human being knows."[16] Nevertheless, Edison confirms having thought "of a machine or apparatus which could be operated by personalities which have passed on to another existence or sphere." Thus, even while ignoring whether these personalities actually exist, he is convinced of knowing the technological preferences of these personalities – if they exist: "But I do claim that it is possible to construct an apparatus which will be so delicate that if there are personalities in another existence or sphere who wish to get in touch with us in this existence or sphere, this apparatus will at least give them a better opportunity to express themselves than the tilting tables and raps and ouija boards and mediums and the other crude methods now purported to be the only means of communication."[17] The modern dead or undead, their existence notwithstanding, prefer to communicate by the most advanced and "delicate" means of communication: 'media' in our post-Second-World-War sense begin to replace 'media' in the sense of around 1900, which were still a necessary component in Edison's multimedia system as portrayed by Villiers. But this is a minor difference, which fits into a narrative of 'technological progress' shared by Edison and Villiers, among almost all their contemporaries. Sometimes, science (or *magic*?) is less far from fiction than one might like.

Notes

[1] Aleister Crowley, *Magick in Theory and Practice* 114 and 107. – While this paper is a completely new text, it relies on material already presented in: Robert Stockhammer, *Zaubertexte: Die Wiederkehr der Magie und die Literatur, 1880-1945*.

[2] See Alexandra Lembert and Elmar Schenkel, *The Golden Egg: Alchemy in Art and Literature*.

[3] Anonymous, "The Society for Psychical Research: Objects of the Society" 4.

[4] See Manfred Schröter, *Philosophie der Technik* 40.

[5] See Max Weber, *Wissenschaft als Beruf* 19.

[6] Villiers de l'Isle-Adam, *L'Ève future*, is henceforth quoted parenthetically within the text; the English translations being my own (RSt).

[7] William Crookes, *Researches in the Phenomena of Spiritualism* (ed. 1874) 14.

[8] William Crookes, "Address by the President" 338.

[9] Crookes, *Researches in the Phenomena of Spiritualism* (ed. 1926) 130.

[10] Crookes, *Researches in the Phenomena of Spiritualism* (ed. 1874) 22, footnote.

[11] See Hugo Ball, *Die Flucht aus der Zeit*, entry 15 June 1926.

[12] See Andreas Höfele, "Raising Tempests: Religion, Science, and the Magic of Theatre," in this volume.

[13] Henri Hubert and Marcel Mauss, "Esquisse d'une Théorie générale de la Magie" 25.

[14] James Huneker, "Villiers de l'Isle Adam," from *Iconoclasts: A Book of Dramatists* (1912), http://www.io.com/~larrybob/villiers.html (12 Jan. 2005).

[15] Austin C. Lescarboura, "Edison's Views on Life and Death: An Interview with the Famous Inventor Regarding His Attempt to Communicate with the Next World" 446. The passage quoted here is taken from the introduction to the interview.

[16] Ibid.

[17] Ibid.

Works Cited

Anonymus. "The Society for Psychical Research: Objects of the Society." *Proceedings of the Society for Psychical Research* 1 (1882-83): 3-6.

Ball, Hugo. *Die Flucht aus der Zeit* (1927). Munich: Kösel and Pustet, 1931.

Crookes, William. *Researches in the Phenomena of Spiritualism* (1870-74). London: J. Burns, 1874. Enlarged ed. Manchester: Two Worlds Publications, 1926.

---. "Address by the President." *Proceedings of the Society for Psychical Research* 12 (1896-97): 338-355.

Crowley, Aleister. *Magick in Theory and Practice* (1929). Secaucus, N.J.: Castle Books, 1991.

Fankhauser, Alfred. *Magie: Versuch einer astrologischen Lebensdeutung.* Zurich: Orell Füssli, 1934.

Hubert, Henri and Marcel Mauss. "Esquisse d'une Théorie générale de la Magie" (1902-03), in Marcel Mauss. *Sociologie et Anthropologie.* Paris: Presses Universitaires de France, 1950. 1-141.

Huneker, James. "Villiers de l'Isle Adam," in *Iconoclasts: A Book of Dramatists* (1912), http://www.io.com/~larrybob/villiers.html (12 Jan. 2005).

Kapp, Ernst (1877). *Grundlinien einer Philosophie der Technik: Zur Entstehungsgeschichte der Cultur aus neuen Gesichtspunkten.* Düsseldorf: Stern Verlag, 1978.

Lembert, Alexandra and Elmar Schenkel, eds. *The Golden Egg: Alchemy in Art and Literature*. Berlin: Galda and Wilch, 2002.

Lescaboura, Austin C. "Edison's Views on Life and Death. An Interview with the Famous Inventor Regarding His Attempt to Communicate with the Next World." *Scientific American*, 30 Oct. 1920: 446 and 458-460.

Schröter, Manfred. *Philosophie der Technik* (1934). Darmstadt: Wissenschaftliche Buchgesellschaft, 1972.

Stockhammer, Robert. *Zaubertexte: Die Wiederkehr der Magie und die Literatur, 1880-1945*. Berlin: Akademie, 2000.

Todorov, Tzvetan. *Introduction à la littérature fantastique*. Paris: Éditions du Seuil, 1970.

Villiers de l'Isle-Adam, Philippe-Auguste. *L'Ève future* (1886). Paris: GF-Flammarion, 1992.

Weber, Max. *Wissenschaft als Beruf* (1919). Stuttgart: Reclam, 1995.

Ghostly Geometry

The Fourth Dimension in Literature

Elmar Schenkel

If you ask people what is meant by the 'fourth dimension,' they may come up – depending on their age – with a vague sort of transcendence, a pop group or a perfume produced in 1953. 'Ecstasy' or 'time' might be further answers. In many ways, the popular ideas surrounding this term share the fate of the word 'alchemy': a pop group, a perfume, transcendence, ecstasy and of course, there is an alchemy of time as well. In this paper, I should like to look at the fourth dimension before its fall, or, as it were, on the eve of the hangover, i.e. the popular inflation and abuse of a term that at one point had a more precise connotation. It is interesting that the term and the concept were rediscovered only in the late 1970s and early 1980s, by such authors as Rudy Rucker, Martin Gardner and Linda D. Henderson. Two reasons might be given for this phenomenon. Between the 1980s and today, great changes in the concept of space have been occurring with the electronic revolution and virtual spaces accessible through computers. Imaginary spaces have become more and more important, vide the internet. The fact that in this period cultural critics began to reconsider a similar conceptual shift in the 19th century demonstrates to what extent perception of historical data is influenced by contemporary issues. It is in the 1980s that 4 D was discovered as an important concept for the 1880s onwards, that was to influence literature, art, philosophy and architecture, physics as well as metaphysics. The second reason for the rediscovery can be found in the pairing of hitherto strange or impossible bed-fellows: modernism and the occult. In a way, modernism before had stood basically for a negation of the 19th century, which in itself was imbued both with progress and religion, whose underside was occultism à la Blavatsky. The famous 1986 Los Angeles County Museum of Art Exhibition on "The Spiritual in Art: Abstract Painting 1890-1985" not only extended the period of the abstract, but focused explicitly on the occult aspects of modernism. The catalogue included contributions on alchemy, Boehme and Linda D. Henderson's essay on the Fourth Dimension. In this context, it became increasingly apparent that the Fourth Dimension as a scientific as well as pseudo-scientific concept served as a cultural bridge between science and religion/occultism.[1]

In my paper, I wish to draw some of these lines connecting literary works to an underground of occult and mathematical ideas, and in particular, I wish to describe the functions and uses to which this concept was put in different cultural settings around 1900.

How did ideas about the fourth dimension (4D) come about in the first place?

Though the term existed before 1800, and was used in a mystical sense by Henry More in the 17th century, its modern fate is bound up with the history of non-Euclidean mathematics.[2] As Henderson, Rucker and other authors have shown, new concepts of space emerged from a mathematics that shifted its interest from planes to curved spaces. Euclidean axioms as to the convergence of parallels in the infinite or the constant sum of angles in a triangle could no longer be maintained.[3] From a cultural and historical point of view, it is important to note that much of this new mathematics can be seen as rising to a challenge: the measurement of the planet, which in itself was not merely a geometrical problem but also one of competition among colonial powers. Jules Verne, as always on the lookout for what was in the political air, associated the scramble for Africa to a scramble for longitude and latitude in his novel *Aventures de trois russes et de trois anglais* (1871-72). In this sense, one could say that non-Euclidean geometry is one of the first results of globalisation. Apart from this geopolitical movement, other cultural and technological factors had an impact on concepts of space in the 19th century. The microscope and the telescope were refined to such a degree that completely new spatial worlds began to unfold in laboratories or observatories. The excessively large and the minutely small called for new languages and expressions, and some of these were found in fantasy stories such as E. T. A. Hoffmann's "Meister Floh," Poe's "The Sphinx" or Fitzjames O'Brien's "The Diamond-Maker." The extension of scales also led to a re-evaluation of dimensions.[4] Three-dimensional space seemed to collapse at the gigantic end of the scale, while by looking through a microscope one could find apparently flat beings endowed with 3D life.

One thinker who was interested and well-versed in both science and literature, in mathematics and the imagination, was the Leipzig psychologist and physicist Gustav Theodor Fechner (1801-1887). As a polymath with a playful sense of the absurd, he was one of the first to toy with the idea of dimensions. In the essay "Der Raum hat vier Dimensionen" published under the pseudonym of Dr. Mises – his quizzical alter ego – he imagines a so-called "*Schattenmännchen*," a little shadow man cut out of paper moving across a screen and having some adventures there. Only when he is moved outside the frame of the screen will a 3D world emerge. The questions raised are basically of a spatial nature, but Fechner is also one of the first to relate the fourth dimension to time. He imagines life as represented by a beam of time from which every single slice corresponds to the state of the human being at a given point in time. It took another half century or so for the concept of time as the fourth dimension to catch on. It was eventually taken up by H. G. Wells and made famous with his *The Time Machine* (1895). With computer visualisations of multidimensional objects, slicing has become one of the most important techniques in this field of spatial representations on the screen. Fechner, who turned increasingly spiritual after a severe psychosomatic crisis, was not the only one in his city to dream, think or speculate about the fourth dimension or strange geometries. One of his contemporaries was August F. Moebius, the inventor of one-sided surfaces, or the Moebius strip. Here again, time is needed to explore topological strangeness. One has to walk or travel around such a twisted strip in order to find out that it has one surface only. Geometry and time are thus becoming more and more indissoluble.[5] Another contemporary of Fechner, the astrophysicist Johann Carl Friedrich Zöllner (1834-1882), was to write a voluminous and polemic work on the fourth dimension as proof of a transcendental world.[6]

Fechner's thought experiments may be seen as the budding ability to imagine four and more dimensions. I should now like to consider the further development under three headings for which I will select suitable examples.

4D concepts play a role in cultural discourses which contribute to the sense of relativity in religion and philosophy, but which also influence narrative techniques, thus preparing modernism to some extent. Most of all, they seem to be scientifically produced evidence of a supernatural world and hence enable religious thought to survive in an agnostic or atheistic environment.

Charles Hinton (London 1853-1907) developed 4D ideas and though he was not motivated by religion, his books and exercises influenced writers and occultists alike, giving them a sense of the transcendental apparently based on science. Hinton's father was a freethinker and once is supposed to have said: "Christ was the Saviour of men, but I am the savior of women, and I don't envy Him a bit."[7] His son, a teacher at Uppingham School, true to his father's spirit, committed bigamy and had to leave England. He spent some years in Japan but left again, this time to the United States. Here he became a lecturer in mathematics at Princeton University and other places. His most famous contribution to culture was the invention of the baseball cannon, which is still used in the US. He died when he was about to make a toast to female philosophers.

When Hinton disappeared from England, he left an obscure London address from which one could order strange geometrical puzzles, as Borges informs us in his introduction to Hinton's stories and essays.[8] The basic idea with these objects, as with Hinton's general approach, is to use analogy when thinking about a 4D geometry. In other words, we should ask ourselves what might a world look like if we were flat, 2D beings. How would we then perceive 3D objects like spheres and cubes? If we were flat, everything would move towards us or from us in lines, like coins seeing other coins moving across a table. Now, taking the analogy from here, what traces would a 4D object leave if it were to pass through our 3D world?

For Hinton, these cogitations were yoga-like gymnastics of the mind. They were meant to show how to reach a higher dimension by exploring our own 3D possibilities: "It is the object of these pages to show that, by supposing away certain limitations of the fundamental conditions of existence as we know it, a state of being can be conceived with powers far transcending our own."[9] The outcome of these exercises is not necessarily a metaphysical one, though it should not be excluded. Secular transcendence may very well carry traces of an earlier religious experience,[10] but Hinton was rather trying to account for the new arising out of the old order of thought. He tried to make the new thinkable. If surfaces constitute 3D objects, solids must bound four dimensional figures.[11] Thus a so-called hypercube would be bounded by eight cubes, an object he would call a "tesseract."

In another essay, Hinton discusses geometrical thought as a way of "casting out the self."[12] Here his methods remind us of approaches to non-European, Indian or Buddhist meditation attained by rigorous mental training. The idea of these exercises is to enumerate the cubes in a given block of 27 cubes and to learn their positions by heart in such a way that every cube is listed from each side adjacent to it. By doing this, Hinton claims, one can gain an objective, i.e. non-perspectival, view of each cube. He called this method also "seeing as a higher child." Some of the clients who bought his puzzles were completely nonplussed and complained about an approaching madness when they exercised; they "are completely mind-destroying," as one former user wrote in a letter.[13]

Hinton was humble about the actual value of these exercises, claiming that they simply demonstrated the functions of the mind and would not constitute a form of higher knowledge.

However, his suggestions were taken up by a number of contemporaries, one of them P. D. Ouspensky, the Russian esoteric philosopher who became immensely influential, not least through his one-time collaboration with the Georgian magus G. I. Gurdjieff. It is through this channel that 4D ideas also entered modern art, particularly through Cubists and abstract artists such as Kandinsky, Malevich, Kupka and Mondrian.[14]

In literature, however, these ideas had a variegated effect. Above all, they helped writers to revise concepts of the absolute and the relative, which eventually was to encompass writing itself. They also enabled writers to discuss spirituality without being outside the scientific worldview. They could talk magic and still remain modern, as it were.

Arguably the most significant literary contribution to the philosophy of the fourth dimension came from the Headmaster of the City of London School, whose specialities lay in Shakespeare and theology, but who also taught Sanskrit, chemistry and the classics. Edwin A. Abbott (1838-1926) published his best-known work *Flatland: A Romance of Many Dimensions* in 1884 and it has remained a minor classic, bridging the gap between the two cultures. As in *Gulliver's Travels*, satire and sense of relativity are connected and both books furthermore are reports about travel in strange lands. Flatland is a plane and we are introduced to its world by the narrator, A. Square. The country is inhabited by geometrical beings – lines, triangles, polygons and circles, and all are representatives of certain social groups. The first part of this slight novel describes the society of Flatland, which turns out to be hierarchical, patriarchal, and to some extent totalitarian, since geometrical misfits are treated in hospitals. Abbott satirises Victorian ethics and society, including the debate on eugenics which started in the 1880s, and above all attacks Victorian prejudices against women. In the second part, he aims at a different level. Here, he wants to introduce the reader to the plurality of worlds, a concept that might hopefully lead to a more relative understanding of one's own position in society. The square visits lineland and pointland and is then initiated to the third dimension by a strange visitor, a sphere. Thus an analogy is built: if it is so difficult for a 2D being to imagine and recognise a three-dimensional world, what are the chances that a 3D reader might be able to imagine a 4D world? Like Gulliver, Square becomes a heretic after his return. He is at odds with his environment and is even imprisoned by the authorities.

Analogy is here meant to produce knowledge of the unknown, or at least acknowledgement of its possibility. *Flatland*, then, has to be seen as a fable about the relativity of knowledge, or should we not rather call it humility?

Abbott's dimensional tale turned out to be the key to all later applications and transformations of 4D concepts, for it covers both a spiritual as well as a philosophical side, linking the satirical tradition from Swift to Orwell.

The Scottish fantasy writer George MacDonald (1824-1905) explored the spiritual implications of a multidimensional world in his last novel, *Lilith* (1895). Even in his earlier fantasies, written in the wake of Romanticism and under the spell of the fairy tale, other dimensions loomed large, as in *Phantastes* (1858). This may serve as a reminder that dimensional thought was rooted, at least to some extent, in Romanticism, in those magical mirrors and blue flowers, those golden pots and miraculous skins that we find in Hoffmann, Balzac, Tieck or Novalis, some of whom clearly influenced MacDonald. The mirror is the pre-eminent door to the supernatural world. As in *Alice Through the Looking Glass* by MacDonald's friend Lewis Carroll, or, later, in Algernon Blackwood's story "The Pikestaff Case," the mirror serves as an entrance in *Lilith* as well. The narrator is led by a phantom from the library to the garret where he finds a strange looking-glass enabling him to enter another world. In this universe, he is confronted with all kinds of riddles and philosophical conundrums treating the question of space and time. The phantom who is a raven (but also a sexton and librarian) is his spiritual mentor and leads him into a spatial and temporal maze. In the following dialogue, it turns out that MacDonald's agenda is a spiritual one, just like Ouspensky's.

When the confused narrator asks the raven where he is, the raven answers: "In the region of seven dimensions." They then discuss space in these terms:

The raven: "Where do you think it [the tree] stands?"
"Why *there*, where you know it is!"
"Where is *there*?"[15]

The tree actually stands at two places at the same time: in a pine wood and on the hearth of a kitchen. To the narrator, it seems he has wandered into a world where "both the material and psychical relations of our world had ceased to hold."[16] The architecture of the strange house is intricate, if not outright paradoxical: "[...]

the more doors you go out of, the farther you get in!"[17]

When he uses the word *where*, the raven reprimands him: "You know nothing about whereness." He considers the possibility that all he is experiencing now is some life beyond death, a postmortal world. Then, again, the house is related to the human body, with the library as its brain. In MacDonald's strange tale, neurology and dimensional thinking are curiously intertwined. Any lack of understanding, however, is due to the narrator's spiritual immaturity, as the raven never ceases to point out. Becoming an individual, according to him, means unfolding oneself in more than three dimensions. Spatial movement here becomes allegorical for the act of self-realisation, which is to say that the physical and the spiritual, the moral and the psychological must be aligned or coordinated for human development to take place. In order for this to happen, they must refer to unseen dimensions. MacDonald's novel is a complex tale about the war between dualism and monism, between polytheism and monotheism, paganism and Christianity. Dimensional thinking is an important element in this unresolved struggle.

For H. G. Wells, in contrast, the fourth dimension has no spiritual function, but serves instead as a useful fictional tool. Though there is no intention to educate the reader spiritually, Wells' entertaining stories always have an eyeopening quality. For Wells, 4D concepts in fiction help him to find formulae for the relativity of perception and of the worlds we live in, which is one of the sustaining themes of his work, as, for example, in "The Country of the Blind." Wells continues the Swiftian theme and satirical note from *Flatland* and thus turns scientific ideas into means of cognition. He is probably the first author to have used 4D concepts in both the spatial and temporal sense.

In "The Plattner Story," the fourth dimension is understood in the spatial sense: the teacher Plattner, who is blown up during a chemical experiment in the classroom, finds himself in a strange world that is only loosely connected to the familiar one. It is actually a very kafkaesque world. His experience can be likened to near-death experiences as they have been reported all over the world and collected by such writers as Raymond Moody. Wells wrote about out-of-body experience as well, as for example in "Under the Knife." The recurrence of this theme may well be due to Wells' own proximity to death when he was dangerously kicked in the kidneys during a football match in the 1880s. For nine days, Plattner stays within this intermediate, both real and unreal world, and is permitted to inspect the houses of the living just as the sphere in *Flatland* was able to see everything that happened in the circles, squares and triangles of Flatland. The themes of invisibility and omniscience, two attributes of the Judeo-Christian god, are thus deeply connected to dimensional fiction. Plattner returns to this world by being dumped from 4D into the 3D garden plot of his boss, the head of the school. Afterwards, it is noticed that Plattner has been reversed spatially, his heart now being on the right side and his left hand taking over the function of his right hand.[18]

In his lesser known fantasy, *The Wonderful Visit* (1895), Wells took up the multidimensional strain again to make his satirical point. A local farmer happens to shoot an angel, who, wounded, is rescued by villagers. While the angel thinks he has fallen into the world of his dreams, the locals are baffled. Ironically, a man of the cloth is the first to refer to science as an explanation: "It is confusing," said the Vicar. "It almost makes one think there may be (ahem) Four Dimensions after all. [...] there may be any number of three dimensional universes packed side by side, and all dimly dreaming of one another. There may be world upon world, universe upon universe."[19] There is a similar satirical reference to this subject in Oscar Wilde's "The Canterville Ghost," which goes to show that 4D concepts must have become fashionable in the mid-1890s.

Though Wells was aware of all this and contributed to the popularisation of the fourth dimension as space, he gave another twist to the concept that was to prove rather revolutionary. Even in spatial fantasies, time is not completely unimportant. Out-of-body experiences, for example, happen on two different timescales, one as perceived by the person experiencing it, the second as seen by the others around him or her. In Wells' "The Man Who Could Work Miracles," in which a man finds he can stop the universe, events are reported that happen below human consciousness or which have never happened at all. If they have, they also must have destroyed the (fictional) reader. In "The Remarkable Case of Davidson's Eyes," a man experiences a split universe: one in which he is in London, another in which he is shipwrecked thousands of miles to the South. For three weeks, he stays in this difficult condition reminiscent of someone caught in cyberspace while pursuing his regular life. Two years later, it turns out that what he saw in the South Seas actually took place. To account for this, the narrator quotes a Professor Wade who suggests a "kink in space," some kind of curved or wrinkled space, that is, which makes two distant spots coincide at certain times and for certain periods.[20] If this holds, we must conclude that Davidson's fusion of two places lasted some three weeks and that space must have changed after this period. Here, possibly for the first time in fiction, space and time are curiously linked and form some kind of third entity. Later on, this was to be called space-time.

In *The Time Machine* (1895), Wells laid out his view on space-time when his time traveller discourses on the four dimensions. His basic tenet is this: "*There is no difference between Time and any of the three dimensions of Space except that our consciousness moves along it.*"[21] His discovery lies in the possibility of intentional travelling, not just drifting, along this line called time. As in the Czech writer Arbes' story "Newton's Brain" (1877), imagination and memory serve to anticipate a more regular machine which can be controlled at any point by humans. Here, for Wells, the question of the fourth dimension turns into a philosophical debate about our existence in time and space. Once human beings have gained control of planetary space, the next step is to control time as an inexhaustible resource. The traveller can do this by denying a fundamental difference between space and time and is thus close to what Einstein was to formulate a decade later.[22] Dimensional thinking here begins to erode an old world picture and to lay the foundation of much that was to be significant in twentieth-century thought. Furthermore, this conjunction of philosophical with physical thinking may well entail a different type of spirituality as well, one that encompasses scientific insight and human experience.

While Wells conducted thought experiments in fictional form, they had no distinctive impact on the narrative techniques themselves. No stream of consciousness emerged from his fictions of space and time and it took the modernism of Joyce, Woolf or Eliot to pick up on this side. However, there are transitions between the late Victorians and modernism and one of the bridges can be found in a much neglected collaborative novel by Joseph Conrad and Ford Madox Ford.

In *The Inheritors* (1901), their first completed collaboration, Conrad and Ford tackled the popular theme of the fourth dimension because they were looking for a bestselling recipe, as both were in desperate need of commercial success. Two influences relating to dimensional ideas may have been significant. Conrad had been impressed by Wells' scientific romances, especially *The Invisible Man*, but he also admired *The Time Machine* and *The Wonderful Visit*. The "immediate catalyst," as Elaine Kleiner put it,[23] was however Conrad's first experience of an X-ray machine one night in September 1898 in Glasgow. Conrad reported about this evening to his editor Edward Garnett and mentioned a discussion of waves and vibrations as "at the bottom of all states of consciousness." He added that horizontal and vertical waves may well coincide and thus make it possible "that two universes may exist in the same place and in the same time – and not only two universes but an infinity of different universes – if by universe we mean a set of states of consciousness."[24]

This concept links up with two developments in dimensional thought: space and time coincide more than even in Wells' *The Time Machine*. While the time traveller moves in time, yet stays in the same space, in Conrad's and Ford's novel time has become simultaneous, at least approximatively. The future is already embedded in the present. The second development relates to Ouspensky's view that the fourth dimension has a psychic quality, which Conrad calls sensation, emotion and thought.[25] The discussion of physical aspects of the fourth dimension is, however, not very extended in the novel.

Granger, an unsuccessful writer, has to take up hack writing for a paper. His first task is to collect 'atmospheres' of famous people, politicians and artists. He is also asked to collaborate with the foreign secretary, Churchill, on a biography of Cromwell. But right from the start he is involved with a different level as well when he meets a stranger, a beautiful woman near Canterbury Cathedral. She claims to come from the fourth dimension and to be one of the first of a great invasion that is about to take place. The "dimensionists" are going to inherit the earth and they will reign with terror and awe. They are free from ethical traditions, indifferent to pain and suffering and will just do their job to replace the weak humanity of the present age. She then goes on to prove her powers by dissolving Canterbury Cathedral in front of his eyes: "One seemed to see something beyond, something vaster – vaster than cathedrals, vaster than the conception of the gods to whom cathedrals were raised. The tower reeled out of the perpendicular. One saw beyond it, not roofs, or smoke, or hills, but an unrealised, an unrealisable infinity of spaces."[26] The phenomenon produced by the dimensionist is a new sense of the sublime; the mode of its description shares features with Nietzsche's elated prose in *Thus Spake Zarathustra*, another contemporary attempt to overcome conventional religion. Indeed, the dimensionists owe a great deal to Nietzsche's *Übermensch*. They also connect up with ruthless social Darwinism. At the same time, the cruel and beautiful lady interferes with a project called "*Le système* Greenland" run by a Belgian nobleman, le Duc de Mersch, a pseudonym for King Leopold of Belgium. With this political figure, the authors refer to King Leopold's scheme for the Congo which is the substance of Conrad's previous novel, *Heart of Darkness*. Thus the novel works on complex levels of self-reference relating both to Conrad and Ford and their collaboration. My point here is that, in addition to being a 4D fantasy about space and time, this novel uses dimensional ideas to collapse narrative conventions. It raises the question to what extent reality is still translatable. As Granger has to confess towards the end: "I found myself in the very position that she had occupied at the commencement of our relations: that of having to explain – say, to a Persian – the working principle of the telegraph."[27] This uncertainty about central events and characters is frequently underlined by elliptical sentences that float around in the air, by the hazy 'atmospheres' Granger is painting and by the enigmatic nature of the dimensionists. Granger learns that he is "at the heart of a machine," as the girl puts it,[28] stressing everyone's automatic and helpless behaviour. *L'homme machine* is a recurring theme not only in science fiction but also in esoteric philosophies close to Ouspensky. In some ways, then, dimensional ideas can also go beyond a discovery of relativism: they may well lead to sectarian reductionism and lots of other 'isms' which pigeonhole humans and see them as machines, thus depriving them of their humanity. It coincides with what Granger, the helpless chap, suspects about the dimensionists: "They had no joy, these people who were to supersede us."[29]

I suspect that this is the first time a Science Fiction concept affects the narrative method. While the magic elements of the fourth dimension disappear in a literal sense, it becomes palpable in language and the reading mind. Thus the fourth dimension turns into an adventure in consciousness. In this sense, the much maligned *The Inheritors* is a bridge not only between Victorianism and Modernism, but also between the unknown and the known, the future and the present. The fourth dimension is no longer a breeding ground for thought experiments, but has gained a direct impact on language and literary expression, which also happens to be the end of a concept.

Notes

[1] Maurice Tuchman, *The Spiritual in Art: Abstract Painting 1890-1985*.

[2] For the history of the term, see, for example, Linda D. Henderson, *The Fourth Dimension: Non-Euclidean Geometry in Modern Art* 17-43.

[3] Henderson, *The Fourth Dimension*, Rudy Rucker, *The Fourth Dimension*. For details about the transition from Non-Euclidean mathematics to a 4D geometry, see Michio Kaku, *Hyperspace* 30-43.

[4] For the impact of the microscope see Catherine Wilson, *Early Modern Philosophy and the Invention of the Microscope*; for the telescope see Richard Panek, *Seeing and Believing*.

[5] For a short story in which Moebius strips figure cf. Martin Gardner, "The No-Sided Professor."

[6] Johann Carl Friedrich Zöllner, *Wissenschaftliche Abhandlungen*.

[7] Quoted in Kaku, *Hyperspace* 68.

[8] Borges, "Preface to Charles Hinton," *Wissenschaftliche Erzählungen* 8.

[9] Charles Hinton, "What is the Fourth Dimension?" in *Speculations* 2.

[10] As, for example, wonderfully captivated in Martin Gardner's short story "The Church of the Fourth Dimension."

[11] Hinton, "What is the Fourth Dimension?" in *Speculations* 10.

[12] Hinton, "Casting Out the Self" in *Speculations* 56-66.

[13] Rucker, "Introduction" in *Speculations* xv.

[14] Henderson, *The Fourth Dimension* 44-116.

[15] George MacDonald, *Lilith: A Romance* 21-22.

[16] MacDonald, *Lilith* 12.

[17] Architecture and the fourth dimension are also treated in Robert Heinlein's short story "And He Built a Crooked House" (1941); see also de Stijl architect Theo van Doesburg's attempts in the 1920s to construct tesseract buildings.

[18] This reversal reminds one of the notorious Slade experiments conducted in Leipzig in 1877-1878 by the astrophysicist Zöllner and the American medium Henry Slade. Slade produced ghostly slate writings and unravelled knots that could only be unravelled if one went through 4D-space. Apparently, he was a great charlatan and tricked Zöllner, and also Conan Doyle, who defended him. The 4D thinker who was not tricked, however, was Zöllner's colleague Gustav Theodor Fechner. See Gardner, "'Dr.' Henry Slade, American Medium" and Christoph Meinel, *K. F. Zöllner und die Wissenschaftskultur der Gründerzeit* 38-57.

[19] Herbert George Wells, *The Wonderful Visit* 25f. Interestingly enough, this theory of many worlds has been taken up again recently, see Gardner, "Multiverses and Blackberries."

[20] This trick also serves as a basis for Madeleine L'Engle's fantasy *A Wrinkle in Time* (1962).

[21] Wells, *The Time Machine* 8.

[22] Einstein may have been influenced by Wells' story, but a more direct influence comes from his readings in the history of science, especially of popular works. One of them was Felix Eberty's *Die Gestirne und die Weltgeschichte* (1846), which Einstein prefaced for a 1923 edition. Eberty prefigures some of Einstein's thought experiments relating to moving on a lightbeam and the space-time correlations. See Jürgen Neffe, *Einstein: Eine Biographie* 78-81.

[23] Elaine Kleiner, preface to *The Inheritors* vi-vii.

[24] Quoted in Martin Ray, *Joseph Conrad: Interviews and Recollections* 95.

[25] Conrad in his letter to Garnett, quoted ibid.

[26] *The Inheritors* 8.

[27] Ibid. 122.

[28] Ibid. 151.

[29] Ibid. 150.

Works Cited

Abbott, Edwin A. *The Annotated Flatland: A Romance of Many Dimensions* (1884). Ed. Ian Stewart. Oxford: Perseus Press, 2001.

Borges, Jorge Luis. "Preface to Charles Hinton," in *Wissenschaftliche Erzählungen*. Ed. Charles Howard Hinton. Stuttgart: Edition Weitbrecht, 1983. 7-11.

Conrad, Joseph and Ford Madox Ford. *The Inheritors: An Extravagant Story* (1901). Liverpool: Liverpool UP, 1999.

Fechner, Gustav Theodor. "Der Raum hat vier Dimensionen," in *Vier Paradoxa*. Ed. Gustav Theodor Fechner. Leipzig: Breitkopf and Härtel, 1875.

Gardner, Martin. "The No-Sided Professor," in *The No-Sided Professor and Other Tales of Fantasy, Humor, Mystery, and Philosophy*. Ed. Martin Gardner. Buffalo, NY: Prometheus, 1987. 45-58.

---. "The Church of the Fourth Dimension," in *Imaginary Numbers: An Anthology of Marvelous Mathematical Stories, Diversions, Poems, and Musings*. Ed. William Frucht. New York: Wiley and Sons, 1999. 169-180.

---. "'Dr.' Henry Slade, American Medium," in *Are Universes Thicker Than Blackberries?* Ed. Martin Gardner. New York: Norton, 2003. 263-272.

---. "Multiverses and Blackberries," in *Are Universes Thicker Than Blackberries*? Ed. Martin Gardner. New York: Norton, 2003. 3-11.

Henderson, Linda D. *The Fourth Dimension: Non-Euclidean Geometry in Modern Art*. Princeton: Princeton UP, 1983.

Hinton, Charles H. *Speculations on the Fourth Dimension: Selected Writings of Charles H. Hinton*. Ed. Rudy Rucker. New York: Dover, 1980.

Kaku, Michio. *Hyperspace: A Scientific Odyssey Through the 10th Dimension*. Oxford: Oxford UP, 1995.

MacDonald, George. *Lilith: A Romance* (1895). Grand Rapids: Eerdmanns, 1981.

Meinel, Christoph. *J. K. F. Zöllner und die Wissenschaftskultur der Gründerzeit*. Berlin: Sigma, 1991.

Neffe, Jürgen. *Einstein: Eine Biographie*. Reinbek: Rowohlt, 2005.

Panek, Richard. *Seeing and Believing: How the Telescope Opened Our Eyes and Minds to the Heavens*. Harmondsworth: Penguin, 1999.

Ray, Martin, ed. *Joseph Conrad: Interviews and Recollections*. Iowa City: Iowa UP, 1990.

Rucker, Rudy. *The Fourth Dimension*. Boston: Houghton and Mifflin, 1984.

Tuchman, Maurice, ed. *The Spiritual in Art: Abstract Painting 1890-1985. Los Angeles County Museum of Art*, 23 Nov. 1986 - 8 March 1987. New York: Abbeville Press, 1986 .

Wells, Herbert George. *The Time Machine* (1895), in *Selected Short Stories*. Harmondsworth: Penguin, 1958. 7-83.

---. *The Wonderful Visit*. New York: Dutton and Company, 1895.

Wilson, Catherine. *The Invisible World: Early Modern Philosophy and the Invention of the Microscope*. Princeton: Princeton UP, 1995.

Zöllner, Johann Carl Friedrich. *Wissenschaftliche Abhandlungen* vols III and IV. Leipzig: Staakmann, 1879.

Disenchanting the World, Enchanting the Reader

The Negotiation of Biochemistry in Sinclair Lewis' *Arrowsmith*

Jarmila Mildorf

Sinclair Lewis' novel *Arrowsmith* (1925)[1] was written at a time when the media had already started to contribute to an increase in lay knowledge of medical problems and biochemistry through a range of newspaper articles, health advice literature, discussion forums, etc. The 1920s in particular saw a wider dissemination of scientific writing in America. Notably the publishers Alexander Macmillan in Britain and Edward L. Youmans in America sought to advance the cause of public education in science, and for this purpose they published the works of young, aspiring scientists in specialist journals and magazines.[2] An American example is *Popular Science Monthly*. The "Science Service," a science news bureau, was founded in 1921 and provided scientific news items for newspapers and magazines.[3] People were encouraged to find out more about science, in opposition to still popular anti-science sentiments that outlawed the teaching of evolution in schools, for example, and by 1927 young people participated in a national essay writing contest in chemistry.[4] Against this background, I aim in this chapter at eliciting the discursive practices by which Lewis' novel portrays medical laboratory work and scientists. How does the novel as literary text make scientific discourse accessible to, and potentially fascinating for, lay readers? In depicting the professional and personal development of Martin Arrowsmith from medical doctor to passionately devoted bacteriologist, Lewis' novel explores the moral and ethical dilemmas involved in walking the tightrope between scientific practice and society's expectations of medical science's healing powers. The conflict becomes most poignant when Arrowsmith is praised for having discovered a 'magic cure' for bubonic plague and subsequently wavers between 'curing' the natives of a plague-ridden Caribbean island with his serum, and strict adherence to scientific procedures in a controlled experiment conducted there. If one considers the narrative text as one possible expression and, at the same time, re-instantiation of the cultural semiotics of biochemistry, then one can ask in what ways the novel negotiates the 'magic' world of medicine and biochemistry within its own artistic spaces of freedom. One strategy, I argue, is the rhetorical employment of magical imagery. Images of magic and magicians constantly mesh with scientific discourse in order to (re)create and make more palatable otherwise inaccessible knowledge of a clearly demarcated and often mythologised professional domain and its potential effects. Thus I pay attention to a set of images frequently used in *Arrowsmith* that no Lewis critic, at least to my knowledge, has yet focused on.

My chapter has the following structure: first, I provide an outline of criticism concerning the novel. I will show that, despite a growing interest in the scientific subject matter of the novel, very little has been said about how this particular subject matter is textually conveyed. Then I will discuss magical imagery on two levels: first, in the depiction of scientists as "magicians," a conceptual interface which, in fact, has had a long tradition in the Western world; secondly, in the usage of scientific discourse, which appears as an almost alien element in the overall economy of the text. Finally, I will investigate to what extent the use of magical imagery correlates with non-literary discourses about science at the time, and how this strategy helps involve the lay reader.

Lewis' *Arrowsmith*: Some Critical Voices

Arrowsmith is one of Sinclair Lewis' most widely read books. By 1959, the book had sold 700,000 copies.[5] From the beginning, Lewis' novels have been rather controversial. While some authors criticised the "shallowness"[6] of Lewis' writing and the "slick magazine language"[7] of certain passages, others praised him for the "fresh new realism"[8] he introduced to the American novel. As Michael Augspurger contends, however, Lewis has been largely "canonized for vividly recording the inanity of provincial American middle-class life in the early twentieth century."[9] Thus, Lewis Mumford, for example, elevates Lewis to "the most effective satirist our country has produced,"[10] and Mark Schorer argues that "the triumph is given to the best qualities of the middle class and that it is its worst qualities that the novel [i.e., *Main Street*, J. M.] castigates: smugness, hypocrisy, a gross materialism, moral cant."[11] In *Arrowsmith*, it is furthermore the alliance between medicine, institutionalised science and capitalism that is uncovered and criticised, as James M. Hutchisson points out: "Arrowsmith analyzes the point in the development of medicine when physicians became empowered through science and when capitalism intervened to determine a political agenda for health care that would be supportive of the new industrial empires of the early twentieth century."[12] The description of the Rouncefield Clinic as a "medical factory" (*Arrowsmith* 271), in which Martin works as a "faithful mechanic" (ibid.) of pathological medicine, exemplifies this criticism, and so does the presentation of the McGurk Institute, which is part of a huge corporation and is, significantly, located in a New York high rise "above a world of stenographers and bookkeepers" (*Arrowsmith* 276).

Yet satire hardly accounts for the popularity of Lewis' novel, especially since one might have expected the American audience to be outraged by the criticism indirectly addressed to them. Augspurger therefore moves the argument further by taking into account the romantic thrust of *Arrowsmith*: "The explanation for this incongruity is that Lewis did provide more than invective. In his most successful decade, a romantic idealism behind Lewis' satire explains at least in part the overwhelming cultural resonance of his novels."[13] The ending of the novel, when Martin leaves behind a life in luxury with his most accomplished and beautiful second wife in order to lead a secluded life as independent scientist with his colleague, Terry Wickett, in the backwoods of Vermont, indeed conveys an image of romantic idealism with its emphasis on the fulfilment of individual dreams and the final retreat to nature. In Augspurger's view it is this offer of a "vision of adversary values"[14] which must have appealed to an audience of professional managerial readers who found themselves torn between professional ideals and the demands of bureaucratic corporate organisations. Earlier critics had already discussed Lewis' romantic side, albeit in more dismissive terms. Martin Light, for example, discredits the character Arrowsmith as "a wanderer, a fancifier, a romantic,"[15] and he calls Lewis a "fantasist,"[16] whose "soap opera episodes"[17] are hardly convincing. Geoffrey Moore considers Lewis "a rebel with too many causes,"[18] and he discerns a "central split" in Lewis, namely "the degree to which he wished to express his appreciation of the prairie people, and the degree, also, to which he felt abashed before the sophisticated, educated Easterners."[19] This split can also be perceived in Martin Arrowsmith, who most of his life dreams of becoming a famous scientist, but at the same time struggles with the social requirements that success imposes on him. As Charles E. Rosenberg succinctly puts it: "His ultimate rejection of society and its demands has been criticized as callow romanticism – and perhaps it is. But it is the logical result of Lewis' desire to depict greatness and his inability to conceive of its being allowed to exist within American society."[20]

While earlier criticism of Lewis' novel was primarily concerned with value judgments concerning the quality of Lewis' writing and the conception of his characters, for example,[21] later critics focused more strongly on the professional and scientific subject matters of *Arrowsmith*, medicine and bacteriology, and situated them in wider sociocultural contexts. I have already mentioned Augspurger's discussion of the conflict between the humanist ideals of medicine and the managerial aspects of medical practice, which came into sharper relief in the early twentieth century, and Hutchisson, who moves in a similar direction by concentrating on the relationship between medical practice and capitalism in the novel. Mary G. Land, by contrast, discusses the scientific debates still dominant at the time of the novel's creation,[22] especially the mechanist-vitalist controversy, i.e., the 18th-century controversy about whether there was a mysterious vital force behind all life, or whether everything could be explained by means of physico-chemical and materialist-energetic processes.[23] By modelling the immunologist Max Gottlieb on Jacques Loeb, head of the Rockefeller Institute's Division of Experimental Biology, who favoured a mechanistic view in his work, Lewis managed to present "a reasonable facsimile of the scientific enterprise," as Land argues.[24] Gert H. Brieger, himself a medical doctor, is equally admiring in his statement that "Lewis and de Kruif [i.e., the Rockefeller scientist who contributed considerably to the conception and elaboration of *Arrowsmith*; J. M.] have succeeded in describing the scene just as medical science was at the take-off stage of its impressive subsequent trajectory."[25]

Jacques Loeb is an important figure for the discussion of Lewis' book, as his philosophy of medical science is a central motif in the novel. Not only is Max Gottlieb equally devoted to mathematical precision in research and to the discovery of truth, but he also becomes an almost God-like mentor for Arrowsmith, who, towards the end of the novel, finally acquires the qualities of a 'real' scientist when he writes "his first important paper, on the effect of X-rays, gamma rays, and beta rays on the anti-Shiga phage" (*Arrowsmith* 406). It is important to bear in mind that the increase in knowledge about bio-chemical processes in the early twentieth century gradually began to demystify people's notions of life. Rosenberg, for example, writes about Loeb's work: "Loeb turned to physiological research in an attempt to prove that animal behavior was simply the sum of inorganic phenomena no different in kind from those studied by the physical scientist. Human behavior, too, he believed, was no more nor less than the product of such physical and chemical forces. The "mystical" aspects of life were to be dissolved in the acid of mathematics and physical chemistry."[26] While the science of biochemistry excludes mysticism, mysticism is far from being absent in Lewis' novel. On the contrary, I will show that 'magical' imagery is frequently drawn upon precisely for the depiction of the scientific enterprise, and I argue that it is this combination which constitutes much of the novel's appeal, as it provides food for the reader's imagination. In the following sections I will delineate three areas that seem to me to be of interest: 1.) The occurrence of the image of the scientist as magician; 2.) the negotiation of medicine, biochemistry, and bacteriology in the novel's 'scientific' discourse; and 3.) the implicit framing of the lay reader as participant-observer in the presented scientific enterprise.

The Scientist as Magician

Research on the Babylonian culture has shown that, from an early time, magic constituted an integral part of both religion and science.[27] As the ethnologist Christoph Daxelmüller argues, magic was part of the religiously legitimised liturgy but it also provided an incentive for the investigation of those phenomena people believed had been discovered in certain constellations of the stars or in intestines.[28] It is perhaps not surprising then that magic and science have frequently been linked. In *Arrowsmith*, the link is first and foremost Max Gottlieb, who appears like an almost mystical figure and who stimulates the young Martin's imagination. Consider the following text passage where Martin observes Gottlieb on his way home from the laboratory: "He was unconscious of the world. He looked at Martin and through him; he moved away, muttering to himself, his shoulders stooped, his long hands clasped behind him. He was lost in the shadows, himself a shadow. He had worn the threadbare top-coat of a poor professor, yet Martin remembered him wrapped in a black velvet cape with a silver star arrogant on his breast" (*Arrowsmith* 10). The imaginary velvet cape with the silver star reminds one of sorcerers and magicians, especially if one interprets the star as a pentagram, the symbol of the quintessential, of power and harmony, and, in a figurative sense, the universal symbol for magic.[29] However, the text does not specify which kind of star Martin imagines, and thus one begins to wonder. The star could, for example, also represent the star of David, which consists of two interlocked triangles and thus symbolises the tripartite composition of the world (animal; astral soul; and spiritual principle).[30] This interpretation is plausible to the extent that the novel also mentions Gottlieb's German-Jewish origins. Another possibility is that the star represents an upside down pentagram, which is the symbol of black magic that is used for negative purposes and in the name of Satan.[31] This reading corresponds with other passages in *Arrowsmith* where students attribute to Gottlieb the powers of a "priest of diabolic mysteries" (*Arrowsmith* 36): "Professor Gottlieb was the mystery of the University. [...] A thousand fables fluttered about him. It was believed that he was the son of a German prince, that he had immense wealth, that he lived as sparsely as the other professors only because he was doing terrifying and costly experiments which probably had something to do with human sacrifice. It was said that he could create life in the laboratory, that he could talk to the monkeys which he inoculated, that he had been driven out of Germany as a devil-worshiper or an anarchist, and that he secretly drank real champagne every evening at dinner" (*Arrowsmith* 8). The semantic field drawn upon in this passage clearly sets up Gottlieb as a practitioner of satanic magic, a "devil-worshiper," who talks to monkeys and makes use of human sacrifice. However, the passage also ridicules the naïve perception of Gottlieb's students, through whom this view is focalised. They attribute to Gottlieb the qualities of a magician because they lack a better understanding of his work and life, and in the same vein they also perpetuate other rumours which reinforce the notion of Gottlieb's deviance, e.g., his anarchic political stance and his excessive life style that allegedly entails drinking "real champagne."[32] Gottlieb's work on the one hand bewilders his students, on the other hand it fills them with admiration and with the same "respect they had for a man who could do card tricks or remove an appendix in seven minutes" (*Arrowsmith* 33). In other words, the text captures both an inherent fear of all the mysteries surrounding immunology and biochemistry in the early twentieth century, and admiration for the precision and techniques of these upsurging sciences. The combination of magical powers and scientific knowledge also reminds one of the image of the magus, i.e., a member of the priestly caste of the Medes and Persians, who were said to have these qualities. In this sense Gottlieb is also comparable to Paracelsus, the Swiss physician and alchemist.

Martin himself succumbs to the "charms of bacteriology" (*Arrowsmith* 242) and the "tragic enchantments of Max Gottlieb's world" (*Arrowsmith* 169). Even while he practises as a general practitioner in the small town of Wheatsylvania, he cannot entirely escape the "mysterious and unreasoning compulsions of his science" (*Arrowsmith* 140). Part of his fascination is kindled by the seemingly empowering new technologies. Thus, Martin is more than pleased with his new practice equipment: "Excitedly he ordered a steel stand, a sterilizer, flasks, test-tubes, and a white-enameled mechanism with enchanting levers and gears which transformed it from examining-chair to operating-table" (*Arrowsmith* 152). When he drives out into the countryside to rescue a young girl who is about to die from diphtheria, he still believes in the power of his medication and his newly bought car, which is supposed to carry him faster to suffering patients: "He thanked the god of science for antitoxin and for the gas motor. It was, he decided, a Race with Death" (*Arrowsmith* 158). Ironically, however, and, one might add, tragically, the "god of science" fails Martin on both counts: first, Martin does not immediately find the way to the farm and thus loses time despite his fast car; then the girl dies at the very moment that Martin injects his antitoxin. Although the girl's mother, Mrs. Novak, "waited for him to do miracles" (*Arrowsmith* 157), and although Martin himself hopes to emerge as a heroic country doctor, he fails miserably. This incident suggests an ambivalent approach towards science and technology: while new scientific insights and tools have the potential of improving health care provision, they are by no means a guarantor thereof. It seems that Lewis' novel not only criticises an overemphasis on the technical side of medical science, but also the unreasonable expectations that patients develop alongside the growing expertise of doctors. The competent doctors at the Rouncefield clinic, for example, are "master technicians" and "captain-generals of medicine," as well as "great priests and healers" (*Arrowsmith* 272), but they are also portrayed as cold and "visionless" (*Arrowsmith* 270).

Martin experiences the same conflict when he discovers what he calls the X-Principle, i.e., a principle whereby bacteria are mysteriously dissolved under certain conditions.[33] What is at stake here is the discovery of bacterial viruses, which was first accomplished by Frederick Twort in 1915.[34] While Martin's initial laboratory work is presented in terms of the magic and mystery of real scientific advance, his subsequent application of his "phage" is clad more in images of charlatanery. Thus, his laboratory work seems to him to reveal "something at the mysterious source of life" (*Arrowsmith* 306), and when he sets out to test his phage on various types of bacteria, even his colleagues cannot escape a sense of mysticism surrounding this enterprise: "To protect himself from infection by the rat-fleas he wore, while he worked with the animals, rubber gloves, high leather boots, straps about his sleeves. These precautions thrilled him, and to the others at McGurk they had something of the esoteric magic of the alchemists. He became a bit of a hero and a good deal of a butt" (*Arrowsmith* 338). The analogy with alchemy is telling here as involvement in alchemist practices was often strongly disapproved of and even led to the persecution of alchemists.[35] Martin is not persecuted but he becomes the laughing stock of the Institute,[36] which again indicates the ignorance of people not initiated to the 'real mysteries' of bacteriology. Instead of being allowed to continue his in-depth research, Martin is sent off to a plague-ridden West Indian island, where he is "expected not only to produce miracles but to explain beforehand how important and mature and miraculous he was" (*Arrowsmith* 349). There is rumour that "he possessed a magic cure for the plague" (*Arrowsmith* 377), and people "came in committees to beg him to heal their children" (*Arrowsmith* 382). In other words, Martin is perceived as a magic healer who has the power to do miraculous things. Hardly anyone understands Martin's real purpose of coming to the island, namely to conduct an experiment concerning the effectiveness of phage by injecting only half of the population and by leaving the other half untreated as a control group. On his return to New York, "the newspapers had reported wonders" (*Arrowsmith* 404), but Martin is dissatisfied with himself as he did not observe the principles of rigid test conditions and administered phage to almost everyone. Thus, while people celebrate the 'magic' of phage made visible in the cure of thousands of people, its 'real' magic in the scientific sense, i.e., the actual principles underlying its functioning, still remain opaque.

To my mind, it is surprising that only a few critics have discussed the West Indies episode in detail. Some critics have rejected this part of Lewis' novel on the grounds of its exotic nature, which seemed unsuitable in a largely realistic novel.[37] Charles E. Rosenberg, for example, mentions the respective chapters only in passing and then only in a footnote.[38] Lisa L. Lynch, on the other hand, discusses the island episode at length and provides a postcolonial reading. She argues that "Lewis's novel suggests the outmoded, arcane nature of European imperialism by associating imperialism with a dangerous failure to understand the concepts of twentieth-century medical science. Modern medical science is then given a specifically American face in the figure of Martin Arrowsmith, the man who steps in to shoulder the "white man's burden" in the midst of a crisis brought on by colonial mismanagement. But Martin's loss of his wife, and his own temporary loss of sanity, hints at possible consequences of such acts of beneficence."[39] I would only partly go along with Lynch's argument. For one thing, the idea of 'being American' is downplayed in the novel when the narrator comments: "Martin was, like most inhabitants of Elk Mills before the Slavo-Italian immigration, a Typical Pure-bred Anglo-Saxon American, which means that he was a union of German, French, Scotch, Irish, perhaps a little Spanish, conceivably a little of the strains lumped together as "Jewish," and a great deal of English, which is itself a combination of primitive Briton, Celt, Phoenician, Roman, German, Dane, and Swede" (*Arrowsmith* 2). Secondly, Martin's American contemporaries lack as much understanding of the importance of his scientific experiments as do the natives on the island. At the same time, one black physician, Dr. Marchand, is supportive of Martin's plans and is generally more scientifically-minded than many of his white American colleagues presented in the novel. Thirdly, Martin's failure on the island is not necessarily indicative of the failure of imperialism, as he also fails several times as a GP back home in the States. I would regard this failure as another example of the ongoing conflict between the goals and requirements of biomedical science on the one hand, and actual medical practice on the other.

This conflict also involves the difference between being a "miracle man" and a "scientist," as Gottlieb points out to Martin (*Arrowsmith* 316). However, as I already mentioned, science itself creates mysteries, and Gottlieb expands on this notion by comparing scientific work to mysticism and to writing poetry. He says to Martin: "To be a scientist – it is not just a different job, so that a man should choose between being a scientist and being an explorer or a bond-salesman or a physician or a king or a farmer. It is a tangle of ver-y [sic] obscure emotions, like mysticism, or wanting to write poetry; it makes its victim all different from the good normal man. The normal man, he does not care much what he does except that he should eat and sleep and make love. But the scientist is intensely religious – he is so religious that he will not accept quarter-truths, because they are an insult to his faith" (*Arrowsmith* 278). It is telling that the scientist's feelings about his work should be the same as, for example, the feelings of a poet when trying to write poetry. It is implied that other people can only feel empathy with scientists if they have themselves experienced this strong desire to uncover the truth, as is also immanent to religious belief and art. The question that arises, then, is: how can these emotions of the real scientist and the principles of science be conveyed to a largely lay audience? As far as the emotional effect of the novel is concerned, the answer is simple: by presenting scientists as human beings with all their flaws and weaknesses, the novel allows us to identify with them to a certain extent and thus to adopt a more sympathetic view. But what about the principles of science that the second part of my question addresses? How can biochemical, and more specifically, bacteriological discourse be presented?

Scientific Discourse in *Arrowsmith*

One possible answer to this question could be: scientific discourse can be presented in analogous images that non-scientists would be able to understand through making inferences from their own life experience and culture. Beside the images of religion and art, one of the most prominent metaphors to describe the life and work of the scientist is magic. Thus, we observe Martin falling into a kind of trance by staring "at a shining flask till it fascinated him like a crystal ball" (*Arrowsmith* 328), and we accompany him when he learns "the involved mysteries of freezing-point determinations, osmotic pressure determinations" (*Arrowsmith* 405) and when he becomes "absorbed in mathematical laws which strangely predicted natural phenomena" (ibid.). From the beginning of the novel, scientific discourse is frequently compared to magic, the *tertium comparationis* being the opacity and complexity of both phenomena. Thus, even as a young boy, when Martin reads about the lymphatic system, "he muttered the long and perfectly incomprehensible words in a hum which made drowsier the duty room" (*Arrowsmith* 2). This muttering with its humming sound reminds one of incantations, an image that is later explicitly mentioned when the narrator describes the scientist Sondelius: "Sondelius was a roarer and a playboy, but he knew incantations" (*Arrowsmith* 181). When Gottlieb tells the young Martin about his own work, he does it "in a vocabulary as heathenish as Tibetan magic: 'Arrhenius and Madsen have made a contribution toward bringing immunity reactions under the mass action law, but I hope to show that antigen-antibody combinations occur in stoic-heiometric proportions when certain variables are held constant'" (*Arrowsmith* 292). Indeed, for a lay readership to hear about "antigen-antibody combinations" and "stoic-heiometric proportions" also sounds like "Tibetan magic." Yet, who would know what Tibetan magic sounded like? Interestingly enough, the imagery of magic that is used to illustrate science is as remote from people's everyday life experience as science itself. Daxelmüller points out that the most intriguing critical and historical problem concerning magic is that magicians and magical practices remain unknown to the present day. All we have is a 'knowledge' that consists of interpretations and discursive traditions of magic which have come down to us over the centuries and have thus been reinforced: "die Täter, die wir einmal «Magier» nennen wollen, sind bis heute ebenso unbekannt wie die Magie, die sie praktizierten. Es bleiben lediglich unser Wissen und unsere Interpretationen von Magie, wie sie über Jahrhunderte hinweg überliefert und dadurch gefestigt wurden" ["the culprits, whom we will call magicians here, have hitherto been as unknown to us as the magic they practised. What remains is simply our knowledge and our interpretations of magic as they have been passed on throughout the centuries and which have thus been reinforced;" my translation].[40]

Just as magic is discursively (re)constructed in our culture, it is also used in *Arrowsmith* to textually reconstruct inaccessible fields of knowledge. This does not solely apply to the depiction of science, however. Occasionally, magical images are evoked in other contexts, e.g., when emotions like fatigue are described: "Martin stood with her at the window, mad with the pride of proprietorship. Suddenly he was so weak, so tired, that the mysterious cement which holds cell to cell seemed dissolved, and he felt that he was collapsing" (*Arrowsmith* 109). This quote is interesting as it combines the semantic domains of magic ("mysterious") and science ("cell to cell") with the rather mundane field of building ("cement"). One forms the impression that even science, although it is generally treated with reverence and in mostly positive terms throughout the novel, need not be taken too seriously.[41] This is probably best illustrated by quoting the following lengthy passage from the novel, which depicts the reception of Martin's first 'serious' scientific paper by his boss and by the scientific community:

Holabird was as much bewildered as Tubbs would have been by the ramifications of Martin's work. What did he think he was anyway – a bacteriologist or a bio-physicist? But Holabird was won by the scientific world's reception of Martin's first important paper, on the effect of X-rays, gamma rays, and beta rays on the anti-Shiga phage. It was praised in Paris and Brussels and Cambridge as much as in New York, for its insight and for "the clarity and to perhaps be unscientifically enthusiastic, the sheer delight and style of its presentation," as Professor Berkeley Wurtz put it; which may be indicated by quoting the first paragraph of the paper: *In a preliminary publication, I have reported a marked qualitative destructive effect of the radiations from radium emanations on Bacteriophage-anti-Shiga. In the present paper it is shown that X-rays, gamma rays, and beta rays produce identical inactivating effects on this bacteriophage. Furthermore, a quantitative relation is demonstrated to exist between this inactivation and the radiations that produce it. The results obtained from this quantitative study permit the statement that the percentage of inactivation, as measured by determining the units of bacteriophage remaining after irradiation by gamma and beta rays of a suspension of fixed virulence, is a function of the two variables, nillicuries and hours. The following equation accounts quantitatively for the experimental results obtained:*

$$k = \frac{\lambda \log e \dfrac{u_0}{u}}{E_0(\varepsilon - \lambda t_1)}$$

(*Arrowsmith* 406-407)

Readers will not fail to notice the irony employed in the juxtaposition of Professor Wurtz's praise of the "clarity" of Martin's paper and the direct quotation from the same paper, which is far from clear. Like other pieces of scientific writing, this paper abounds in nominal constructions and complicated technical terms, which are incomprehensible to lay persons. In addition, a mathematical equation is presented to the reader to demonstrate not only the knowledge behind Martin's scientific work, but also the apparent uselessness of this knowledge for common people. Even more educated readers will not necessarily be able to verify or falsify this formula unless they have been trained in biophysics. In other words, Lewis could have used any equation, even a nonsensical one, without loss or harm.[42] Thus, the reader is indirectly aligned with "the lay visitors to the laboratory, who believed that sanguinary microbes would leap upon them from the mysterious centrifuge, from the benches, from the air itself," and for whom Martin feels "professional scorn" (*Arrowsmith* 32). In the following section, I will consider the reader in more depth.

Involving the Reader as Participant-Observer

With reference to Hugh Holman's comments on Lewis' novels,[43] Augspurger contends that "satire's primary influence resides not in its ability to reform those it criticizes, but in its power to consolidate and reaffirm the sensibility of the criticizer."[44] Thus, Augspurger continues, "satire can also have an affirmative power. By 'inviting' readers to share a way of looking at the world, the satirist creates a divide between a perceptive 'us' and a misguided 'them'."[45] Put differently, if the narrator in *Arrowsmith* ridicules ignorant patients for expecting too much from their doctor, or visitors to the laboratory for reacting in an almost hysterical manner, or narrow-minded bureaucrats for lacking an understanding of real science, then readers are indirectly also invited to identify all these shortcomings and to side with the narrator in their own judgements and perceptions of such characters. On the other hand, more self-critical readers may well discern some of their own characteristics in such presentations, e.g., their fear of new sciences, their attitudes towards doctors, etc.[46] It is interesting to note in this context that the serial version of *Arrowsmith*, which preceded the book publication and appeared in the *Designer and the Woman's Magazine*, was considerably altered in order not to be "too repulsive for its readers."[47] Especially references to certain medical diseases and the description of the dissecting room at the medical college are left out, for example. I would argue, by contrast, that it is exactly the medical and biochemical parts of the novel that add a special flavour to it and thus attract the interested reader.

James Owen Drife sees the advantage of surgery as a subject matter for novels in that "it is physical and it makes people faint. Gritty truth has always been fashionable in narrative from Charles Dickens to Irving Welsh, and surgeons are in regular contact with blood and guts. The operating theatre is a place of drama."[48] Likewise, it is the scenes where medical students dissect rats and rabbits, where Martin injects patients with medication and grapples with death in the face of the bubonic plague epidemic, that readers get goosebumps and shudders are sent down their spines. The island scenes in particular must have been gripping as the described epidemic possibly reminded readers of the massive flu pandemic that had swept the globe in the autumn of 1918.[49] Moreover, as I pointed out at the beginning of my chapter, there was a strong movement in favour of public education in science. People's imagination concerning science and scientists was also stimulated by Einstein's visit to America in 1921. As Paul A. Carter maintains, Einstein appealed through his "unworldliness,"[50] and for many Americans he became a "guide to the new century's *mysterium tremendum*."[51] One can perhaps draw a parallel between Einstein and Gottlieb, who both, as foreigners and devoted scientists, at the same time alienate and fascinate their contemporaries.

What is particularly interesting when considering the discursive presentation of science in 'magical' terms in Lewis' novel, is the fact that many of the non-literary discourses concerning science at the time used a similar language. Thus, for example, Robert L. Duffus wrote in *Collier's* in 1924: "The scientists have gone off all by themselves and made a magic that even they didn't foresee. [...] This magic is transportation and great cities and wars and Ford factories and much else."[52] The conflict between pure and applied science is depicted by George Horace Lorimer in a 1926 *Post* editorial in the following way: "Applied in [sic] the starry-eyed goddess, the wonder worker, who gets on the first page of newspapers and captivates the imagination of men. She speaks the language of the people, gives them what they want, and every year perfects new gifts to make life easier or longer or more amusing. [...] Pure Science is the wallflower, the elder sister who lives secluded and remote, unknown and unpraised. She does not advertise her astounding feats, and could not if she would, for the only language she knows is a jumble of Latin, Greek, calculus, and mathematical formulas."[53] This passage strongly reminds one of the way that pure and applied science are depicted in *Arrowsmith*: Pure science as represented by Gottlieb and by Martin towards the end of the novel is marked by an incomprehensible discourse ("Tibetan magic"), and a lack of recognition by the outside world, whereas applied scientists are presented as "witch doctors" (*Arrowsmith* 29) and "miracle man" (*Arrowsmith* 316), in analogy to the "wonder worker" in the quoted passage above. By using magical imagery to describe the scientific enterprise in his novel, Lewis tapped into contemporary discursive practices, and thus he managed to endow science with a more popular language that would help it speak for itself, even if that was only made possible in the artistic realm of the novel's storyworld. Readers thus won over to the side of science, could then benevolently listen to the 'mysteries' of science displayed to them, even though they may not have fully understood them. In a way, we assume the position of Martin's first wife, Leora, in "listening to" Lewis' narrative: "Martin was off again, and if Leora did not altogether understand the relation of the synthesis of antibodies to the work of Arrhenius, yet she listened with comfortable pleasure in his zeal" (*Arrowsmith* 58).

Conclusion

The image of magic appears constantly in contexts where the scientific enterprise is depicted, both from the viewpoint of the scientists themselves, who seek to dig deeper in the 'mysteries' of bacteriology and immunology, but also from the perspective of lay people who look upon science and its application in medical practice as incomprehensible, 'magical' endeavours. As I demonstrate, magical images are not unique to literary discourse but can be found in other, non-literary discourses surrounding the natural sciences in the early twentieth century. This meshing of discourses of science and discourses of magic has had a long tradition in the Western world and has perhaps contributed substantially to the ways that we perceive of science. In my opinion, this is not surprising if one considers that both magic and science seem to converge in one important point: they both create enigmas and thus stimulate outside observers' imagination. Ironically, science's goal to get to the 'truth' of life and to ultimately explain its underlying laws and functionings, has not clarified the world for us but keeps mystifying us by ever raising new questions. As Ingo Berensmeyer maintains with reference to Hans Blumenberg's concept of reality as an "open context": "Science is one instance in which this modern concept is exemplified: its horizon is open insofar as it is constantly modified by new observations, new discoveries, so that scientists have to be trained to expect the unexpected,"[54] and he continues by saying that "the increase of knowledge will necessarily bring with it an increase of non-knowledge, of blank spaces and uncertainty."[55] The narrator in *Arrowsmith* hints at the same problem by describing science as "the unswerving quest for causes which, as it drove through layer below layer, seemed ever farther from the bottom-most principles," and he comments on Martin's "intolerable strain of learning day by day how much he did not know" (*Arrowsmith* 110-111).

People watching magic tricksters perform also become accustomed to expecting 'the unexpected.' Yet magic, i.e., 'real' magic, is even more opaque. One cannot even be sure whether magic and magicians ever really existed. Nevertheless, most people have no difficulty conjuring up images of magicians and of magical scenarios in their minds, as such images have come down to us through folklore and fairytales, and nowadays also through fantasy films and role play games, for example. It is this stock of preconceived images and models that Lewis used for his novel. By presenting biochemical discourse to a largely lay readership through images of magic, Lewis succeeded in making this at first sight alien and incomprehensible knowledge domain palatable and accessible, albeit not necessarily more comprehensible, and thus he managed to turn a scientific topic into an interesting subject matter for a novel. This must also be regarded within the wider context of attempts in the 1920s to popularise science, and to make scientific results available to a larger audience, as I showed earlier in this chapter. While the magical vocabulary and images in *Arrowsmith* often draw upon clichés concerning both magic and science, it is exactly this recognisability that helps readers identify their own (mis)conceptions and thus perhaps overcome them in a process of critical self-reflection. In this sense, Lewis' novel parallels science in that it also constitutes "work whose end is satisfying because there is never an end" (*Arrowsmith* 300).

Notes

[1] Subsequent references to the book are placed in the text, including title and page numbers.

[2] Cf. James Strick, "Darwinism and the Origin of Life: The Role of H. C. Bastian in the British Spontaneous Generation Debates, 1868-1873" 58.

[3] Cf. Paul A. Carter, *Another Part of the Twenties* 71-72.

[4] Cf. ibid. 75.

[5] Cf. Michael Augspurger, "Sinclair Lewis' Primers for the Professional Managerial Class: Babbitt, Arrowsmith, and Dodsworth" 94.

[6] Geoffrey Moore, "Sinclair Lewis: A Lost Romantic" 163.

[7] Martin Light, *The Quixotic Vision of Sinclair Lewis* 90.

[8] Alfred Kazin, "The New Realism: Sherwood Anderson and Sinclair Lewis" 119.

[9] Augspurger, "Sinclair Lewis' Primers for the Managerial Class: Babbitt, Arrowsmith, and Dodsworth" 73.

[10] Lewis Mumford, "The America of Sinclair Lewis" 102.

[11] Mark Schorer, "Sinclair Lewis" 51.

[12] James M. Hutchisson, "Arrowsmith and the Political Economy of Medicine" 123.

[13] Augspurger, "Sinclair Lewis' Primers for the Professional Managerial Class: Babbitt, Arrowsmith, and Dodsworth" 74.

[14] Ibid. 94.

[15] Light, *The Quixotic Vision of Sinclair Lewis* 86.

[16] Ibid. 89.

[17] Ibid.

[18] Moore, "Sinclair Lewis: A Lost Romantic" 161.

[19] Ibid. 158.

[20] Charles E. Rosenberg, "Martin Arrowsmith: The Scientist as Hero" 458.

[21] Cf. Lyon N. Richardson, "*Arrowsmith*: Genesis, Development, Versions," and Mark Schorer, *Sinclair Lewis: A Collection of Critical Essays.*

[22] Cf. Mary G. Land, "Three Max Gottliebs: Lewis's, Dreiser's, and Walker Percy's View of the Mechanist-Vitalist Controversy."

[23] This debate had been revived, among others, in Thomas Huxley's Biogenesis and Abiogenesis address (Strick, "Darwinism and the Origin of Life: The Role of H. C. Bastian in the British Spontaneous Generation Debates, 1868-1873" 68ff.), which discusses the question of the origin of life. "Biogenesis" refers to the creation of life only from pre-existing life, while "abiogenesis" means the creation of life from non-living matter, also called "spontaneous generation."

[24] Land, "Three Max Gottliebs: Lewis's, Dreiser's, and Walker Percy's View of the Mechanist-Vitalist Controversy" 314.

[25] Gert H. Brieger, "*Arrowsmith* and the History of Medicine in America" 36.

[26] Rosenberg, "Martin Arrowsmith: The Scientist as Hero" 454-455.

[27] Cf. Christoph Daxelmüller, *Zauberpraktiken: Die Ideengeschichte der Magie* 46ff.

[28] Ibid. 51.

[29] Cf. Mária Szepes and Wictor Charon, *Die geheimen Lehren des Abendlandes: Academia Occulta* 329.

[30] Ibid. 328.

[31] Ibid. 329.

[32] This rumour incidentally also conveys criticism concerning the middle classes, who have traditionally opposed the eccentric and luxurious ways of the aristocracy or higher social classes.

[33] Readers may identify an ironical cross-reference to the so-called X Club here, i.e., a club of scientists who engaged in evolutionary theory. Famous members of this club were Thomas H. Huxley and John Tyndall (Strick, "Darwinism and the Origin of Life: The Role of H. C. Bastian in the British Spontaneous

Generation Debates, 1868-1873").

[34] American Society for Microbiology, "Significant Events Of The Last 125 Years." Lewis' book follows largely the historical events by making Martin discover his principle by accident, much as Twort had made his discovery while growing viruses and noticing that the bacteria on his plates became transparent. Likewise, Felix d'Herrelle's independent discovery of what he coined "*bactériophage*" in 1917 is used in the novel for the sake of verisimilitude but also to prevent Martin from gaining fame for his discovery.

[35] Cf. Szepes and Charon, *Die geheimen Lehren des Abendlandes: Academia Occulta* 37; Daxelmüller, *Zauberpraktiken: Die Ideengeschichte der Magie* 116; 147-148. Daxelmüller describes, for example, cases in the Medieval period where monks were occasionally accused of dealing in alchemy or Templars such as Gilles de Rais (1404-1440) became the epitome of evil alchemist practices and ritual magic.

[36] That doctors are presented in comical terms has had a long tradition in literature. In Molière's comic dramas or in the medieval *commedia dell'arte*, for example, one frequently encounters the stock character of the quack (Anne Hudson Jones, "Images of Physicians in Literature: Medical Bildungsromans" 734).

[37] Robert Morss Lovett, for example, contends: "When, however, Arrowsmith in fighting the plague in St. Hubert is bidden by his scientific conscience to divide the population into two parts, [...] we have either an example of scientific fanaticism or a piece of pure fiction. [...] This air of unreality hangs over the latter part of the book as Mr. Lewis becomes more absorbed in his purpose" (Lovett, "An Interpreter of American Life" 34).

[38] Cf. Rosenberg, "Martin Arrowsmith: The Scientist as Hero" 456n.

[39] Lisa L. Lynch, "*Arrowsmith* Goes Native: Medicine and Empire in Fiction and Film" 207.

[40] Daxelmüller, *Zauberpraktiken: Die Ideengeschichte der Magie* 19. For Daxelmüller, this problem leads to the difficulty to determine whether magic in the Western world was simply an ethical fiction that helped stabilise social and cultural systems, whether and how it became a model for action and a mental principle in folk culture, or if it even represents a basic cognitive structure and thus functioned as a major catalyst for the development of Western culture (cf. ibid. 44-45).

[41] The fact that both laypersons and scientists are ridiculed to a certain degree points towards a Quixotic presentation of science. For a discussion of this aspect in Lewis' work see Light's *The Quixotic Vision of Sinclair Lewis*.

[42] In fact, as Gregor Schiemann and Dirk Vanderbeke pointed out to me, the equation is non-sensical and faulty in a number of ways. This, however, does not deflect from the ironical implications of the use of such a formula in a non-scientific text, quite on the contrary.

[43] Cf. Hugh Holman, "Anodyne for the Village Virus."

[44] Augspurger, "Sinclair Lewis' Primers for the Professional Managerial Class" 86.

[45] Ibid.

[46] For example, many people today are sceptical towards or even scared of new developments in genetics and gene technology. This may well be analogous to people's fears about bacteriology and immunology in the early twentieth century.

[47] Richardson, "*Arrowsmith*: Genesis, Development, Versions" 239.

[48] James Owen Drife, "Narrative in Surgery" 116. This is probably one of the main reasons why Robin Cook's medical thrillers and TV series such as *ER*, *Casualty*, etc. are so popular today.

[49] Cf. American Society for Microbiology, "Significant Events Of The Last 125 Years."

[50] Carter, *Another Part of the Twenties* 66.

[51] Ibid. 71.

[52] Cited ibid. 76.

[53] Cited ibid. 77.

[54] Ingo Berensmeyer, "No Fixed Address: Pascal, Cervantes, and the Changing Function of Literary Communication in Early Modern Europe" 624.

[55] Ibid.

Works Cited

American Society for Microbiology. "Significant Events Of The Last 125 Years." http://www.asm.org/MemberShip/index.asp?bid=16731 (18 Oct. 2004).

Augspurger, Michael. "Sinclair Lewis' Primers for the Professional Managerial Class: Babbitt, Arrowsmith, and Dodsworth." *Journal of the Midwest Modern Language Association* 34.2 (2001): 73-97.

Berensmeyer, Ingo. "No Fixed Address: Pascal, Cervantes, and the Changing Function of Literary Communication in Early Modern Europe." *New Literary History* 34.4 (2003): 623-637.

Brieger, Gert H. "*Arrowsmith* and the History of Medicine in America." *Möbius* 2.3 (1982): 32-38.

Carter, Paul A. *Another Part of the Twenties.* New York: Columbia UP, 1977.

Daxelmüller, Christoph. *Zauberpraktiken: Die Ideengeschichte der Magie.* Düsseldorf: Albatros, 2001 [1993].

Drife, James Owen. "Narrative in Surgery," in *Narrative Based Medicine: Dialogue and Discourse in Clinical Practice.* Ed. Trisha Greenhalgh and Brian Hurwitz. London: BMJ Books, 1998. 110-117.

Holman, Hugh. "Anodyne for the Village Virus," in *Comic Imagination in American Literature.* Ed. Louis D. Rubin. New Brunswick, NJ: Rutgers UP, 1973. 247-258.

Hutchisson, James M. "Arrowsmith and the Political Economy of Medicine," in *Sinclair Lewis: New Essays in Criticism.* Ed. James M. Hutchisson. Troy, NY: Whitson, 1997. 110-124.

Jones, Anne Hudson. "Images of Physicians in Literature: Medical Bildungsromans." *Lancet* 348 (1996): 734-736.

Kazin, Alfred. "The New Realism: Sherwood Anderson and Sinclair Lewis," in *Sinclair Lewis: A Collection of Critical Essays.* Ed. Mark Schorer. Englewood Cliffs, NJ: Prentice-Hall, 1962. 119-128.

Land, Mary G. "Three Max Gottliebs: Lewis's, Dreiser's, and Walker Percy's View of the Mechanist-Vitalist Controversy." *Studies of the Novel* 15.4 (1983): 314-331.

Lewis, Sinclair. *Arrowsmith.* New York: Signet, 1998 [1925].

Light, Martin. *The Quixotic Vision of Sinclair Lewis.* West Lafayette, Indiana: Purdue UP, 1977.

Lovett, Robert Morss. "An Interpreter of American Life," in *Sinclair Lewis: A Collection of Critical Essays.* Ed. Mark Schorer. Englewood Cliffs, NJ: Prentice-Hall, 1962. 32-35.

Lynch, Lisa L. "*Arrowsmith* Goes Native: Medicine and Empire in Fiction and Film." *Mosaic* 33.4 (2000): 193-208.

Moore, Geoffrey. "Sinclair Lewis: A Lost Romantic," in *Sinclair Lewis: A Collection of Critical Essays.* Ed. Mark Schorer. Englewood Cliffs, NJ: Prentice-Hall, 1962. 151-165.

Mumford, Lewis. "The America of Sinclair Lewis," in *Sinclair Lewis: A Collection of Critical Essays.* Ed. Mark Schorer. Englewood Cliffs, NJ: Prentice-Hall, 1962. 102-107.

Richardson, Lyon N. "*Arrowsmith*: Genesis, Development, Versions." *American Literature* 27.2 (1955): 225-244.

Rosenberg, Charles E. "Martin Arrowsmith: The Scientist as Hero." *American Quarterly* 15 (1963): 447-458.

Schorer, Mark, ed. *Sinclair Lewis: A Collection of Critical Essays*. Englewood Cliffs, NJ: Prentice-Hall, 1962. 46-80.

---. "Sinclair Lewis," in *Seven Modern American Novelists: An Introduction*. Ed. William van O'Connor. Minneapolis: U of Minnesota P, 1964. 46-80.

Strick, James. "Darwinism and the Origin of Life: The Role of H. C. Bastian in the British Spontaneous Generation Debates, 1868-1873." *Journal of the History of Biology* 32 (1999): 51-92.

Szepes, Mária and Wictor Charon. *Die geheimen Lehren des Abendlandes: Academia Occulta*. Munich: Orbis Verlag, 2001.

'Science is magic that works'

The Return of Magic in Literature on Science[1]

Dirk Vanderbeke

The relationship between magic, science and knowledge in general has for centuries been marked by the dichotomy of right vs. wrong, and even if recent science studies in the wake of constructivist concepts attempt to show that science's truth claims are ill-founded, this has hardly led to a revival of academic belief in pre-scientific modes of knowledge. In contrast to some attacks (and a few esoteric interpretations) Paul Feyerabend in his influential and controversial book *Against Method* never demanded that "Equal weight [...] should be given to competing avenues of knowledge such as astrology, acupuncture and witchcraft."[2] Instead, he rather argued that reasonable and useful knowledge can exist on the basis of false premises and that some of that knowledge was unfortunately discarded with the advent of modern science.[3] Over the last decades quite a few studies have explored the explanatory power and functionality of early modes of knowledge, e.g. Georgio de Santillana's and Hertha von Dechend's *Hamlet's Mill: An Essay Investigating the Origins of Human Knowledge and Its Transmission Through Myth*.[4] Moreover, it may be useful to bear in mind that the almost universal discredit of magic may be based on a distorted image of pre-scientific knowledge and on the unreserved acceptance of the present paradigms.[5] Keith Thomas has pointed out that: "At all times most men accept their basic assumptions on the authority of others. New techniques and attitudes are always more readily diffused than their underlying scientific rationale. [...] Most of the millions of persons who today would laugh at the idea of magic or miracles would have difficulty in explaining why. They are victims of society's constant pressure towards intellectual conformity."[6]

The true/false dichotomy informs the perspective of James George Frazer's *The Golden Bough*, unquestionably one of the most influential books on magic in the early 20th century. Frazer argues that magic is founded on the same basic presuppositions as science: "[I]ts fundamental conception is identical with that of modern science; underlying the whole system is a faith, implicit but real and firm, in the order and uniformity of nature. The magician does not doubt that the same causes will always produce the same effects, that the performance of the proper ceremony, accompanied by the appropriate spell, will inevitably be attended by the desired result [...]"[7] The problem, then, is not one of the premises but of their validity: "The fatal flaw of magic lies not in its general assumption of a sequence of events determined by law, but in its total misconception of the nature of the particular laws which govern that sequence." Thus the difference is ultimately dependent on the outcome: "It is therefore a truism, almost a tautology, to say that all magic is necessarily false and barren; for were it ever to become true and fruitful, it would no longer be magic but science. From the earliest times man has been engaged in a search for general rules whereby to turn the order of natural phenomena to his own advantage, and in the long search he has scraped together a great hoard of such maxims, some of them golden and some of them mere dross. The true or golden rules constitute the body of applied science which we call the arts; the false are magic."[9] Frazer's assumption that magic is indeed founded on basic natural laws has been challenged, and Stephen Sharot points out that "the ethnographic evidence is overwhelming that the personal idiom is predominant in the mode of thought of those societies ('primitive,' 'preliterate,' 'traditional') that are said to have most magic."[10] But then magic is a very broad term covering the most diverse traditions of knowledge and practices in equally diverse cultures and societies,[11] and a look at one of the major texts of European magic indeed presents us with a perspective that confirms Frazer's view of a belief in universal laws. In his first book of *De occulta philosophia*, Agrippa von Nettesheim lays down the foundations of magic and their relationship with the other branches of knowledge:

"Magick is a faculty of wonderfull vertue, full of most high mysteries, containing the most profound Contemplation of most secret things, together with the nature, power, quality, substance, and vertues thereof, as also the knowledge of whole nature, and it doth instruct us concerning the differing, and agreement of things amongst themselves, whence it produceth its wonderfull effects, by uniting the vertues of things through the application of them one to the other, and to their inferior suitable subjects, joyning and knitting them together thoroughly by the powers, and vertues of the superior Bodies. This is the most perfect and chief Science, that sacred and sublimer kind of Phylosophy [philosophy], and lastly the most absolute perfection of all most excellent Philosophy. [...] Whosoever therefore is desirous to study in this Faculty, if he be not skilled in naturall Philosophy, wherein are discovered the qualities of things, and in which are found the occult properties of every Being, and if he be not skilful in the Mathematicks, and in the Aspects, and Figures of the Stars, upon which depends the sublime vertue, and property of every thing; and if he be not learned in Theologie [theology], wherein are manifested those immateriall substances, which dispence [dispense], and minister all things, he cannot be possibly able to understand the rationality of Magick. For there is no work that is done by meer Magick, nor any work that is meerly Magicall, that doth not comprehend these three Faculties."[12]

Even if we take into account that the mathematics of Agrippa is quite different from our understanding of mathematics, as it is chiefly concerned with geometry and celestial observation,[13] the account indeed presents us with a "rationality of magic." Of course, there is the firm belief in astrology and the principle of "As above, so below," and it is this branch of esoteric superstition that is most severely under attack whenever irrationalism and pre-scientific beliefs are scrutinised. However, it is also particularly this principle that occasionally resurfaces in modern science. When Niels Bohr described the structure of the atom in his Nobel Prize in Physics Award Address 1922, he added: "In this picture we at once see a striking resemblance to a planetary system, such as we have in our own solar system."[14] His colleague Max Born quickly grasped the analogy and pointed out that this concept performed "a great magic; indeed its form is rooted in the superstition (which is as old as the history of thought) that the destiny of men could be read from the stars."[15] Of course, Bohr's model of the atom had to be modified and the similarity with the solar system vanished, but the example shows how easily analogies can be forged between scientific and pre-scientific concepts – I will return to this point later.

In Kurt Vonnegut's novel *Cat's Cradle* we find the suitable abbreviation of Frazer's dictum that serves as the title of this paper: "Science is magic that *works*,"[16] another version of the same idea is Arthur C. Clarke's Third Law: "Any sufficiently advanced technology is indistinguishable from magic."[17] And indeed, if we look at the flights of fancy offered as possible achievements of magic there is little doubt that they have by now become part of our daily life – and in some suitably equipped modern housing the command "Let there be light" may well bring forth the desired result.[18] What seems to have been lost – at least according to some authors – is the enchantment of reality, and in a famed passage Robert Musil wrote:

> Wenn es die Verwirklichung von Urträumen ist, fliegen zu können und mit den Fischen zu reisen, sich unter den Leibern von Bergriesen durchzubohren, mit göttlichen Geschwindigkeiten Botschaften zu senden, das Unsichtbare und Ferne zu sehen und sprechen zu hören, Tote sprechen zu hören, sich in wundertätigen Genesungsschlaf versenken zu lassen, mit lebenden Augen erblicken zu können, wie man zwanzig Jahre nach seinem Tode aussehen wird, in flimmernden Nächten tausend Dinge über und unter dieser Welt zu wissen, die früher niemand gewusst hat, wenn Licht, Wärme, Kraft, Genuß, Bequemlichkeit Urträume der Menschheit sind, – dann ist die heutige Forschung nicht nur Wissenschaft, sondern ein Zauber, eine Zeremonie von höchster Herzens- und Hirnkraft, vor der Gott eine Falte seines Mantels nach der anderen öffnet, eine Religion, deren Dogmatik von der harten, mutigen, beweglichen, messerkühlen und -scharfen Denklehre der Mathematik durchdrungen und getragen wird. Allerdings, es ist nicht zu leugnen, daß alle diese Urträume nach Meinung der Nichtmathematiker mit einemmal in einer ganz anderen Weise verwirklicht waren, als man sich das ursprünglich vorgestellt hatte. Münchhausens

> Posthorn war schöner als die fabriksmäßige Stimmkonserve, der Siebenmeilenstiefel schöner als ein Kraftwagen, Laurins Reich schöner als ein Eisenbahntunnel, die Zauberwurzel schöner als ein Bildtelegramm, vom Herz seiner Mutter zu essen und die Vögel zu verstehen schöner als eine tierpsychologische Studie über die Ausdrucksbewegung der Vogelstimme. Man hat Wirklichkeit gewonnen und Traum verloren."[19]

However, the complementary view may also be valid, and while the disenchantment with science and technology is certainly recognisable, there are also the "Enchantments of Technology" – a book with this title has recently been published by Lee Bailey.

The idea that 'science is magic that works' is, of course, based on a pragmatic approach which evaluates practical results rather than a cognitive or explanatory potential. This bias needs to be questioned, as it neglects at least three important points: a) Within cultures based on a belief in magic the conviction in the practical effects of magic were probably equally strong as ours in the useful application of science.[20] b) Wrong theories can lead to useful knowledge and practical results. c) Science itself is historically contingent. The last point, in particular, is important when we look at the true/false dichotomy. Even if one rejects the radical conclusions as voiced by the proponents of the 'strong programme of social constructivism' – which I most certainly do – the history of science is not simply a march from one truth to a better truth, it is also the history of the falsification of scientific theories, i.e. of theories that yielded experimentally verifiable results but nevertheless had to be discarded when faced with so-called anomalies or more comprehensive theories. And, as Stephen Sharot points out, "[i]t makes little sense for scientists' hypothesis or theories that have been found false to be automatically recategorized as magic."[21] Moreover, it happens occasionally that concepts that had been discarded are revived when new facts arise: Einstein's cosmological constant is probably the best-known recent example. He had introduced it in his equations as a counter-gravitational force to maintain a stable universe, but when the expansion of the universe was discovered he called it his greatest mistake. However, in the 1990s it was shown that the expansion of the universe accelerates instead of slowing down and so the cosmological constant was re-introduced to explain this phenomenon – together with a hitherto undetected dark energy and dark matter.

There are, of course, various other suggestions of how to mark the distinction between magic and science as different approaches to knowledge. Among those are the distinction between 'closure' and 'openness' of the respective theories,[22] the role of tradition and the wisdom of the ancients vs. the rejection of authorities, the irrationality vs. the rationality of the approach, or the belief in the supernatural vs. the empirical foundation in experience and reality. However, none of these suggestions can really serve to distinguish the totality of all magical approaches from the scientific enterprise:

a) "In the 1890s Professor John Trowbridge, chairman of the physics department at Harvard University, was urging graduate students to pursue their studies in other fields, since in his opinion there was no further important research to be done in his specialty,"[23] and recent suggestions of a 'theory of everything' at least indicate a belief in the possiblitity of closure in modern physics.

b) Within times of 'normal science' as described by Thomas S. Kuhn[24] scientists work on the basis of the prevailing paradigm, and a failure to explain a puzzle does not immediately call for a new paradigm – occasionally this leads to a rejection of useful, but diverging, research (for an example cf. Kevles, passim[25]).

c) Theories are occasionally accepted on an intuitive basis even though concluding evidence has not been offered or adverse facts have not been accounted for.[26]

d) The question of what is to be regarded as supernatural is in itself historically contingent, and scientific research has validated forces or phenomena that would have been rejected as fantastic or supernatural at earlier times. And here I want to supply some examples.

One of the most pressing problems is as old as modern science itself, it is the question "What is gravity?". Newton already faced the embarrassment of introducing a hypothetical force with properties that clashed with his own demands on rational explanations of physical phenomena. In a letter from 1692 or 1693, admittedly a time of crisis for Newton, he wrote: "That one body may act upon another at a distance through a vacuum, without the mediation of any thing else, by and through which their action and force may be conveyed from one to another, is to me so great an absurdity, that I believe no man, who has in philosophical matters a competent faculty of thinking, can ever fall into it."[27] This embarrassment was obviously also felt by Roger Cotes, the editor of the second edition of Newton's *Principia Mathematica* [1713], and in his introduction he ferociously defends the work against the accusation of introducing an 'occult' quality or property.[28]

Looking at modern science one cannot fail to notice that the problem has not really been solved. In Einstein's theory of relativity, gravity no longer acts as a force but appears as a property of spacetime itself,[29] but in quantum theory it is still a force – and like all the other forces it acts by the exchange of particles.[30] The two concepts of gravity are famously incompatible, and physicists have not yet reached a Grand Unified Theory that would embrace both views or replace them with a new paradigm. Thus, one of the cornerstones of physics is still in an undecided state.

However, gravity is part of our everyday life and thus the uncertainty about its precise nature does not affect our immediate concepts of reality. But there are aspects in quantum theory that are not only beyond our present scope of understanding but also evoke analogies with pre-scientific concepts. One of the basic principles underlying many magical world views and practices is contagious magic. Frazer explains it as follows: "Contagious magic proceeds on the notion that things which have once been conjoined must remain ever afterwards, even when quite disserved from each other, in such sympathetic relation that whatever is done to the one must similarly affect the other. [... I]ts physical basis, if we may speak of such a thing [...] is a material medium of some sort which, like the ether in modern physics, is assumed to unite distant objects and to convey impressions from one to the other."[31] As an example Frazer offers the anointing of a weapon to heal or irritate the wound it has made.[32] What is involved here is a form of instantaneous action over distance,[33] an action that disobeys one of the most basic laws of physics which states that nothing can move or act faster than the speed of light.

It was exactly this law which was invoked by Einstein, Podolsky and Rosen in their famed thought experiment, intended to show that quantum theory must have some basic flaws and further variables need to be introduced to arrive at a complete theory. The argument ran as follows:

Heisenberg's Uncertainty Principle states that it is impossible precisely to determine the position and momentum of a particle. If one attempts to measure one, the other will necessarily remain unknown. Now, according to quantum theory, the states of two photons or particles which are emitted from a single source are correlated. In consequence, it should be possible to measure the momentum of one and the position of the other and thus arrive at a precise knowledge of the respective states in both of them. Bohr rejected this argument and pointed out that in quantum theory the two particles remain in an inseparable quantum system, which means that any measurement performed on one of them instantaneously affects the other. The behaviour is described in a wave function which in the unobserved state is a probability wave. If a measurement is carried out, the probability wave collapses and we get a defined state, but, as the particles are correlated, the probability wave function also collapses for the other particle, no matter how far distant it is. Einstein rejected this explanation because it introduced a "ghostly action at a distance."[34] In 1966 John Bell worked out his theorem, according to which "an explicit causal mechanism exists whereby the disposition of one piece of apparatus affects the results obtained with a distant piece."[35] He also pointed out that "the Einstein-Podolsky-Rosen paradox is resolved in the way which Einstein would have liked least."[36] Since then the appropriate experiments have been successfully carried out – for example by Alan Aspect in 1982 – and the results confirmed the predictions of quantum theory. The world we live in is indeed non-local, the term for the strange correlation is quantum entanglement, and it may serve as one more indication that "[n]ot only is the universe stranger than we imagine, it is stranger than we can imagine" (this dictum is variously attributed to Sir Arthur Eddington and to J. B. S. Haldane). By now physicists are working on a practical use of the phenomenon for so-called quantum-computers.[37]

The concept of quantum entanglement can easily be radicalised – however I should like to point out first that this is not 'state of the art' theory but rather an isolated view, even though it is being discussed by serious physicists. The universe in all probability came into existence with the Big Bang, i.e. a situation when all its particles were densely packed and everything interacted with everything else. If interacting particles remained correlated ever after, this would lead to the conclusion that in this universe everything is connected to everything else, a very magical concept indeed.[38] David Bohm suggested that the universe should not be seen as constituted of localised elements but as an undivided whole. He illustrates this concept with the image of a hologram: "The key new feature of the record is that each part contains information about the *whole object*. [...] That is to say, the form and structure of the entire object may be said to be *enfolded* within each region of the photographic record. When one shines light on any region, this form and structure are then unfolded to give a recognizable image of the whole object. We propose that a new notion of order is involved here, which we call the implicate order [...]. In terms of the implicate order one may say that everything is enfolded into everything."[39] Quantum entanglement, the quantum property of a non-local and non-causal relationship of distant elements, then results from the implicate order which governs the undivided whole. Bohm argues that "we may regard each of the 'particles' constituting a system as a projection of a 'higher dimensional' reality, rather than as separate particles, existing with all the others in a common three dimensional space."[40]

Bohm's theory argues that in this universe the seemingly separate parts and the whole are intricately interwoven, that one reflects the other and that each part contains information about the whole. It is a particular form of symmetry across scale: the part resembles the whole. As such it shows some analogy to the principle of magic already mentioned: "As above so below," and indeed this principle re-appears in very different guises and on very different levels in modern physics; another example would be the symmetry across scales in non-linear dynamics a.k.a. chaos theory and in fractals.[41]

Of course, any scientist would be quick to point out that quantum entanglement and Bohm's hypothetical 'implicate order' act on a vastly different level from contagious magic or the astrological influence of stars on human lives. And I certainly do not want to suggest that quantum physics simply validates magic and superstition or that contagious magic and quantum entanglement are inherently the same. This is emphatically not the case, even if the interconnectedness of distant phenomena may appear similar – a similarity that, as I will argue later, has possibly been taken up in literature. However, what I do want to suggest is that concepts that would have been ridiculed in science for a long time, i.e. concepts like non-locality or an interconnectedness between the part and the whole, are now accepted or seriously discussed – the same goes for 'mind over matter' theories or, once more on a quite different level, for the possibility of a particle being at two places at the same time.

The weirdness of the quantum world is, of course, stressed in every text dealing with the properties of the sub-atomic realm. Indeed, in the texts on modern science and quantum physics in particular, magic and mysteries are frequently evoked to express the strangeness of the phenomena and the counter-intuitive theories and concepts that try to explain them. Sir Arthur Eddington wrote about "Science and Mysticism;" in Roger Penrose's book *The Emperor's New Mind*, the chapter on quantum theory is entitled: "Quantum Magic and Quantum Mystery;" there are books titled *The Ghost in the Atom* or *The Dark Side of the Universe* – and all these are texts written by serious scientists. A quick survey of the words "quantum magic" on the Internet yielded 434 hits ranging from "Quantum Magic – Use the Fourth Force to Transform Your Life" (www.geocities.com/716948/framehh/quantum.html, last visited December 6, 2004) or "Deepak Chopra; Ancient Wisdom and Quantum Magic" (logosresourcepages.org/deepak.htm, last visited December 6, 2004) to "Quantitative Multiple-Quantum Magic-Angle-Spinning NMR Spectroscopy of Quadrupolar Nuclei in Solids" (web.mit.edu/fbml/cmr/griffin-group/pdf-publications/wu.jacs118.2.pdf, last visited December 6, 2004). According to its best known advocate Ed Witten, the term 'M Theory,' one of the variants of string theory discussed in physics at present, can stand for "Mother of All Theories," Membrane Theory, Murky Theory, Matrix Theory or Magic Theory. Moreover, in recent theories we also find some other paraphernalia that come with notions of magic like dark matter or dark energy.[42]

Of course, it is fallacious to read any deeper significance into such arbitrary terms. It cannot be stressed often enough that in the sciences words are not of supreme importance, as the foundations of a theory lie in the mathematical apparatus. New terms are frequently needed to denote, but not explain, a particular phenomenon, and the choice of words – for example flavour, colour or strangeness for the properties of quarks (itself an arbitrary term taken from James Joyce's *Finnegans Wake*) – does not necessarily reflect any deeper analogy. However, the choice of terms also indicates that we are in the presence of phenomena that are, at least for the moment, beyond our understanding. Moreover, the recourse to a vocabulary of magic and the supernatural has facilitated the esoteric approaches that came under the umbrella term 'New Age.' Nor has literature been slow to grasp the narrative potential of the scientific imagery, and there are a multitude of texts that exploit or explore the relationship between modern physics and earlier modes of knowledge. The topic lends itself to epistemological questions, and two major approaches can be distinguished in this regard:

a) The discoveries of modern science are used to validate pre-scientific and frequently mystical concepts, and

b) the similarities are taken as an indication that there is no progress in science but rather a perennial recirculation between various theories that have been around since antiquity.

Both of these positions are problematic. The attempt to validate mysticism does not take into account that recent theories frequently have a short half-life. Jeremy Bernstein wrote in a review on Fritjof Capra's New Age Philosophy: "[T]he one thing I am sure of, beyond any doubt, is that the science of the present will look as antiquated to our successors as much of nineteenth-century science looks to us now. To hitch a religious philosophy to a contemporary science is a sure way to its obsolescence."[43] The relativistic approach that rejects notions of progress fails to acknowledge the predictive value of recent theories; indeed, over the last decades most of the experimental discoveries of quantum physics were first suggested by theoretical models and then realised in the various colliders and bubble chambers.

The first approach validating previous modes of knowledge can be found in Jeanette Winterson's *Gut Symmetries*, which explores, among other topics, similarities between Paracelsus's alchemy, the cabala and quantum physics.[44] Another example is Jim Dodge's *Stone Junction*; it is a rather strange text, which begins as a celebration of various forms of counterculture but then changes its perspective and presents the reader with a magical or mystical world. It ends with the protagonist staring into a bowling-ball-sized diamond, realising that he is Hermes, Thoth, Mercury, Hermes Trismegistos reborn – and then he vanishes into the diamond light.[45] The text is riddled with references to modern physics – occasionally no more than bits of pseudo-scientific jargon,[46] and America's leading physicist is presented as "an alchemist at heart":[47] "He wanted to explore the other side of the equation, the conversion of energy into mass, and ultimately, he supposed, the obliteration of the distinction. He confided [...] that he had made a fundamental scientific error in his career – he'd viewed the universe as a machine instead of a thought."[48]

The relativistic perspective which questions the possibility of scientific progress on the one hand and the explanatory value of recent theoretical models on the other hand is evoked in Peter Ackroyd's *First Light*. In this novel a modern astronomer challenges notions of an advancement of science when the grave of one of his Neolithic precursors is discovered: "Ah yes. Science. But who is to say that our science is any better than the science of the astronomer buried in Pilgrim Valley? [...] Science is like fiction, you see. We make up stories, we sketch out narratives, we try to find some pattern beneath events. We are interested observers. And we like to go on with the story, we like to advance, we like to make progress. Even though they are stories told in the dark."[49]

The first paragraph of the first chapter– appropriately entitled "The Uncertainty Principle" – rushes the reader through the various astronomical theories since antiquity: "Once there were creatures of light leaping across the firmament, and the pattern of their movement filled the heavens. But the creatures soon fled and in their place appeared great spheres of crystal which turned within each other; their song vibrating through all the strings of the world. These harmonies were too lovely to last. A clock was ticking in the pale hands of God, and already it was too late. Yes, the wheels of the mechanism began to turn. What was that painting by Joseph Wright of Derby? I saw it once. Was it called "The Experiment?" I remember how the light, glancing through a bell-jar, swerved upwards and covered the whole sky. But this too went out: the candle flame was blown away by the wind from vast furnaces, when the electrical powers swept across the firmament."[50] On the following page the astronomer arrives at the conclusion: "I am on a storm-tossed boat out at sea, the dark waves around me. This was what the earliest men saw in the skies above them – an unfathomable sea upon which they were drifting. Now we, too, talk of a universe filled with waves. We have returned to the first myth."[51]

All of these texts discuss modern physics in relation to previous modes of knowledge, but then there are also those which lift the principles of quantum physics from the sub-atomic level to the world of our experience. Of course, this presents some problems, as we do not encounter quantum effects in our lives, and thus a literature that introduces such phenomena leaves the realm of realism and takes up features that are usually associated with fantastic literature. In his stories about the dreams of Mr. Tiompkins, the physicist George Gamov depicted a macro-world governed by the laws of quantum physics and some aspects of relativity theory adapted to our level of experience, and a very strange world it turned out to be.

In his play *Hapgood*, Tom Stoppard links quantum physics with the world of espionage, and quantum entanglement seems to be at the core of some of the bewildering aspects of the text. Objects and characters frequently appear as doubles, even though it is impossible to observe both at the same time, and they are correlated in their properties or behaviour.[52] And then there is Thomas Pynchon's *Gravity's Rainbow* and its protagonist's inexplicable connection with the V2-Rockets which strike at London in a random pattern. It happens that a map of the city on which Tyrone Slothrop marks the places of his erotic adventures exactly mirrors the future sites where the rockets will hit. Quantum entanglement is never directly mentioned in the novel, but there is a connecting link between Slothrop and the V-weapons, as the scientist who performed Pavlovian experiments on Slothrop in his infancy later also works on a particular device used for the rocket. As Pynchon's interest in particle physics is among the few documented facts about the life of this enigmatic author, and as frequently the most bewildering aspects of his novels are indeed firmly grounded in science, I find it hard not to see Bell's theorem and quantum entanglement at the root of Slothrop's interconnectedness with the rocket. In the novel it serves to emphasise the paranoid notion that "*everything is connected*, everything in the Creation."[53] In addition, the text is also riddled with references to magic, alchemy, astrology, spiritism and other forms of pre-scientific knowledge, and thus contagious magic may well serve as a complementary explanation for the strange phenomenon.

An immediate action over distance figures strongly in Umberto Eco's *The Is-*

land of the Day Before, in which magically informed scientists of the seventeenth century try to use an *unguentum armarium* or weapon salve to create an instant link between a wounded dog on a ship in the Pacific Ocean and the weapon that cut the wound in England. The dog's wound is kept open during the whole journey, and at particular times an irritating ointment or powder is applied to the weapon in order to provoke a reaction from the dog. The cruel procedure is one more attempt to synchronise time between the distant locations and thus to determine the degrees of longitude, one of the pressing problems for navigation at that time. As the text is set in the seventeenth century, the analogy with Bell's theorem is never expressly stated. However, in this book Eco introduces various parallels between pre-scientific concepts and modern physics, and also a host of anticipations of later scientific discoveries and theories,[54] and thus it seems to be valid to read the weapon salve and the practice of contagious magic as one more similarity, in particular as it is the most prominent among such premonitions. Moreover, not only recent scientific discoveries and theories are anticipated in this novel, there are also references to aspects of psychoanalysis,[55] Max Black's 'interaction view of metaphor,'[56] an anecdote from the adolescence of George Washington,[57] the mystery of the Mary Celeste,[58] the story of Captain Bligh and the Bounty,[59] and a book written by the protagonist in the solitude of an abandoned ship in the Pacific shows similarities with various novels by Alexandre Dumas, the elder. In addition, the arts and sciences of the past are also constantly evoked, and thus in the novel – and in particular on the ship – all knowledge from antiquity to the present coincides.

In this respect, the location in the Pacific which gave the book its title may be read as serving two distinct aspects. In the novel, the protagonist is shipwrecked and stranded right next to the 180th meridian, and he can look across and see an island on the other side of the geographic date line, the 'island of the day before.' On the one hand this image fuses time and space, on the other hand the present and the past are juxtaposed and the protagonist muses: "[I]f at that same instant I were on the line of the meridian, it would be midnight on the dot, but if I looked to the west, I would see the midnight of Friday and if I looked east, I would see the midnight of Thursday."[60] And just as yesterday is contemporaneous with today, many ideas and perspectives of the past are in principle interchangeable with recent theories. Knowledge appears as an endless succession of stories, and frequently these stories are variants of the same basic patterns. But most important of all, the period in which the novel is set, the middle of the seventeenth century, may be read as analogous to the position on the 180th meridian. It is a time looking equally into the past as into the future, still influenced by the knowledge of the ancients and bearing the full weight of religious dogma, but also in expectation of new discoveries and theories, many of which are prefigured in the works of the natural philosophers. And, as the choice of the main motif of contagious magic and its counterpart in quantum entanglement indicates, the world was and is a strange place, governed by inexplicable phenomena which may well be seen as mysteries and magic.

Notes

[1] When I began to write this paper I realised that it would have to begin just like another paper I gave at the 2004 ESSE conference in Zaragoza. It tackled the problem from a complementary angle – the title was "Language, Lore, and Learning in *The Lord of the Rings*."
[2] William J. Broad, "Paul Feyerabend, Science and the Anarchist" 534. Broad's inclusion of acupuncture with astrology and witchcraft associates a serious, and by now widely accepted, medical treatment with the epitomes of superstition, a nice rhetorical trick which makes it just so much harder to fully contradict the statement.
[3] Cf. Paul Feyerabend, *Against Method* 78fn.
[4] Giorgio de Santillana and Hertha von Dechend point out that "[t]he science of astrophysics reaches out on a grander and grander scale without losing its footing. Man as man cannot do this. In the depth of space he loses himself and all notions of his significance. He is unable to fit himself into the concepts of today's astrophysics short of schizophrenia. Modern man is facing the inconceivable. Archaic man, however, kept a firm grip on the conceivable by framing within his cosmos an order of time and eschatology that made sense to him and reserved a fate for his soul" (Santillana and von Dechend, *Hamlet's Mill: An Essay Investitgating the Origins of Human Knowledge and Its Transmission Through Myth* 5-6).
[5] It may be necessary to stress here that this is not intended as an argument for scientific relativism. I merely want to point out that people of all times have been able to achieve useful knowledge within their respective world views, even if those world views were founded on erroneous premises. Thus any condescension towards previous modes of knowledge is ill-founded, in particular if one takes into account that at present the gap between those who are well informed about the most recent scientific paradigms and those who are not may well be wider than at earlier times.
[6] Keith Thomas, *Religion and the Decline of Magic* 774.
[7] James George Frazer, *The Golden Bough* 56.
[8] Ibid. 57.
[9] Ibid.
[10] Stephen Sharot, "Magic, Religion, Science, and Secularization" 264.
[11] It is hardly possible to find a common denominator for concepts and practices as different as animism, totemism, spiritism, shamanism, witchcraft, or the astrology and alchemy as practised in the late Middle Ages and the Renaissance, all of which are included under that giant umbrella 'magic.'
[12] Heinrich Cornelius A. von Nettesheim Agrippa, *Three Books of Occult Philosophy (De occulta philosophia)* bk. 1 ch. 2.
[13] "Mathematicall Philosophy teacheth us to know the quantity of naturall Bodies, as extended into three dimensions, as also to conceive of the motion, and course of Celestiall Bodies" (ibid.).
[14] Niels Bohr, "Nobel Prize in Physics Award Address" 316.
[15] Quoted from Arthur I. Miller, "Erotica, Aesthetics and Schrödinger's Wave Equation" 114-115.
[16] Kurt Vonnegut, *Cat's Cradle* 137. Italics in the original.
[17] Quoted from Brain Aldiss with David Wingrove, *Trillion Year Spree: The History of Science Fiction* 281.

[18] Actually, several of the perennial objectives of magic were also mentioned as programmatic goals of science in Bacon's New Atlantis or the list added to this text, e.g. flight, the prolongation of life, the curing of diseases counted incurable, the making of new species, the altering of features etc. (cf. Francis Bacon, *The Advancement of Learning and New Atlantis* 245 and 249), they resurface in early science fiction and have, of course, since then been realised.

[19] Robert Musil, *Der Mann ohne Eigenschaften* 39.

[20] "[M]agic is not chronically vulnerable to disproof; failure to achieve compensators can always be explained within the magical worldview by such factors as countermagic, a mistake in the charm of spell, or the absence of 'correct' relationships with the supernatural beings involved" (Stephen Sharot, "Magic, Religion, Science, and Secularization" 273).

[21] Ibid. 264.

[22] Ibid. 266. Sharot also points out that Robin Horton who suggested this distinction discarded it in a later article (cf. Sharot 266).

[23] Jefferson Hane Weaver, *The World of Physics* vol. 1 31.

[24] Thomas S. Kuhn, *The Structure of Scientific Revolutions* 23-42.

[25] Daniel J. Kevles, "Forschungen gegen den Strom."

[26] "[D]espite the weakness of the early experimental evidence for general relativity, Einstein's theory became the standard textbook theory of gravitation in the 1920s and retained that position from then on, even while the various eclipse expeditions of the 1920s and 1930s were reporting at best equivocal evidence for the theory" (Steven Weinberg, *Dreams of a Final Theory* 77).

[27] Newton, Letter to Richard Bentley ca. 1692-93, quoted from Delo E. Mook and Thomas Vargish, *Inside Relativity* 148.

[28] Cf. Roger Cotes, "Preface to the Second Edition" of Isaac Newton's *De Philosophiae Naturalis Principia Mathematica* XXII-XXIII.

[29] "The force of gravity, in general relativity, becomes a fictional consequence of the geometric properties of spacetime and is not a separate phenomenon of nature the way it is in Newton's model of the world. As we warned [earlier], Einstein's model of gravity still fails to "explain" the phenomenon of gravity in the sense desired by Newton. Just as Newton's model was unable to "explain why" masses should attract one another, Einstein is unable to "explain why" masses should affect spacetime. He can describe the effect, but he provides no reason for its dependence on matter" (Mook and Vargish, *Inside Relativity* 168).

[30] "The particle associated with gravitational waves is called the graviton. No one has ever observed a graviton. It is hard enough even to detect a gravitational wave, as they interact only weakly with matter. But as long as quantum theory applies to gravitational waves, gravitons must exist. We know that gravitons must interact with matter, for when anything massive oscillates it produces gravitational waves. Quantum theory says that, just as there are photons associated with light, there must be gravitons associated with gravitational waves" (Lee Smolin, *Three Roads to Quantum Gravity* 151).

[31] Frazer, *The Golden Bough* 43.

[32] Ibid. 47-48.

[33] Francis Bacon states that the ointment "may be applied to the weapon, though the party hurt be at great distance" (Francis Bacon, *Sylva Sylvarum* 76). However, he also points out that he is not fully inclined to believe in the experiment (cf. ibid.).

[34] Quoted from Paul C. W. Davies and James R. Brown, *The Ghost in the Atom* 15; my explanations are also indebted to their account of the EPR experiment.

[35] John Bell, "On the problem of hidden variables in quantum mechanics" 11.

[36] Ibid. It is quite interesting that Frazer in his description of contagious magic mentioned the ether as an analogy of the hypothetical material medium uniting distant objects (see above). Thus Frazer refers to a concept that was taken to be true at his time but rejected with the advent of the theory of relativity. However, in an interview about the experiments of Alan Aspect and their theoretical impact, John Bell suggested that a return to the concept of an ether might not be a bad idea (cf. Davies and Brown, *The Ghost in the Atom* 49-50). As far as I know this has not been taken up, and John Bell's suggestion seems to be based on the hope "to be able to take a realistic view of the world, to talk about the world as if it is really there, even when it is not being observed" (ibid.).

[37] Cf. Michael A. Nielsen, "Spielregeln für Quantencomputer" passim.

[38] When I talked about this aspect with a physicist he pointed out that in order to work on this premise it would be necessary to know the wave function of the universe – probably an even larger task than the notorious one faced by Laplace's demon.

[39] David Bohm, *Wholeness and the Implicate Order* 177.

[40] Cf. ibid. 188; on recent conceptualisations of the holographic principle cf. also Jacob D. Bekenstein, "Das holographische Universum" passim.

[41] Cf. Dirk Vanderbeke, "Of Parts and Wholes: Self-similarity and Synecdoche in Science, Culture and Literature" passim.

[42] James Trefil, *The Dark Side of the Universe* passim, and Wayne Hu and Martin White, "Die Symphonie der Schöpfung" passim.

[43] Jeremy Bernstein, "A Cosmic Flow" 8.

[44] Cf. Dirk Vanderbeke, *Theoretische Welten und literarische Transformationen: Die Naturwissenschaften im Spiegel der 'science studies' und der englischen Literatur des ausgehenden 20. Jahrhunderts* 272-289.

[45] Jim Dodge, *Stone Junction* 406.

[46] One of the major features of the protagonist is his ability to vanish. His teacher describes the mystical process as "becoming light moving through space," in the final state "[y]ou can lift that electrical field you call a hand and scratch that whirling constellation of energy you call your head, but you are not flesh and bone, ashes or dust. You are released from the constraints of matter [...]" (ibid. 217). The teacher's description of the diamond mixes elements of alchemy with the Big Bang theory: "the diamond is an interior force given exterior density, the transfigured metaphor of the prima material, the primordial mass, the Spiritus Mundi. I'm assuming you are familiar with the widely held opinion that the entire universe was created from a tiny ball of dense matter which exploded, sending pieces hurtling into space, expanding from the center. The spherical diamond is the memory, the echo, the ghost of the generative cataclysm; the emblematic point of origin. Or, as some astrophysicists believe, the universe will reach some entropic point in expansion and begin to collapse back into itself, in that case the diamond may be the homing point, the seed crystal, to which it will all come hurtling back together – and perhaps through itself into another dimension entirely. Or it might be the Philosopher's stone, we alchemists speak of [...]" (ibid.).

[47] Ibid. 32.

[48] Ibid.

[49] Peter Ackroyd, *First Light* 159.
[50] Ibid. 3. The "song vibrating through all the strings of the world" probably also forges a link between the 'music of the spheres' and the string theory in quantum physics.
[51] Ibid. 4.
[52] Cf. Vanderbeke, *Theoretische Welten und literarische Transformationen: Die Naturwissenschaften im Spiegel der 'science studies' und der englischen Literatur des ausgehenden 20. Jahrhunderts* 289-302.
[53] Thomas Pynchon, *Gravity's Rainbow* 703. Italics in the original.
[54] Among these anticipations are the discovery of the law of gravity by the fall of a fruit (363), Thomas Burnet's *Telluris theoria sacra* (Sacred Theory of the Earth) published 1681-1690 (ch. 21, in particular 261-268), the discovery of the bubonic plague bacillus that actually took place in 1894 (248), aspects of relativity (66 and 433-434), forces that act by the exchange of particles (167), the 'many worlds' theory (66 and 267), the creation of the universe by energy and a giant explosion akin to the Big Bang (256), eddies of atoms in the interstellar void as the origin of stars (318), fractals and self-similarity (423-424), and some basic principles of thermodynamics, i.e. heat as the result of the movement of atoms (476).
[55] Umberto Eco, *The Island of the Day Before* 81-82 and 134.
[56] Ibid. 90.
[57] Ibid. 24.
[58] Ibid. 36-37.
[59] Ibid. 206.
[60] Ibid. 265.

Works Cited

Ackroyd, Peter. *First Light* (1989). London: Penguin, 1993.

Agrippa, Heinrich Cornelius A. von Nettesheim. *Three Books of Occult Philosophy (De occulta philosophia* (1510), trans. John French. London, 1651), http://www.esotericarchives.com/agrippa/home.htm.

Aldiss, Brian Wilson with David Wingrove. *Trillion Year Spree: The History of Science Fiction* (London: Gollancz, 1986). Thirsk: House of Stratus, 2001.

Bacon, Francis. *The Advancement of Learning and New Atlantis* (1605/1627). Ed. Arthur Johnston. Oxford: Clarendon Press, 1986.

---. *Sylva Sylvarum, or a Naturall Historie in Ten Centuries,* in *The Works of Francis Bacon* vol. II. London: Baynes and Son, 1824.

Bear, Greg. "Schrödinger's Plague" (1982), in *The Norton Book of Science Fiction*. Ed. U. K. LeGuin and B. Attebery. New York: Norton, 1993. 477-484.

Bekenstein, Jacob D. "Das holographische Universum." *Spektrum der Wissenschaft* (November 2003): 34-41.

Bell, John. "On the Problem of Hidden Variables in Quantum Mechanics," in *Speakable and Unspeakable in Quantum Mechanics*. Ed. John Bell. Cambridge: Cambridge UP, 1988. 1-13.

Bernstein, Jeremy. "A Cosmic Flow." *The American Scholar* (Winter 1978/79): 6-9.

Broad, William J. "Paul Feyerabend, Science and the Anarchist." *Science* 206 (February 1979): 534-537.

Bohm, David. *Wholeness and the Implicate Order* (1980). London: Routledge, 1995.

Bohr, Niels. "Nobel Prize in Physics Award Address" (1922), repr. in *The World of Physics* vol. 2.

Ed. Jefferson Hane Weaver. New York: Simon and Schuster, 1987. 315-338.

Cotes, Roger. "Preface to the Second Edition" of Isaac Newton's *De Philosophiae Naturalis Principia Mathematica* (1687). Trans. Andrew Motte (1729). Berkeley: Harvard UP, 1960. XX-XXXIII.

Davies, Paul C. W. and James R. Brown. *The Ghost in the Atom*. Cambridge: Cambridge UP, 1986.

Dodge, Jim. *Stone Junction* (1990). Edinburgh: Rebel Inc., 1999.

Eco, Umberto. *The Island of the Day Before (L'isola del giorno prima* (1994)), trans. William Weaver. London: Minerva, 1996.

Eddington, Sir Arthur. "Science and Mysticism," in *The Nature of the Physical World*. Cambridge: Cambridge UP, 1929. 316-342.

Feyerabend, Paul. *Against Method* (rev. 3rd edition, first edition 1975). London: Verso, 1993.

Frazer, James George. *The Golden Bough* (abridged edition, 1922). New York: Touchstone, 1996.

Gamow, George. *Mr. Tompkins in Paperback*. Cambridge: Cambridge UP, 1967.

Hu, Wayne and Martin White. "Die Symphonie der Schöpfung." *Spektrum der Wissenschaft* (Mai 2004): 48-55.

Kevles, Daniel J. "Forschungen gegen den Strom," trans. P. Knecht, in *Verborgene Geschichten der Wissenschaft* (*Hidden Histories of Science* (1995)). Ed. R. B. Silvers. Berlin: Berlin Verlag, 1996. 73-109.

Kuhn, Thomas S. *The Structure of Scientific Revolutions (1962)*. 3rd. ed. *Chicago: U of Chicago P, 1996.*

Mook, Delo E. and Thomas Vargish. *Inside Relativity*. Princeton: Princeton UP, 1987.

Miller, Arthur I. "Erotica, Aesthetics and Schrödinger's Wave Equation," in *It Must Be Beautiful: Great Equations of Modern Science*. Ed. Graham Farmelo. London: Granta, 2002. 110-131.

Musil, Robert. *Der Mann ohne Eigenschaften* (1930/33). Ed. Adolf Friese. Reinbek: Rowohlt, 1981.

Nielsen, Michael A. "Spielregeln für Quantencomputer." *Spektrum der Wissenschaft* (April 2003): 49-56.

Penrose, Roger. *The Emperor's New Mind* (1989). London: Vintage, 1990.

Pynchon, Thomas. *Gravity's Rainbow* (1973). London: Picador, 1975.

Santillana, Giorgio de, and Hertha von Dechend. *Hamlet's Mill: An Essay Investigating the Origins of Human Knowledge and Its Transmission Through Myth* (1969). Boston: Nonpareil, 1999.

Sharot, Stephen. "Magic, Religion, Science, and Secularization," in *Religion, Science, and Magic: In Concert and in Conflict*. Ed. Jacob Neusner, Ernest S. Frerichs, and Paul Virgil McCracken Flesher. New York: Oxford UP, 1989. 261-283.

Smolin, Lee. *Three Roads to Quantum Gravity*. London: Phoenix, 2001.

Thomas, Keith. *Religion and the Decline of Magic* (1971). London: Penguin, 1991.

Trefil, James. *The Dark Side of the Universe*. New York: Charles Scribner's Sons, 1988.

Vanderbeke, Dirk. "Of Parts and Wholes: Self-Similarity and Synecdoche in Science, Culture and Literature." *Reconstruction* 4 (2004) (*Postmodern Science*). Ed. C. Jason Smith and Robert Froemke. http://www.reconstruction.ws/044/vanderbeke.htm.

---. *Theoretische Welten und literarische Transformationen: Die Naturwissenschaften im Spiegel der 'science studies' und der englischen Literatur des ausgehenden 20. Jahrhunderts*. Tübingen: Niemeyer, 2004.

Vonnegut, Kurt. *Cat's Cradle* (1963). Harmondsworth: Penguin, 1988.

Weaver, Jefferson Hane, ed. *The World of Physics*, vol. 1. New York: Simon and Schuster, 1987.

Weinberg, Steven. *Dreams of a Final Theory*. London: Vintage, 1993.

Winterson, Jeanette. *Gut Symmetries* (1997). New York: Vintage, 1998.

Media Differences

Magic and Literary Fascination

Hans Ulrich Seeber

The Return of Magic in Art and the Aesthetics of Fascination

My starting-point in this section is two philological findings that have attracted little literary-critical attention up till now. First, it can be shown that since the romantic movement itself, writers have often called on the analogy of magic to indicate the special and mysterious structure of a literary text and the influence it exercises on the reader. There seems to me to be a very close correlation between this, and the growing importance and indeed inflationary diffusion of the concept of fascination. An 'aesthetics of fascination' has tacitly taken foothold, unaccompanied by any theoretical underpinning.

I begin with a collection of quotations all of which, metaphorically or quite literally, speak of art in the context of magic, thereby suggesting that the two are closely related, or even the same.

"The poet, described in *ideal* perfection, brings the whole soul of man into activity, with the subordination of its faculties to each other, according to their relative worth and dignity. He diffuses a tone and spirit of unity, that blends, and (as it were) *fuses*, each into each, by that synthetic and magical power, to which we have exclusively appropriated the name of imagination. This power [...] reveals itself in the balance or reconciliation of opposite or discording qualities: of sameness with difference, of the general with the concrete; etc." (Samuel Taylor Coleridge).[1]

"[...] The grand power of poetry is its interpretative power; by which I mean, not a power of drawing out in black and white an explanation of the mystery of the universe, but the power of so dealing with things as to awaken in us a wonderfully full, new, and intimate sense of them, and of our relations with them. [...] The interpretations of science do not give us this intimate sense of objects as the interpretations of poetry give it; [...] Eminent manifestations of this magical power of poetry are very rare and very precious" (Matthew Arnold).[2]

"Man attempted to compel the gods, to win power over them by rhythm; he captured them in poetry as in a magic snare" (Friedrich Nietzsche).[3]

"All art, therefore, appeals primarily to the senses, and the artistic aim when expressing itself in written words must also make its appeal through the senses, if its high desire is to reach the secret spring of responsive emotions. It must strenuously aspire to the plasticity of sculpture, to the colour of painting, and to the magic suggestiveness of music – which is the art of arts" (Joseph Conrad).[4]

"If magic is a hidden power, springing we do not know whence, then the quality that distinguishes *The Ancient Mariner* from mere verse must be called magic. If verse is memorable speech, then *The Ancient Mariner* is magical speech. [...] poetry is verse plus magic" (James Reeves).[5]

In the critics here quoted, the following characteristics of poetry suggest it to be close to or identical with magic:

(1) The relationship with magic seems to be more than the casual result of linguistic need, *more than a conventional illustrative flourish*. Reeves simply equates poetry with magic. Coleridge seems to evoke ancient occult traditions when he attributes to poetry a divine synthesising power.

(2) All the quotations bear witness to the presence, in both poetry and magic, of *a mysterious power* which overwhelms the recipient. Nietzsche speaks of poetry as a magical linguistic technique that makes the gods themselves obey the poet-sorcerer. The inexplicability of this power compels us to distinguish scholarly-scientific from magical utterance.

(3) As the sorcerer purveys insight into the nature of things such as is denied the average mortal, just so does the poet sharpen the *perceptions* of the reader by representing and interpreting the world so as to put us in a completely new, sensual and vivid relationship with it.

(4) While the sorcerer with the help of techniques of fascination is able to exercise therapeutic or destructive power over his fellow men, *the fascinating and suggestive power of the artist's shaping activity* also casts its spell on the recipient. This mode of reception is characterised by something very different from the coolly observant, uninvolved enjoyment of classical aesthetics, Kant's disinterested delight. Rather does art exercise on the recipient a power that tends to annul his rational autonomy. The transcendental otherness of art transforms the encounter between artefact and recipient into a dynamic event.[6]

(5) When poetry is approximated to magic, a philosophical shift has taken place. The power wielded by magic is spiritual in nature, and it is implicitly just such a spiritual power that Robert Crawford has in mind when, in his introduction to *The Penguin Book of Poetry*, he points out that in the age of electronic communication the "numinous vibrancy of poetry"[7] is very important. The dictionary meaning of "numinous" is "arousing spiritual or religious emotions."

(6) Against this thesis of the 'remagicisation' of art, it might be objected that some major writers of the twentieth century like George Bernard Shaw and Bertolt Brecht explicitly conceived dramatic art as an instrument at the service of enlightenment. The creation of an illusory world which, like a drug or a spell, transports the recipient out of the world of everyday reality and offers relief through escapist distraction, is exactly what those two writers meant to avoid. Yet the counter-objection might be, that such rationally-enlightening theatre cannot achieve its effect unless it succeeds in developing its own kind of magic.

(7) In fact the 'aesthetics of fascination,' though not up till now the object of explicit theorising, has played an ever-greater role since the aesthetics of the sublime asserted itself. Burke's investigations into the sublime approach what could be called the hypnotic effect of both art and nature. The aloof, disinterested plea-

sure in beauty, the observer-role of the recipient, gives way to that rapture which Burke identifies as passionate recognition and intense admiration.

(8) The passion caused by the great and sublime in *nature*, when those causes operate most powerfully, is Astonishment; and astonishment is that state of the soul, in which all motions are suspended, with some degree of horror.[8]

This intensity of aesthetic experience, which simultaneously threatens, and (as will be shown) transports the percipient subject into a paradoxically split condition of psychic immobility and imaginative release, recalls that category of experience nowadays denoted by the fashionable term *Faszination*. In everyday German, and in literary-critical and culture-critical language, the word fascination (fascinate, fascinating) has become an essential, indeed unavoidable term indicating maximal consent to something. The word and the concept have had a sensational career. Whereas in Shakespeare's time (when it was, however, seldom employed, not once by Shakespeare himself in his writings) it generally meant, still, magical practices of evil enchantment or healing, it has now gone through an astonishing process of revaluation, freeing itself from its magical diabolical origins to become a universally-employed term denoting a particular subject-object relationship. An evidently essential expression has come into being whose history has not been written, and which demands a theory that has not yet been provided. Such a theory would have to have its basis in a psychologically elaborated affective aesthetics, not primarily focussed upon meaning, but upon, for example, the relationship between cultural or poetic alterity and fascination, surface effects of fashion and art and fascination. The possible cognitive potential of this concept has nowhere been plumbed.

The history of the concept and of its role in literary history offers, at first sight, four semantic aspects:

(1) *Bewitchment and destruction*. This is particularly true of the early modern period which stresses the connection of the term with sorcery and witchcraft "fascination or bewitching" (1615, *OED*), "witchcraft and fascination" (1681, *OED*). This magical technique of fascination as known since antiquity is transformed, in a philosophically sophisticated romantic mode, into material for aesthetic play aimed at effects of horror (in e.g. *The Monk*, when the monk's rhetoric is credited with fascinating effects).

(2) Fascination as occult *healing powers and methods* is rendered scientific and is practically applied in research into hypnosis in the medicine of the 19th century. In 1882 Charcot secures official recognition of his method of hypnosis which operates by influencing the psyche by suggestion, both verbally and non-verbally (an example of the latter being to have the patient gaze at a shining object). At a time when mesmerisation or Mesmerism still assumed the existence of a fluidum filling all space and thus making such influence possible, medically accepted research into hypnosis had already abandoned this unprovable assumption.

(3) The currently prevalent meaning of fascination has *secularised* the term,

applying it to an experience that can be produced by almost any object. The nineteenth century was fixated principally on erotic fascination.

(4) In applying the term to *aesthetic experiences*, the analogy between aesthetic and magical influence is particularly stressed. Freud had already established "words were originally magic, and the word still retains today much of its ancient magical power [...] Words evoke emotions."[9] And recent literary-critical studies, without reflecting on the history of the concept of fascination, pose with ever-increasing frequency decisive questions such "What is it that makes literature fascinating?"[10] or "bewitched by pictures?".[11]

Then again, the history of ideas provides convincing reasons for the present-day relevance of the aesthetic meaning of the term:

(1) The occultists of the sixteenth century (Paracelsus and others), including indeed Bacon (*Novum Organum*), stress the relationship between fascination and imagination. The powers of the fascinator or magician are due to the strength of his imagination. In terms of language this means that literary artists, just like magicians or magi, create a virtual reality, the illusion of a world, with the help of words which do not point to an actually existing referent.

(2) The fascination exercised by the devil's rhetoric and that of literature is the same. The orthodox theology of the early modern period implies that only God can have immediate access to man's soul. If the devil desires such access he must needs make a detour and operate through imagery and rhetoric – arts with appeal to the lower faculties of the soul. These, moreover, being figurative and ambiguous, can deceive.

(3) The amoral concept of fascination seems to have played a decisive role in the emergence of modern aesthetics. For it provides an escape-route from traditional moral social and aesthetic hierarchies, as also from the binary restraints of rhetorical discourse. Fascination can be attributed equally to the beautiful, the ugly, the good and the evil.

Fascination seems to be first and foremost the sign of successful literary communication. Otherwise, words would not function as advertising slogans. The fact that texts and pictures attain the power they do means that they captivate the recipient, but at the same time liberate his imagination. For fascination is only conceivable if the text succeeds in stimulating the reader's imaginative participation and creativity. A paradoxical situation is thus created. Rapt, or bound and shackled by the seductive power of the text (a power showing many aspects and functions including the production of illusions, beauty and rhetoric, but also truth experienced as novelty, and the anticipation of better things), the reader or recipient experiences simultaneously the 'lightness of being,' his imagination being liberated yet engaged. It is as though, in frozen immobility, he danced. This paradoxical double structure of captivity and stimulated imaginative indeed reflexive mobility, seems to me a basic characteristic of the poetics of fascination – this simultaneity of silence, stammering and yet liberated speech, this captivity which is also free movement, this emotional ensnarement which also allows lucidity.

Oscar Wilde's novel *The Picture of Dorian Gray* throws a significant light on the above remarks. In this novel, a key work of English aestheticism, Wilde carries out, programmatically and thoughtfully, a demoralising of art which at the same time restores to art its magical power. This 'remagicisation' manifests itself not least in the motif of the magic picture which renders, as a visual sign, the psychic disintegration of the sinful, indeed criminal, hedonist Dorian. Still more important is the conspicuously frequent employment (unnoticed by literary research to date) of the term "fascination," which turns the novel into a medium for reflection (implicit for the most part) on aesthetic fascination. For in the story this category of influence (anticipating that exercised by the novel itself on its real-world readers) regularly appears when the fascinating influence of beautiful art, of beautiful objects, or of beautiful personalities, is the subject of reflection. "For the wonderful beauty that had so fascinated Basil Hallward and many others beside him, seemed never to leave him."[12] As is disturbingly underlined by Dorian's rejection of the actress Sibyl Vane after her lamentably inferior performance, aesthetic fascination, that is, here, the subject's enamouredness in aesthetic appearance as the only value, strips him of moral feelings and renders him cruelly cold. It is no wonder that the philosophical stance expounded with rhetorical force by Lord Henry setting aesthetic-hedonistic pleasure as the highest principle of life, provoked cries of outrage from the contemporary public, even while Wilde and others were able to assert, that because of the outcome of the plot, the work was profoundly moral. As "fascination" can be exerted not only by the beautiful but also by what is evil, it is a category of influence that adequately accommodates the double appeal of this novel.

Wilde's novel ends up as a reflexion, both implicit and explicit, on aesthetic fascination. The decisive factor in the present context is the fact that the desired and achieved effect is described in terms deriving from an occult tradition of knowledge (a former 'science' now becomes 'unscientific'), terms such as "fascination," "charm," "spell." When Lord Henry wishes to fascinate Dorian, this is reminiscent of the 'fascinating' (i.e. 'bewitching') practices of sorcerers, shamans and mesmerists; it reminds us of non-rational forms of 'knowledge' that were just then, around 1900, being intensively and scientifically investigated from an ethnographic and psychological perspective[13] and many others. And Dorian is in fact described as somebody bewitched by the influence of Henry's rhetoric, staring at the latter as if in the grip of magical powers, and surrendering his identity (for Henry "charmed his listeners out of themselves").[14] Both terms "fascinate" and "charm" mean originally to bewitch (fascinare, bind) especially by means of song (charm, carmen), and such literal meanings are precisely what the text activates, here and elsewhere. But this 'de-metaphorisation' of expressions generally employed only metaphorically in a modern context ('charm,' 'fascination,' 'spell') is nothing other than the turning into magic, or rather, *back* into magic, of aesthetic communication. This is what was being deliberately practised by the fin-de-siècle adherents of symbolic art.[15] The libertine dandy and aesthete Henry overpowers Dorian by a speech which deliberately annuls scientific, moral and realistic standards and stakes all on provocation and the liberated play of phantasy. The effect of this speech on its recipient can be conceived only in terms of paradox – it is a frozen captivity, yet also a liberation. On the one hand the philosophic-poetic speech spellbinds the listener as the snake does the rabbit, but on the other hand, as is visible elsewhere too, the victim's imaginative activity is stimulated ("wonder"). Fascination as a productive aesthetic experience describes precisely that combination of spell-bound perception and liberated phantasy called forth by an encounter with something extraordinary both in form and content.

In other words, works of art cannot be reduced to cognitive insights, concepts and moral norms. They contain something mysterious ("mystery," *The Picture of Dorian Gray* 45; 76) and at the same time powerfully affecting, something which will not fit into the above-listed categories, and which at the same time shows the source of that power to lie in fascination. This distance from morality and from scientific discourse is precisely the reason why, in the text, Henry's musical voice, and indeed music in general, is explicitly named as triggering fascination.

Around 1900, under the influence of medical research into hypnosis and suggestion, the latter, as well as fascination itself, became indistinguishable categories in art-criticism even though the concept of suggestion replaced fascination in scientific discourse. As an artist interested in the aura-like mystery and power of the artwork, Wilde did not welcome the tendency to turn magic into orthodox science.[16]

"The rapt attention of my revolted Padma" or Literary Fascination and the Annihilation of Time

The magic or quasi-magic experience of fascination in a sense lifts the reader out of space and time. Dorian has this excuse for coming late: "'I am so sorry, Harry', he cried, 'but really it is entirely your fault. That book you sent me so fascinated me that I forgot how the time was going'" (*The Picture of Dorian Gray* 156).

The experience is made possible by a piece of narrative literary art which also transcends time by fixing fleeting memories, thoughts, perceptions in its attempt to create the illusion of life. Overcoming the limitations imposed upon natural man by the power of space and time is indeed the chief function of magic, technical inventions and art, telepathy, space travel, telecommunications and above all language and literature which are the most obvious examples. They allow human beings, just as fantasy does, to transcend their natural limitations. In what follows, I wish to explore the function of overcoming time by the closely interrelated devices of magic, technology and language in two novels which explicitly comment on this function, Lord Dunsany's fantasy *The King of Elfland's Daughter* (1924) and Salman Rushdie's famous contribution to the tradition of magic realism, his postmodern and post-colonial novel about India, *Midnight's Children* (1981).

Characteristically, Lord Dunsany was, with Yeats and others, an active member of the Celtic literary revival which revived romantic traditions of spiritualism, anti-rationalism and cultural criticism. The story of his fantasy novel culminates in the triumph of Elfland, the timeless, unchanging realm of magic and fantasy, over the Earth-oriented society of Erl which is subject to "perpetual movement and perpetual change."[17] With the help of a magical sword Alveric, the son of the lord of Erl, elopes with the Elf King's daughter Lirazell only to discover that she cannot adapt to the culture of Erl, is lured back to Elfland and finally overcomes the division and the "ruin of Time" (*The King of Elfland's Daughter* 39) by persuading her father to incorporate Erl into the sphere of magic. Whether we have to read this as a poetological allegory propagating the superiority of magic, beauty and fantasy over realism is not quite clear, but the hallucinatory descriptive passages evoking the spell Elfland's beauty casts over the inhabitants of "Earth," Alveric and his son Orion in particular, seem to confirm this reading. The recurring motif of the fascinated gaze or fascinated listening of those observers who are exposed to a magical, beautiful and timeless alterity of unicorns, blowing horns, enchanted forests, intense colours and "pale-blue mountains" (*The King of Elfland's Daughter* 81) is further evidence of the book's thematic linkage of magic and aesthetics.

What is particularly important in the present context is the novel's emphasis on language and signs as the source of magic power and technology. Both the witch of Erl, who makes a magical sword for Alveric, and the king of Elfland derive their supernatural power from runes. The word 'rune,' apart from referring to the letters of the Germanic alphabet, can also mean a "similar character or mark having mysterious magical powers attributed to it" or a "charm denoted by magic signs" (cf. *OED*). As the witch makes the sword from thunderbolt-metal in the heat of fire she not only shouts her runes (*The King of Elfland's Daughter* 5) but subsequently also sings a ravishing song full of poignant memories of the past. It is as if the sword were imbued with the magic essence of nature and the past with the help of poetic words. Both the runes and poetic language stir memories which Lirazel, the fascinated reader of the King's rune (*The King of Elfland's Daughter* 61), cannot resist. The privileged magic or poetic sign, one must conclude, both freezes and releases time.

In Lord Dunsany's novel the marvellous in Todorov's sense is such a dominant mode that technology is hardly present but in its unashamedly magical version. In Rushdie's version of magic realism there is, however, a complex coexistence and fusion of technology and magic, history and magic, reality and myth, realism and fantasy where everything seems to interconnect as if the novel's language and structure intended to flesh out the idea of a global syncretism of cultures *and* the magic idea of cosmic unity at the same time. In line with this basic ambivalence

Saleem, the narrator of the story, dons both the mask of the craftsman or technician and the mask of the magician.

(1) On the level of historical discourse embedded into the novel, India is depicted as a society in transition where magical and modern world views coexist and clash in various ways. Snake-charmers and the use of love potions play quite a significant role in the novel's depiction of Indian culture. Adam Aziz, Saleem's grandfather, who studied medicine at Heidelberg unsuccessfully battles against the belief in magic and superstition represented by his wife who, for example, refuses to be photographed as if she were a member of a primitive tribe ("cameras could steal your soul"[18]) The ancient "boatman Tai" (*Midnight's Children* 15), who embodies the timeless traditionalism of Indian life and who fascinates Adam by his "magical talk" (*Midnight's Children* 15) bitterly rejects the "alien thing" (*Midnight's Children* 21) of modern technology preferred by Adam. His "magical talk" is the unmeditated expression of a magical world view which does not acknowledge the Western notions of space and time. In a sense Saleem, Adam's grandchild, continues his grandfather's tradition by ending up in a pickle factory and describing his craft of preserving memories of India's postcolonial history by using the image of pickling. Preserving pickles in jars is a craft, a technology, which allows us to outwit time and nature: "To pickle is to give immortality [...]" Yet the "symbolic value of the pickling process" (*Midnight's Children* 459) to illustrate the process of artistic creation is surely overburdened when pickling is also meant to illustrate those transforming activities by which a work of art achieves "shape and form – that is to say meaning" (*Midnight's Children* 461). The image of alchemy (suggesting the transformation of crude material into a highly valuable metal) would be much more appropriate.

(2) Not surprisingly, therefore, Saleem the artist identifies himself much more with magic than with technology. Poetic technique, however, is a mediating concept which can be easily accommodated with magic's careful use of language. To begin with, his name Sinai, connecting him with the Sufi "master magician" Ibn Sina (*Midnight's Children* 304), determines his fate as it has to according to the magic view of language and naming ("our names contain our fates," *Midnight's Children* 304). In magic practice language is not merely an instrument of communication but an agent of tremendous power which can create or destroy. Like Scheherezade Saleem, expecting the "Black Angel of death" (*Midnight's Children* 447), strives both to defer and to complete his autobiography to preserve memories for his son Aadam and the nation at large. His early magic ability to enter other people's minds via telepathy is presented as a literal fact which also serves to emphasise the narrator's omniscience and omnipotence. At times he views himself as the God-artist: "[...] the bodies I occupied acted at my command; [...] I was somehow making them happen [...] which is to say I had entered the illusion of the artist, and thought of the multitudinous realities of the land as the raw unshaped material of my gift" (*Midnight's Children* 174). However, his omnipotence

is only artistic and verbal, since Saleem the utopist and would-be saviour of India fails miserably. His art is only that of the tricksters and illusionists of Delhi's magicians who conjure up visions of the past and the future. Similarly, Saleem's words conjure up vivid memories of the past while his metafictional reflections reveal the tricks of the narrator's trade. Thus passages of hallucinatory illusion-making, particularly those referring to the enchanted jungle of Sunderban ("the jungle of phantasms and retribution," *Midnight's Children* 424; "magical jungle," *Midnight's Children* 364) are superseded by illusion-breaking comic and ironic reflections.

The most intriguing aspect of the relationship between magic and art is unquestionably the use of rhetorical tropes. Fundamentally there is a close affinity between the magic world view and the logic of tropes, since the latter connect and unify different parts of nature and culture through the principle of sympathetic similarity. Indian culture is obsessed with drawing parallels and creating correspondences (*Midnight's Children* 300). And Saleem's exuberant rhetoric and imagination constantly undermine familiar oppositions between reality and fantasy, history and myth, technology and magic by metaphorically suggesting an affinity between, for example, the radio, the telephone, the railway and magic practices. As Saleem, his son Aadam and Picture Singh travel by railway to the scene of Picture Singh's showdown with another snake-charmer, Aadam is caught in "the rhythm of the magic" (*Midnight's Children* 450) and Saleem hears the liberating "secret word: abracadabraabracadabra sang the wheels as they bore us back-to-Bom" (*Midnight's Children* 450). The metaphoric vehicle culled from magic lore integrates technology, magic and aesthetic incantation. The typical sounds of railway travelling are transformed into a magic charm which in a sense also sums up the message and the artistic credo of the novel. The first word uttered by the baby and future magician Aadam are "Abracadabra," a "cabbalistic formula" (*Midnight's Children* 459).

Since the words and the interpretation reflect the narrator's point of view the instance must be distinguished from the passages when magic metaphors reflect the superstitious mind's understanding of technology. They must also be distinguished from the passages which deconstruct the borderline between symbolic and magic objects, sign and supernatural reality. Symbols are properly speaking signs which tacitly determine the meaning of the whole and the fate of characters. They exercise a concealed magic power over the events.[19]

As soon as this power is marked and becomes explicit, the symbolic object is transformed into a magic object. This is the reason why the silver spittoon ("a talismanic spittoon," *Midnight's Children* 448), the clock magically determining the fate of the midnight children, and the perforated sheet, function both as signs and as magical objects. The perforated sheet does not only symbolise the subjectivity and fragmentary nature of Saleem's narrative construction of his life. It is, at the same time, a "talisman," a magical object guiding the narrator's efforts and determining the fate of the family ("the magic of the sheet," *Midnight's Chil-*

dren 26). Again, the metaphor "tyranny," modified by the adjective "occult" ("occult tyrannies", *Midnight's Children* 9), transforms "clocks" into active, person-like magic objects which do not merely mark time but also interfere with the course of history.

(3) Finally and most importantly: unlike science and technology which aim at doing away with natural language altogether, magic and literary art rely on the power of evocative, poetically organised language to overwhelm and enchant the recipient. Rushdie's novel constantly focuses upon this effect of "mesmerizing" ("ears burning with fascination," "I watched, mesmerized," *Midnight's Children* 454) the listener or reader or observer. De-metaphorising the word 'fascination,' the text elaborates and reflects upon the literal, magic implications of the aesthetic experience. Invariably the effect upon the recipient of literary, technical and magic art defeating time ("in words and pickles, I have immortalized my memories," *Midnight's Children* 459) is fascination, an experience which catapults the reader out of space and time. Just as the old sailor on the picture in Buckingham Villa fascinates "young Raleigh" with his "liquid tales," just as Padma, the narrator's listener, is "frozen into immobility" (*Midnight's Children* 122) by the narrator's "glimpses of myself" (*Midnight's Children* 122), so the reader is expected to be captivated by the truth and magic of an incredibly rich tale of Indian life. Yet since cultures in a globalised world interfuse and interact and since the power of magic and art are universal, the mesmerising effect of successful art invites intercultural references for purposes of illustration. Saleem, who has an English father and an Indian mother, alludes to the Greek myth of Orpheus to give the reader an idea of the "sublime magic" of Jamila Singer's "angel's voice" (*Midnight's Children* 312): "[...] on that first night, as so often afterwards, Jamila sang to Major Alauddin Latif. Her voice wafted out through the window and silenced the traffic; the birds stopped chattering and, at the hamburger shop, the radio was switched off; the street was full of stationary people, and my sister's voice washed over them [...]" (*Midnight's Children* 313). Evidently the artist Rushdie and his narrator celebrate the demise of conventional realism and the advent of a newly-won imaginative freedom. The passage in fact sems to fuse magic and rhetoric since we do not know whether the magic events, apart from the clearly metaphorical and hyperbolic phrase "angel's voice," are produced by a metaphor ("sublime magic") or by a real magic agent (again "sublime magic").

To sum up: Fascination is a characteristic effect of magic and of art. The texts analysed activate its literal, original meaning. During the spell the recipient is released from the everyday awareness and influence of time and space; restlessness gives way to Padma's "rapt attention" ("the rapt attention of my revolted Padma," *Midnight's Children* 320). Furthermore, magic, artistic technique and technology aim at creating structures which overcome the power of space and time. In Dunsany's novel the magic beauty of Elfland symbolises the timelessness and the power of art. Saleem's consciously crafted jars of pickles (the chapters of his novel) preserve, as does Elfland, conscious and unconscious memories, which, from the point of view of psychology, explain the power of suggestive magic appeal. This appeal also uses verbal and non-verbal techniques known from charms and fairy tales – incantation, the evocative power of names, oral narrative, repetitions, magic events. But it would be absurd to claim that literary art's fascination is only triggered by its fantastic elements. Mimetic illusion-making reminds one of the function of magic mirrors which disclose secrets unknown to the average person. As George Eliot points out at the beginning of her realist novel *Adam Bede*, the narrator, just like the Egyptian sorcerer, conjures up visions of the past nobody has ever seen. The magician's "magic mirror" has been replaced by the imagination and the verbal craft of the writer.

Notes

[1] Samuel Taylor Coleridge, *Biographia Literaria* 649.

[2] Matthew Arnold, "Maurice de Guérin" 157-158.

[3] Friedrich Nietzsche, *Die Fröhliche Wissenschaft* 99-100 (translation by John Fowler).

[4] Joseph Conrad, "Preface," *The Nigger of the Narcissus*.

[5] James Reeves, *Understanding Poetry* 34; 38.

[6] On the relationship between event and aura see Dieter Mersch, *Ereignis und Aura: Untersuchungen zu einer Ästhetik des Performativen*.

[7] Simon Armitage and Robert Crawford, *The Penguin Book of Poetry from Britain and Ireland since 1945* xix.

[8] Edmund Burke, *A Philosohpical Enquiry into the Origin of our Ideas of the Sublime and the Beautiful* 57.

[9] Sigmund Freud, *Vorlesungen zur Einführung in die Psychoanalyse* 43.

[10] Evelyne Keitel, *Von den Gefühlen beim Lesen: Zur Lektüre amerikanischer Gegenwartsliteratur* 9: "Was aber ist das Faszinierende an Literatur?"

[11] Heinz J. Drügh und Maria Moog-Grünewald, *Behext von Bildern? Ursachen, Funktionen und Perspektiven der textuellen Faszination durch Bilder.*

[12] Oscar Wilde, *The Picture of Dorian Gray* 156.

[13] Vgl. Marcel Mauss, *Soziologie und Anthropologie*, vol. 1: *Theorie der Magie, Soziale Morphologie*; Edward Burnett Tylor, *Primitive Culture*; James George Frazer, *The Golden Bough: A Study in Magic and Religion*; Alfred Lehmann, *Aberglaube und Zauberei.*

[14] Wilde, *The Picture of Dorian Gray* 66.

[15] Cf. Richard Taylor, "Die mystisch-okkulte Renaissance: Rituelle Magie und symbolistische Dichtung" 100-114.

[16] On fascination cf. my article "The Fascination of Beauty and of Evil in Oscar Wilde's *The Picture of Dorian Gray*" 321-338.

[17] Lord Dunsany, *The King of Elfland's Daughter.*

[18] Salman Rushdie, *Midnight's Children* 100.

[19] Cf. on this Uwe Durst, "Begrenzte Wunder: Ansätze zu einer Theorie des Wunderbaren in realistischen Erzähltexten" 161-175.

Works Cited

Armitage, Simon, and Robert Crawford, eds. *The Penguin Book of Poetry from Britain and Ireland since 1945*. London: Viking, 1998.

Arnold, Matthew. "Maurice de Guérin," *Essays in Criticism*, quoted in *Selected Prose*. Ed. Peter John Keating. Harmondsworth: Penguin, 1970.

Burke, Edmund. *A Philosophical Enquiry into the Origin of our Ideas of the Sublime and Beautiful*. London: Routledge and Kegan Paul, 1967.

Coleridge, Samuel Taylor. *Biographia Literaria*, quoted acc. to *The Oxford Anthology of English Literature: Romantic Poetry and Prose*. Ed. Harold Bloom and Lionel Trilling. New York: Oxford UP, 1973.

Conrad, Joseph. "Preface," *The Nigger of the Narcissus*. London: Dent, 1964.

Drügh, Heinz J. and Maria Moog-Grünewald, eds. *Behext von Bildern? Ursachen, Funktionen und Perspektiven der textuellen Faszination durch Bilder*. Tübingen: Niemeyer, 2001.

Dunsany, Lord. *The King of Elfland's Daughter* (1924). New York: Ballantine Books, 1977.

Durst, Uwe. "Begrenzte Wunder: Ansätze zu einer Theorie des Wunderbaren in realistischen Erzähltexten." *Neohelicon* (2002): 161-175.

Frazer, James George. *The Golden Bough: A Study in Magic and Religion*. London: Longmans, Green, 1901.

Freud, Sigmund. *Vorlesungen zur Einführung in die Psychoanalyse*. Frankfurt am Main: Fischer, 1969.

Keitel, Evelyne. *Von den Gefühlen beim Lesen: Zur Lektüre amerikanischer Gegenwartsliteratur*. Munich: Fink, 1996.

Lehmann, Alfred. *Aberglaube und Zauberei*. Stuttgart, 1898.

Mauss, Marcel. *Soziologie und Anthropologie*, vol. 1: *Theorie der Magie, Soziale Morphologie*. Munich: Hanser, 1974.

Mersch, Dieter. *Ereignis und Aura: Untersuchungen zu einer Ästhetik des Performativen*. Frankfurt am Main: Suhrkamp, 2002.

Nietzsche, Friedrich. *Die Fröhliche Wissenschaft*. Stuttgart: Kröner, 1986.

Price, Martin, ed. *The Restoration and the Eighteenth Century*. New York: Oxford UP, 1970.

Reeves, James. *Understanding Poetry*. London: Pan Books, 1972 [1965].

Seeber, Hans Ulrich. "The Fascination of Beauty and of Evil in Oscar Wilde's *The Picture of Dorian Gray* (1981)," in *To Vindicate the Ways of God to Man: Theodicy and Literature*. Ed. Rudolf Freiburg and Susanne Gruss. Tübingen: Stauffenburg, 2004. 321-338.

Taylor, Richard. "Die mystisch-okkulte Renaissance: Rituelle Magie und symbolistische Dichtung," in *Die 'Nineties': Das englische Fin de siècle zwischen Dekadenz und Sozialkritik*. Ed. Manfred Pfister and Bernd Schulte-Middelich. Munich: Francke, 1983. 100-114.

Tylor, Edward Burnett. *Primitive Culture*. London, 1871.

Wilde, Oscar. *The Picture of Dorian Gray* (1891). London: Penguin, 1988.

Magic and the Film

The Enchantments of a Modern Medium

Johann N. Schmidt

Modernity has traditionally been associated with enlightenment, rationality, and the demise of the occult. But meanwhile we know better: the Freudian "experience of the uncanny" not only arises irrepressibly on the reverse side of reason, it is also deeply inscribed in the double nature of what, at first sight, appears to be the triumph of modern science and rationalised technology over the forces of magic and mystery. It is the cinema, above all other advanced technical media, that gives such strong proof that the phantasmagoric is far from being laid to rest for good.

The affiliation of science to magic ritual can be demonstrated by the example of historical precedents. Thus, the inventors of the telephone showed a strong interest in spiritualistic séances, where they hoped the new apparatus would help to promote communication between those present and the dead. As Tom Gunning shows, the seemingly obsolescent spiritualist movement and results of modern technology had a good deal more in common than is generally believed.[1] In 1856 David Brewster (in his book *Stereoscope*) thought it was quite possible for the photographer "to carry us even into the realm of the supernatural" and "give a spectral appearance to one or more of the figures and to exhibit them as 'thin air' amid the sordid realities of the stereoscopic picture."[2] Photography helped to make the hidden visible: ghostly superimpositions made their appearance where the bamboozled viewer would have never expected it.

Today we are too ready to disavow such a "primitive" response to the magic image. But instead of relating it to times long gone, we should rather acknowledge its inherent modernity. Far from being a mere duplicate, it floats in a realm beyond pre-figured reality and hypostatised meaning. What seems unequivocal as a mirror-like reflection of reality, is interspersed by associations, ambiguities, dreams, and a sense of wonder that undercuts the crude mechanism of ideological certainties. It is precisely this state of an imaginative "in between" that has long challenged the cinema's status as an art form, especially amongst the more conservative devotees of literature and drama for whom neither the photographic image (as a technological replica of reality) nor the mystifying deception of the camera's "eye-trickery" can claim to offer a seriously artistic view of our world.[3] As Parker Tyler puts it: "In striving through imaginative works to create the illusion of reality the movie screen must constantly transcend its own mirror nature of literal reflection."[4]

Many photographic as well as filmic effects had their precursors not in the scientific lab, but in phantasmagoria and magic shows, the pantomime, and sensational exhibitions of new technologies all of which left the impression of fraudulence and counterfeit. While there were also highly serious science lectures and demonstrations, most of them were a curious mixture of popular entertainment and pseudo-scientific presentation which laboured to keep the workings of their instruments under the cover of secrecy.[5] The combination of educational objectives and spiritualist phenomena is strikingly evident in a notice of 1877 that advertised a "Polytechnic Séance." Quite symptomatically, cinema had thus its origin at the crossroad of education and diversion, of instruction and the attraction of curiosa.

The categories of the marvellous and the fantastic underwent an ongoing renegotiation in "respectable" demonstrations of optical illusion and visual metamorphing, but also in the "trickery" of variety programmes (denounced as "infamous magic" by positivist scientists). At the heart of magic lies the dialectic between what can be seen and what is invisible. Objects that exist only in imagination (and are hidden from the eye,) become visible as phantasmagoric appearances by means of optical instruments, magnetism, mechanics, and chemistry. The dialectic also becomes evident in cinematic concepts that have oscillated and will continue to oscillate between the opposite poles of a "cinema of attraction" and those films that try to involve the spectator into "reality as it is." In the mode of reception, finally, cinema functions both as a mass experience and as a very intimate event where the viewer is left alone in darkness.

In a study on special effects Michele Pierson gives a vivid description of such scientific demonstrations in the 1860s which combined lecturing with startling feats of technological novelties.[6] Again we realise that the vocationally minded expert was never far from the conjurer and the virtuoso magician. Both performed in a dark theatre with the help of mirrors, spirit lamps and other gimmicks of stage magic. Their "wonder apparatuses" not only projected dangerous adventures but made of the showings themselves somewhat dangerous events: oil lamps, coal gas, oxygen-ether limelights and early electricity were considered thoroughly detrimental to public safety.[7]

Precisely at the intersection of science and magic was the *image* that functioned as the mediator between a reflection of reality and illusionistic deception. Spectral pictures could display the iconic accuracy of photographic likeness and at the same time the transparency and insubstantiality of a ghost-like apparition. Matter not only materialised but also de-materialised when the focussing and the distribution of light helped to extinguish the visible character of a given object. Quite often optical devices were developed by eccentric inventors and self-appointed visionaries who dreamt of the downright impossible such as pictures that were moving, talking, taking on colours and three-dimensional shape. Only a short time later the apt term "cinemagicians" was coined for all those pioneers who made big steps in the development of electricity, mechanics, and optical illusion.

Perhaps the most famous amongst them, Georges Méliès, started out by turning a defunct theatre into an updated theatre of magical illusions, including spiritualist numbers, "in which devices of lighting, careful control of point of view, and an elaborate optical shutter [...] recreated the effects of a materialization séance."[8] Thus the mediating apparatus took over the medium's rôle. About one century later when the actor Brandon Lee was killed while shooting *Crow*, the film could be completed also without his presence – or more precisely: his presence was generated with the help of computerised images. Méliès would have also been delighted to see Tom Hanks shaking hands with the late President Kennedy in *Forrest Gump* which, in a way, recaptured much of the "spectral spectacle" cinema of early days. Schwarzenegger's opponent robot in *Terminator 2* is a man consisting entirely of fluid metal that can dissolve in every direction. Special effects owe more to the iconography of magic than the technical expertise that made them possible might suggest. In fact, technology not only enables these effects, it is itself imbued with the early popular discourse about the practice of illusory tricks.

In the past magical performances, when produced on the stage of a variété, were unique in the sense that they were non-recurring; they only "happened" at the very moment they were presented, with all the risks of failure. But once they could be mechanically reproduced they were endlessly repeatable and thus suited for serial fabrication. The cinema as we have understood it for more than a century (projecting images on a screen), originated at roughly the same time as the car engine, the radio and other products that became viable only with the mechanisation of production. The mechanical and automaton-like configurate also as a principle of what is preferably shown in moving pictures: fast cars, trains, elevators, revolving doors. This principle is also to be referred to the medium itself when we just think of how many copies of a film are produced for its first showing, or the importance that has been attached to remakes since the early days of cinema.

Today moving pictures have taken one further step forward, and yet they increasingly look back to transformational processes of the 18th and 19th centuries. Due to video and DVD, watching a movie is no longer a unique experience with a beginning and an end. What happens is rather a dissection and dispersion of the linear narrative to achieve a new temporality of reception. The highly suggestive effect on the part of the user is complemented by the transformations to which the image can be subjected: "morphing" (a good playing card changes to a bad one in the hand of a trickster), "warping" (dirty dishes melt away and disappear in the kitchen sink), "enhancing" (colours begin to glow, humans sparkle like stars), etc. In many ways they bring back, though technically on a far more advanced level, the early superimpositions and the stop-action camera once introduced by Méliès.

There is general assumption that 28 December 1895 was the birthday of cinema, and that the Grand Café near the Boulevard des Capucines was its birth place.

But this is not even half the truth. As Tom Gunning aptly states: "Origins are slippery things, figuratively and literally, and the cinema is possibly the most slippery medium that has ever existed."[9] *In sensu stricto* Paris was the scene of the first cinematographic performance, arranged by Louis and Auguste Lumière for a paying audience.

However, the cinema was not "invented" – if that term has any relevance at all – at a certain place and at a certain time.[10] In fact, there existed many competitive, non-linear options quite independently from each other, which makes it difficult to construct a teleological course of evolution. Even in 1895-96 various systems of projection were spread all over the world: In Berlin (the Bioscop double projector of the brothers Max and Emil Skladanowsky), in London (Robert William Paul's theatrographer) or in New York, where the first parlour with a kinetoscope was opened by Thomas Alva Edison in a former shoe shop. Edison's device was very much like a peep show box for individual use, giving the impression of lifelike motion; the corresponding camera, called "kinetograph," weighed over a ton and had 48 pics per second. Of the American Jean Aimé Le Roy we know that he performed a programme to a paying audience one week before the brothers Lumière. Other sources tell us that they first presented their invention of a gripping device for transporting the film to selected members of the *Société pour l'encouragement de l'industrie* as early as in May 1895. It gives the impression of a race towards the same goal but at different places and with different equipment. William Friese-Greene and George Eastman were the ones who brought in celluloid for roll films. Celluloid owed its existence for filmic use to William Kennedy Laurie Dickson who had studied the early transparent celluloid of John Carbutt. In early times cellulose served as a substitute for ivory in the production of billiard balls. These "pioneers" stood not at the beginning but almost at the end of the long road that led to the cinema. This road was not the wide vista of myth but it was rather crooked and it opened up into many byways and cul-de-sacs. Innumerable are the names of apparatuses to produce and demonstrate moving pictures, initially non-projective, such as Joseph Plateau's phenakisticope (literally "deceptive view"), Dr John Paris' thaumatrope ("wonder-turner"), Stampfer's strotoscope (a disc rotating before a mirror), Emile Reynaud's praxinoscope (spinning mirrors), Ottomar Anschütz's electrical tachyscope ("Schnellbildseher"), Cecil Wray's kineoptoscope, the mutoscope, the zoetrope, the viviscope, and Etienne-Jules Marey's photographic gun.[11] Perhaps one of the key discoveries was made (or at least re-formulated) by Plateau in 1836: the laws of the stroboscopic effect were founded on the observation that visual motion no longer appeared to the human eye as divided into individual pictures as soon as these pictures met the eye in quick succession. William George Horner made use of this inertia of optical perception when he fastened images on the inside of a cylinder that was set into motion. The observer who looked through the narrow apertures of the cylinder did not see one picture after the next, but "moving pictures," combining to sequences of dancers, acrobats, jugglers, and boxers.

Perhaps the best-known proof of the mechanism of viewing, though from an opposite standpoint, was delivered by Eadweard Muybridge who, as a result of a bet, was commissioned in 1872 to photograph the galloping movement of horses. It was not known whether all four legs came off from the ground at once in what was called "animal locomotion." Five years later Muybridge installed a series of cameras whose mechanism was released at very short intervals by the horses touching thin threads in full gallop. When dissected into a series of images, the "flying horses" became objects of scientific demonstration, but the reverse process of projecting the pictures backwards or of using the technique of the stop-action camera brought them back into the realm of the uncanny where the body disobeys time and space. Muybridge's tremendous importance lies in his demonstrating how magic triumphs over science by using the apparatus for creating a world beyond systematic knowledge.

The dream of "living photographs," however, had been cherished a long time before all these apparatuses and the invention of photography came into being. Certainly one of the most influential predecessors of cinema, both in terms of the optical apparatus and the modality of performance, is the magic lantern or *laterna magica*. This invention, principally attributed to the Dutch humanist Christiaan Huygens, proved a popular showpiece of magicians and electrical engineers alike. It founded nothing less than the whole tradition of projected images as an important strand of visual culture throughout the centuries. This culture was concerned with the character of "natural vision" but, even more so, with the transformation and deformation of vision by means of optical refraction, luminous intensity, and mechanical succession. The magic lantern gave the impression of motion through a movable glass slide, and its cognomen "lantern of fear" was owed to rather scary illusions like that of an animated skeleton removing its skull and other *diableries*. The close-up of the cinema had its predecessor in the effect of combing the magic lantern and the microscope which displayed eyes and legs of tiny insects or the pores of the human skin (reminiscent of Gulliver's voyage to Brobdingnag).

What elevated the inventions of the magic lantern and the camera obscura above being mere proto-forms of cinema was, on the one hand, the aspect of popular spectacle presented by showmen, and, on the other hand, the narrative structure of mini-tales of mythical, historical, comical and sometimes obscene character. The subtle blending of one image into another anticipated the "dissolve" of cinematic story-telling; it also resembled the mechanical moving slides which were on the market after 1870. In the mid-nineteenth century the diorama, projecting light through semi-transparent painted canvases, was some sort of lantern to be entered by the viewer in order to be immersed in the pre-designed illusion of a sunset, a rainbow, of lightning, snow, or fire. This illusion, incidentally, had a far greater visual impact and a higher degree of perfection than the earliest films with their grainy surface and their chopped-up sequential structure.

In our contemporary rhetorics of sincerity and authenticity a common reproach made to visual media is that of manipulation. It rings of misuse, deception, propaganda, and fakery. Only an image that approaches verisimilitude in representation is considered as able to convey truth and enlightenment (the latter to be understood quite literally as a state where our psychic scene is lit into the last corner). From early on, however, directors revelled in the manipulating force of pictures for the sake of magic effects. In the 1913 production of Stellan Rye's *The Student of Prague* the unearthly double is portrayed through the old photographic trick of multiple exposure, producing a wraith-like image. The phantom-like quality of cinema was again and again authenticised by "true" phantoms (*Fantômas, Judex, Les Vampires*) who transformed substantial social space into an eerie realm for reveries, nightmares, and mysteries. In fact, overexposures and superimpositions were considered as part of "spiritual photography." Josef von Sternberg gleefully deceived our senses, as he thought that trickery was at the origin of all art that helped us to perceive reality in a less habitualised manner. "What interests me," Sternberg remarked, "is the expression by exploring technology":[12] In *The Scarlet Empress* the clapper of a bell is transformed into a human being with his head downwards – a ghostly expression of the cruelties under the tsarist régime.

We know of another pioneer of international cinema, Friedrich Wilhelm Murnau, that he was a follower of alchemist disciplines, and so it is not surprising that he tried to reveal the hidden side of life and elucidate its existence in an intricate playing together of light and shade (it should be remembered that light presupposes the darkness of the cinema or of an auditorium in which magical trickery is performed). Murnau also gives us an idea of life as a gigantic labyrinth where everything is fleeting, transitory, and illusory. His films abound with places of transition, mirrors and reflecting surfaces. Thus water becomes translatable into the fluctuating spirit that is free of solid material substance, and yet it can be perceived as a surface reality.[13] In a way this is a paradigm for the moving pictures. The vibrant filmic image seems to be so much more powerful than the reality it depicts because of the link between natural and mechanical representation on the one hand and poetically "fugitive" dimensions on the other hand. When Alfred Hitchcock declared that he wouldn't allow reality to rear its ugly head, we hear not only the skilled manipulator but also someone who is highly conscious of the tradition in which he stands. Thus the title of his film *The Lady Vanishes* (1937) is a clear reminiscence of Méliès's first substitution film *Escamotage d'une dame chez Robert-Hondin* (1898).[14]

The double nature of the filmic image (stability and fluidity) has been described by a number of theorists, possibly because it seems to lie at the heart of what constitutes film as both "art" and "non-art." Here it should be remembered once more that the debate whether cinema could be called "the seventh art" has been taken up again and again without leading to a definite result, which hardly seems to be surprising when we think of its origins. As Pierre Klossowski states, the plain filmic image reflects, at first sight, no idiosyncrasy (as in art) but only a raw reality, even if this "reality" is of a fictional nature. Mostly it is perceived as nothing more than a bearer of resemblance, a means of verification. By way of chemical and physical operation we are in good faith of getting a positivist idea of reproduction, centered firmly in the realm of technology. But as soon as empirical reality breaks away from the contexts and the modes in which we see things "as they are," i.e., as soon as it becomes subjected to the laws of mechanical discontinuity and the intrinsic qualities of images, this very reality presents itself (in Klossowski's words) as "its phantasmic double."[15] Méliès's devices of stop- action, superimposition, of accelerating and slowing down, show an imaginative interplay of time and space, or what the art historian Panofsky observed in cinema as the "spatialization of time" and the "temporality of space."[16]

Right from the beginning of film history, there has been a distinct consciousness of recreating an illusion of reality and, at the same time, of conveying the aleatory play of fantasy and the marvellous. It is not surprising that classical Hollywood cinema laid great importance on the principle of continuity cutting which should make us forget that we actually are watching a movie. But what we might understand as a strict avoidance of any devices that could draw our attention to cinema's artificiality, the effect of magic – of what is behind the visible – has by no means disappeared. As Karen Beckman points out, even continuous narrative in cinema is no less a trick, however less self-declaratory, than any of the special effects.[17]

Films are lessons about the relationship between material and movement. The combination produces an effect both "realistic" and "dematerializing" in regard to the representational quality of the material. It is the demarcation line where magic profits from technology and technology unveils its portion of fantasy, affectivity, and visual magic. This lies at the heart of the sense of wonder which has played such an important role ever since the early times of cinema. It is still present, though greatly professionalised, in a culture of computer-generated images (CGI) where connoisseurship and fandom are the modern attributes of the magician. "New Magic" was the title of one of Douglas Trumbull's showscan movies in the 1980s which were shot at frame rates up to 96 frames per second. Enchantment and technophilia are still just two sides of the same coin in the creation of a poetic universe out of the scientific spirit.

Notes

[1] See Tom Gunning, "Phantom Images and Modern Manifestations: Spirit Photography, Magic Theater, Trick Films, and Photography's Uncanny" 42-71.

[2] Quoted ibid. 47.

[3] "This doesn't hold true for the avant-garde at the outset of the 20th century which felt sure that novel technological processes like photography and film not only add new forms to the existing ensemble of the arts, but [that] these reproductive media have far-reaching influences in reconfiguring the field at every level" (Malte Hagener, *Avant-Garde Culture and Media Strategies: The Networks and Discourses of the European Film Avant-Garde, 1919-39* 3). Interestingly, Hagener adds that the self-definition of an avant-gardist at the time was also that of an engineer and scientist, practical joker, magician, and mugger.

[4] Parker Tyler, "From 'Magic and Myth of the Movies'" 727.

[5] See Iwan Rhys Morus, *Frankenstein's Children: Electricity, Exhibition, and Experiment in Early Nineteenth-Century London* passim.

[6] Michele Pierson, *Special Effects – Still in Search of Wonder* 11-51.

[7] Deac Rossell, *Living Pictures: The Origins of the Movies* 10.

[8] Gunning, "Phantom Images and Modern Manifestations: Spirit Photography, Magic Theater, Trick Films, and Photography's Uncanny" 62.

[9] Tom Gunning, "Introduction" xix.

[10] See Thomas Elsaesser, *Filmgeschichte und frühes Kino: Archäologie eines Medienwandels* 47.

[11] A comprehensive summary is given in Albert A. Hopkins, *Magic Stage Illusions, Special Effects and Trick Photography* 490ff.

[12] Quoted in Frieda Grafe, "Josef von Sternberg – Technik als Kunstmittel" 46.

[13] See Charles Jameux, "Alchemie des Bildes" 529.

[14] See also Karen Beckman, *Vanishing Women: Magic, Film and Feminism*; Lucy Fischer, "The Vanishing Lady: Women, Magic and the Movies" 339-354.

[15] Pierre Klossowski, "Pierre Klossowski über Film" 544.

[16] Erwin Panofsky, "Stil und Stoff im Film" 345.

[17] Beckman, *Vanishing Women: Magic, Film and Feminism* 131.

Works Cited

Barnouw, Erik. *The Magician and the Cinema*. Oxford: Oxford UP, 1981.

Beckman, Karen. *Vanishing Women: Magic, Film and Feminism*. Durham: Duke UP, 2003.

Ceram, C. W. [Kurt Wilhelm Marek]. *Archäologie des Kinos*. Reinbek: Rowohlt, 1965.

Crary, Jonathan. *Techniques of the Observer. On Vision and Modernity in the Nineteenth Century*. Cambridge, MA: MIT Press, 1992. Trans.: *Techniken des Betrachtens: Sehen und Moderne im 19. Jahrhundert*. Dresden: Verlag der Kunst, 1996.

During, Simon. *Modern Enchantments: The Cultural Power of Secular Magic*. Cambridge, MA: Harvard UP, 2002.

Elsaesser, Thomas. *Filmgeschichte und frühes Kino: Archäologie eines Medienwandels*. Munich: edition text + kritik, 2002.

Faure, Elie. "Mystik des Films." *Filmkritik* 5 (1969): 329-337.

Fischer, Lucy. "The Vanishing Lady: Women, Magic and the Movies," in *Film before Griffith*. Ed. John Fell. Berkeley: U of California P, 1983. 339-354.

Grafe, Frieda. "Josef von Sternberg – Technik als Kunstmittel." *Filmkritik* 1 (1967): 43-50.

Gunning, Tom. "Phantoms, Images and Modern Manifestations: Spirit Photography, Magic Theater, Trick Films, and Photography's Uncanny," in *Fugitive Images: From Photography to Video*. Ed. Patrice Petro. Bloomington: Indiana UP, 1995. 42-71.

Hagener, Malte. *Avant-Garde Culture and Media Strategies: The Networks and Discourses of the European Film Avant-Garde, 1919-39*. Amsterdam, 2005.

Hankins, Thomas and Robert Silverman. "The Magical Lantern and the Art of Scientific Demonstration," in *Instruments and the Imagination*. Ed. Thomas Leroy Hankins. Princeton: Princeton UP, 1995. 37-71.

Hopkins, Albert A., ed. *Magic- Stage Illusions, Special Effects and Trick Photography*. New York: Munn, 1976.

Jameux, Charles. "Alchimie des Bildes." *Filmkritik* 9 (1967): 527-530.

Klossowski, Pierre. "Pierre Klossowsky über Film." *Filmkritik* 9 (1969).

Mannoni, Laurent. *The Great Art of Light and Shadow: Archaeology of the Cinema*. Exeter: U of Exeter P, 2000.

Moore, Rachel O. *Savage Theory: Cinema as Modern Magic*. Durham: Duke UP, 2000.

Morus, Iwan Rhys. *Frankenstein's Children: Electricity, Exhibition, and Experiment in Early Nineteenth-Century London*. Princeton: Princeton UP, 1998.

Musser, Charles. *Before the Nickelodeon: Edwin S. Porter and the Edison Manufacturing Company*. Berkeley, CA: U of California P, 1991.

Panofsky, Erwin: "Stil und Stoff im Film." *Filmkritik* 6 (1967): 343-355.

Pierson, Michele. *Special Effects – Still in Search of Wonder*. New York: Columbia UP, 2002.

Rossell, Deac. *Living Pictures: The Origins of the Movies*. Albany, N.Y.: State U of New York P, 1998.

Smith, Grahame. *Dickens and the Dream of Cinema*. Manchester: Manchester UP, 2003.

Sobchack, Vivian, ed. *Metamorphing: Visual Transformation and the Culture of Quick-Change*. Minneapolis: U of Minnesota P, 2000.

Tyler, Parker. "From 'Magic and Myth of the Movies'" (1947), in *Film Theory and Criticism*. Ed. Gerald Mast et al. New York: Oxford UP, 1992. 725-729.

A Magic Lantern Of Sound?

The Aeolian Harp between Cheap Showmanship and Spiritual Mystery, here shown as a Catalyst within the History of Music

Walter Windisch-Laube

I

"It's Magic" (Sarah Vaughan)

In the 1696 'musical tract' *Phrynis Mitilenaeus* by Wolfgang Caspar Printz, which actually is, in part, a kind of musicians' *roman à clef*, the following passage can be found:

"Als Herr *Eumenes*, den ich nebst etlichen andern begleitete / in das erwehnte *Museum* hinein trat / empfieng ihn ein Bild oder *Statua*, mit einer kurtzen / doch schönen und artlichen Rede / welches uns alle erstaunend machte / sintemahl dieses Bild nicht nur die Augen verwendete / sondern auch im Reden den Mund nicht anderst / als ein lebendiger Mensch bewegte. Da wir beschäfftiget waren allerhand *Mathematische*, und sonderlich wunderbahre *Optische* Kunst=Stücke zu besehen / öffnete Herr *Katharinus Asicherus* ein Fenster / da höreten wir eine frembde und artliche *Harmoni*, und wusten nicht wo sie herkahm. Es ist aber mein Vorhaben nicht alle dergleichen wunderbahre Dinge zu erzehlen: massen ich sonst der *Music* gantz darüber vergessen müste: Doch muß ich noch dieses erwehnen / daß wir viel nicht für natürlich gehalten hätten / wenn uns nicht dieser Wunder=Mann die Ursachen derselben entdecket und erwiesen hätte."[1]

The Rome visitors Eumenes and Phrynis, after having received a welcome from a talking statue, got to hear mysterious 'sounds that give delight;' so they may have asked themselves and each other using Ferdinand's words in Shakespeare's *The Tempest* 1.2.391: "Where should this music be? i' the air, or the earth?"[2] Six years before *Phrynis Mitilenaeus*, Printz himself in his *Beschreibung der Edelen Sing- und Kling-Kunst*[3] had given the explanation of the phenomenon by means of which Athanasius Kircher (1602-1680), i.e. "Katharinus Asicherus" in the quoted text, used to surprise and amaze his visitors within the famous Museum Kircherianum by opening a window; what they heard was the sound of an Aeolian harp. Kircher, the master of several 'free arts,' thus celebrated a sort of miracle or magic which on the one side was founded in baroque science, on the other was ideologically or religiously based (and even applied with missionary intentions[4]).

Appropriately an exhibition in Fulda on the occasion of Kircher's 400th birthday was given the title: "Magie des Wissens,"[5] 'the magic of knowledge.' Fulda, not too distant from Kircher's native place (Geisa), had in those days been the district capital, where he had spent several years of his youth.

Athanasius Kircher is widely reputed to have invented not only the Aeolian harp (the wind harp, a string instrument played by the wind only), but the laterna magica[6] as well, and even the camera obscura.[7] In fact, although he cannot really be called the inventor of any of these, yet in each instance he was one of the innovators or, in the case of the Aeolian harp, he was rather the *practical* inventor and, as in several other matters, the one who effectively presented it to a wider public.

Athanasius Kircher, Aeolian harp[8]

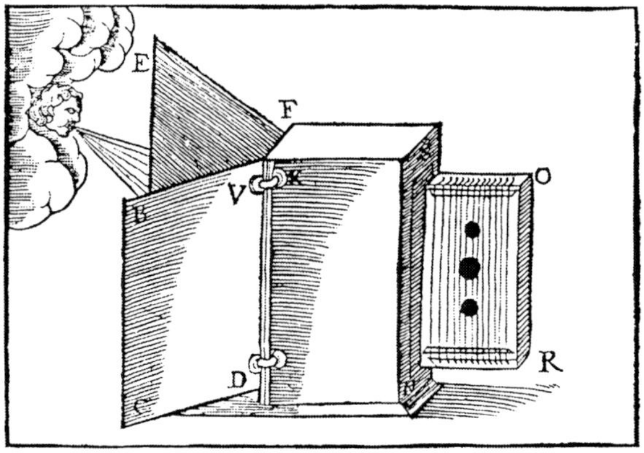

That's what this essay is about: the correlation of quasi magical 'show effects' and innovative impulses within musical (and also literary) history, especially concerning the Aeolian harp symbol.

One may argue over which of two analogies (though both are weak) is the stronger: that between the laterna magica and the Aeolian harp or that between this instrument and the camera obscura. The first pairing emphasises the amazing effect on the viewer or listener, and, moreover, the illuminating and conspicuous role of the Aeolian harp in musical history, the second pairing accentuates rather the representative function of this very special instrument of *musica instrumentalis* as a mediator between the 'music' of the spheres (*musica mundana*) and the human mind or soul (*musica humana*). However, Kircher's source, Giovan Battista della Porta's *Magia naturalis*, without actually using either term, describes (in its second edition) both the camera obscura and laterna magica,[9] in full, and as connected to each other, and in the sequel refers to the Aeolian harp phenomenon, too – the first mention of it in scientific literature at all.[10]

As in his *Musurgia universalis* (1650) and *Phonurgia nova* (1673),[11] Kircher probably explained to his Rome visitors afterwards what they had heard or rather what phenomenon they had had to do with. Likewise, explanations for some corresponding magical sound effects were supplied subsequently by Jean Paul in *Hesperus* and by E. T. A. Hoffmann in his *Kater Murr*. The narrator of *Hesperus* (1793[12]) enthusiastically describes those supernatural strains and the protagonist's fascination in their presence, whereupon the reader is told about the terrestrial, the 'Aeolian' reason for such synaesthetic enchantment and is also guided (in a metaphorical way) to their metaphysical implication.[13] In Hoffmann's novel the analogous procedure appears ironically refracted: As the parts or pages of the 'waste paper' manuscript follow each other in reverse order (irregularly interrupted by the tomcat Murr's *Lebensansichten*/'memoirs'), the reader is informed much earlier[14] than the protagonist, Kapellmeister Kreisler, himself,[15] where the here more terrifying acoustic events come from – in this case from a huge adaptation of the 'Aeolian' instrument, from the 'weather harp.'[16] Thus, we are shown a connection between certain baroque and Romantic modes of thinking and writing: A metaphysically orientated enthusiasm is conveyed by means of scientifically well founded craftsmanship and exact, intellectual, nevertheless speculative construction. Meister Abraham, Kreisler's fatherly confidant, by uniting (or even merging) inventor, magician and philosopher, on several occasions reminds us directly of Athanasius Kircher.

Several hundred years before Kircher, kites including a wind-string were constructed in China[17] (and the allround scholar probably knew of them[18]). Nevertheless Kircher was the one who 'invented' the Aeolian harp[19] in the sense that he experimented on and gave instructions for making a string instrument[20] especially to generate sounds evoked by no human agency, nothing but the wind. In fact the analogous phenomenon had, even in a scholarly way (see above, regarding Porta), been described earlier: initially in myths and legends[21] as well as in literary works (cf. a Chinese Tang-epoch lyric text, and especially Rabelais' *Gargantua et Pantagruel* IV/55).[22] The instrument maker's results Kircher arrived at through observations and experiments, i.e. empirically (later described in his great works *Musurgia* and *Phonurgia*), were first and foremost practicable, but not only this: his number-based constructions with several strings all of equal length also reflect Kircher's view of the world – he saw all real things as members of an integrated unity, numerically harmonised, a concept not too far from those of Leibniz, Kepler, Marin Mersenne and Spinoza. In Kircher's philosophy the wind appears as a symbol, or rather as a medium of God's spirit;[23] music has to be interpreted as an ideal way to praise Him,[24] and numbers, in the form of proportions physically responsible for all 'Aeolian' sound effects, represented to Kircher the universal principle of creation and nature. At least in this respect the amazing and sensorial side of Kircher's demonstrations, which often seem more cheap showmanship, is not separable from the innovative impulses consistent with his way of thinking. Thus a baroque world view takes some single – and primarily practical – steps towards the 'new physics' and at last towards the Enlightenment.[25]

If music is, per se, already a kind of magical science, regarded of old (*Artes liberales*, Leibniz[26] etc.) as a complement and even another facet of mathematics, the sounds of the Aeolian harp are to be considered so even more. To none of all sound producers (except perhaps electronic equipment) does coincidence play – interacting with mathematical proportions – a comparably important role. Therefore the reign of higher rule seems to be present at all times, either, in the baroque view, as an anticipation of the next world and of eternity, or as a sudden idea of totality (of the absolute or of the universe), revealed to the mind of the Romantic artist. By consulting ethereal beings[27] to describe 'Æolian visitations'[28] and 'Eolian magic,'[29] Romantic poetry shows its roots both in direct experience (Wackenroder's and Tieck's)[30] of baroque art and in the various continuations of Renaissance and baroque intellectual beginnings.[31] However thoroughly the phenomenon may have been investigated by science, and indeed explained, the generation of the 'Aeolian' sound remains somewhat mysterious (to describe the vibration of a string alone is most complicated, all the more so if it depends on the wind). Of course it is the magical and also enigmatic effect of the Aeolian harp in connection with its metaphysical dimension which has fascinated generations of poets and musicians.

II

"A Kind of Magic" (Queen)

As far as the history of *music* is concerned, the next innovations in Aeolian harp symbolism took place not earlier than the turn to the 19th century, when that 'wind instrument' had already been widely distributed in English gardens, castles, ruins et cetera. From then on, auditory experience both of real Aeolian harps and of earlier musical compositions showed an influence on musical works dealing with the Aeolian harp (as subject and symbol), mostly settings of poems about this instrument, or about the analogous acoustic phenomenon when taking place on a traditional chordophonic (stringed) instrument. We can say the magical effect of the wind harp itself, and of its echo in the arts, leads to a more and more ingenious translation of the concept into artistic practice. So magic and a kind of science proceed hand in hand, while the aspiration to find or even produce the absolute, pure sound has become an urgent issue.

Charles Dibdin (1745-1814), a British composer, dramatist, poet, novelist, actor, singer and entertainer, still fairly well-known for his songs[32] and for his 30 or so "Table entertainments,"[33] created the first musical composition dedicated in its title to the Aeolian harp. It was part of the Table entertainment "Most Votes Carry It," which had its premiere on 9 October 1802.[34] In its separately published song *The Aeolian Harp* Dibdin's lyric speaker sings the praises of plucked string instruments, beginning with "Amphion's lute and Orpheus['] lyre." "But more the senses to ensnare give me the soft melodious strain, [...] Sounds that the ear so smoothly greet from the celestial self play'd sweet Æolian harp."[35] This passage in both the piano and the added solo part is directly followed by a dominant chord held through six bars. Its tones are overlapped and completed by a repetition of the latter part of the cited sentence, too – set to a new melody, with a seventh and a ninth whose resolution is postponed.

About two or, at the most, six years later Christian Schreiber's likewise vocal composition *Die Aeolische Harfe*[36] includes a passage which must have appeared magical to contemporary listeners, for it is probably the earliest whole tone scale to be found:

Christian Schreiber, *Die Aeolische Harfe*[37]

III

"Magic Pie" (Oasis)

Some twenty years later Hector Berlioz began to enrich the art of instrumentation. In no preceding century had the history of music been determined and pushed forward in nearly every way by industrial and technical progress, e.g. the development of wind and keyboard instruments. Thus it is not surprising that orchestration moved from sheer craftsmanship to a sublime science – serving the purpose of enchanting the audience. Jean Georges Kastner and Hector Berlioz, being friends, were the two (French)[38] composers who established the *study of instrumentation* as a subject in itself, and they did this not least with regard to the Aeolian harp (which may also – as by Athanasius Kircher[39] and others[40] – be called one of the first automatic musical instruments). Berlioz' cantata *La Mort d'Orphée* (1827) contains, as its final movement, an orchestral miniature here called "Tableau musical,"[41] dealing with the wind in the lyre of Orpheus who is now dead. This short piece (29 bars) was not only the first pure instrumental work exploiting the Aeolian harp effect, but also the first example of purely instrumental programme music in Berlioz's Œuvre. In addition, it played an important personal role for its composer;[42] he was busy with it through more than three decades.[43] Finally it appeared in the 'lyrical monodrame' *Lélio ou Le Retour à la vie* with the title *La Harpe Éolienne – Souvenirs*.[44] The surprise Berlioz gave his auditors was due, in the first place, to the fact that he devoted even an orchestral piece, not a vocal one, to the wind-played instrument; a second surprise was the shortness of the setting and a third the orchestration effects[45]: partly imitating a wind harp, the orchestra (including a harp, another novelty[46]) provides a steady foundation for the intense clarinet solo of an eloquent, and emotional character. As a detail concerning performance Berlioz reports in his *Grande Traité d'Instrumentation et d'Orchestration Modernes*[47] that he told the soloist to wrap the clarinet in a leather bag – to intensify the impression of hearing merely a 'shadow' of music, and to increase the evocation of (as the writer-musician calls it) "spleen"[48] which means (a poetical) melancholy. As a particular compositional detail the orchestral miniature ends in an unresolved seventh chord, not the dominant, but the tonic with added seventh. The surprise or magical elements the composer presents here are not so much show and sensation as the result of typically romantic reflections and considerations. Most of the components determining the short movement refer to the fragment principle working all through the opus as a whole;[49] moreover *La Harpe Éolienne* appears to be the pioneer composition for a new quality of instrumentation which is now treated as a compositional feature of determining importance in itself, and not merely as an illustrative addition.[50]

Less evidently consistent is Jean Georges (or Johann Georg)[51] Kastner's[52] "Grand monologue lyrique avec chœurs" *Stephen, ou La Harpe d'Éole*, which he published in 1856 as the artistic part of his "Livre-Partition" *La Harpe d'Éole et la musique cosmique. [...]*.[53] Kastner's importance in the history of music is due to the fact that he generated or invented two genres.[54] Even earlier than Berlioz, and not without an influence on him, Kastner founded what we call the study of orchestration; the first, 1837 edition was titled: *Traité Général d'Instrumentation*. In it the Aeolian harp sounds are described as "des sons vraiment magiques," seeming to come from a foreign world.[55] Later, from 1849 on, Kastner developed those books, the "Livre-Partitions,"[56] in which he combined science and art: an extensive scholarly study with a rather large-scale composition, in most cases a text-generated one. So *Stephen*, including the attempt to imitate the sound of an Aeolian harp in a refined way as naturally as possible. For this in particular, Kastner makes abundant use of a new orchestral effect, producing (natural as well as artificial) harmonics, on harps, and mainly on bowed strings, mostly *con sordini* from *pp* up to *pppp* and in combination with syncopations, tremolos at different speeds, and further with echo and lontano (faraway) effects from flutes and clarinets. The first of these passages,[57] all of which play the role of leading sound, is headed: "Effet d'orchestre imitant la harpe d'éole," and its tone (or should we say magic) exercises an edifying influence on the protagonist. In this respect, and in that of orchestration Kastner's 'Aeolian harp' sections can be understood as a continuation of Berlioz' *La Harpe Éolienne*.

Kastner's tendency to reproduce the 'Aeolian' sounds as truly as possible (*ars simia naturae*?) may remind present-day listeners of electronic instruments like the synthesizer (of course able to imitate wind harp sounds, too), which in the last decades have been 'corrupted' – as their pioneers would probably express it – to the sound-imitating so-called "keyboard."[58]

Quite different was – five decades after Kastner – Franz Schreker's starting point within the field of instrumentation, in which he, too, employed the idea and symbol of the wind harp. Schreker's technique of composition generally aims at a languishing sonority. In his opera *Der ferne Klang* ('The distant sound;' first performed in 1912),[59] the largest musical work dealing with the Aeolian harp (even as a main subject), he uses celesta, harp and strings – in cleverly thought-out sound mixtures – as favourite instruments[60] to create an equivalent for (not an imitation of) the bewitching and impalpable 'sound-shape' of that wind-played 'automaton.'

IV

"Every Little Thing She Does Is Magic" (Sting)

Franz Liszt surely was one of the most innovative musicians during the 19th century, not only for the technique of piano playing, but also in questions of musical composition.[61]

On Western ears the Hungarian cimbalom has a similarly strange and somehow exotic effect as the Aeolian harp. Liszt, in 1860, brought the two instrumental spheres together, inspired by the famous Nikolaus Lenau poem about the 'Three Gipsies' (*Die drei Zigeuner*) – which in Germany is probably the most popular echo of the Aeolian harp phenomenon in the arts. Liszt's 'grand idea' of music in an "alliance plus intime avec la poésie"[62] seems to be concentrated in a few bars of this Lenau setting; its accompaniment was also set for orchestra by the composer himself and, in 1864, he paraphrased the song and its wind harp effect in his only important composition for violin and piano, then arranged by Jenö Hubay for violin and orchestra in 1931).[63] Very likely Franz Liszt of all former composers was the one whose connections to the Aeolian harp symbol were the most frequent and varied.[64]

What takes place at the centre of Lenau's poem as well as in its musicalisation[65] could be called a magic parallelism. The two verses containing the Aeolian harp phenomenon are situated exactly in the middle of the seven-stanza text, and are, moreover, emphasised by the metre: the verse "Und sein Zimbal am Baum hing," is in this poem the first containing two successive accentuated syllables – and even ending as well as rhyming with them; the following line, "Über die Saiten der Windhauch lief," is the only nine-syllable verse within the scheme of four alternating with three accentuated syllables in one line. It is not difficult to realise that just these two verses allude to the biblical Psalm 137[66]: the wind touches the strings of a cimbalom hung up in a tree, while a dream touches the heart of the sleeping third Gipsy. Lenau has put this into the rhetorical or poetic figure of a *parallelismus membrorum*, which is constitutive for the Psalter texture anyway. Lenau's parallelism is a triple one: "Und der dritte behaglich schlief, / Und sein Zimbal am Baum hing,", these two lines form one affinity; "Über die Saiten der Windhauch lief, / Über sein Herz ein Traum ging" another one, and each of the crosswise rhyming verse-pairs constitutes one more parallelism.

The cimbalom passage in Liszt's musicalisation combines several sorts of parallelism, too: between solo voice and piano part (motivic and intervallic) as well as among the piano chords to be played with a tremolo and replacing one another from bar to bar in a mediant relation[67]: F#^{7}, A, F#^{7}, A, A^{7}, C, A^{7}, C, E, E, C#^{7}, C#^{7}. The composer's main intention therefore seems to be that of providing sonorous, harmonic surprise. But more than this, the special succession of sounds marks the centre of the composition, both the formal and the intellectual. Music, dream (on which word the culminating point of the melody is set) and yearning, as well as distant times and cultures, coincide in these 20 bars of magical effect. Not least among the causes of this effect is the tonal reproduction and stylisation of the cimbalom tone, too, which offers the compositional starting point for this stanza, but also for Liszt's whole setting of the poem. Even its beginning not only alludes

Peter Cornelius, *Auftrag*[68]

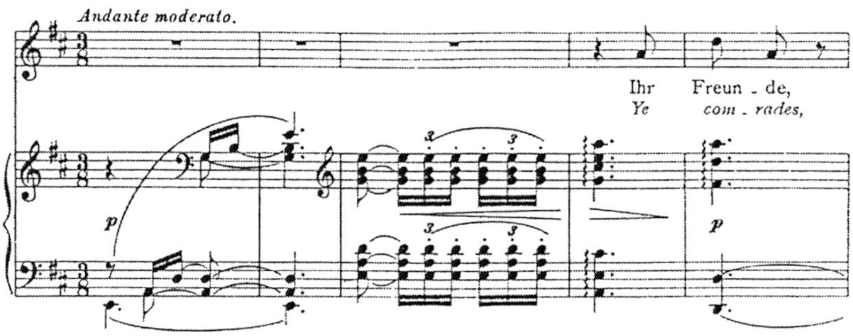

to the composer's 13th Hungarian Rhapsody and to his own piano concerto in E*b* major, but as well includes – at least from the fourth full bar on – diverse pianistic adaptations of the above-mentioned Hungarian or Gipsy folkloristic instrument. In various of his Rhapsodies Franz Liszt integrated similar cimbalom imitations, extensively in the 10th, 11th and 14th, but *Die drei Zigeuner* is probably the first *lied* almost completely constructed out of those sounds, and furthermore follows the czardas-form in a singular way, too. The close correlation between notes and words gives Liszt the opportunity of conveying a meaning beyond colouristic images: putting the cimbalom sound-magic at the service of the Romantic idea. Some catchwords regarding it are, in this case, the antiphilistine attitude or at least yearning, the longing for sound per se, the unfolding of the unknown, the art of sublime citation and fragmentation. With the radiance of the sound motif throughout the whole of this work the composer gives both an interpretation of Lenau's poem and a glance at the importance of the Aeolian harp symbol for the intellectual history of the nineteenth century.

The effect of imitating a traditional musical instrument on the piano in the name of the 'self-played' Aeolian harp is also to be found in an *lied* by the Liszt and Wagner follower Peter Cornelius (1824-1874). The composer takes a guitar's open strings as a leading motif or central idea for his art song *Auftrag* ('mission;' op. 5.6, 1862). As the underlying text, a poem by Ludwig Christoph Heinrich Hölty (its third stanza, the most magical, was added by Johann Heinrich Voß), speaks of a *harp*, the musical shape is to be interpreted as a stylization – in accordance with Romantic tendencies to the imprecise naming of string instruments.[69] The open string sound therefore points to the tuning of a human soul, and correspondingly that guitar-string fourth chord of the pure instrumental *ritornello* is resolved each of the four times in a different way. The magical as well as symbolic effect of this obviously influenced Richard Wagner when he wrote the Beckmesser serenade in his *Meistersinger* opera.[70]

V

"Magic Carpet Ride" (Steppenwolf)

Max Reger did not choose the 'magic' of an instrumental characteristic transferred to another instrument. Instead he took the magical charm of one single sonority, just two tones (d flat and f in the two-line octave) forming the interval of a major third and maintained in syncopated movement throughout a whole composition; its title: *Aeolsharfe* (op. 75.11), simply 'Aeolian harp.'[71] Written in 1903, it is based on a German poem of the same title – and of not particularly high quality – by Hermann Lingg (1820-1905)[72] and seems to be an example of Art Nouveau in music.[73] The (somewhat importunate) recurrent third gives Reger the opportunity to use a kind of free tonality in which the ductus of the parts or voices plays a distinguished role, in the vocal part as well as in the instrumental parts below those steadily-recurrent tones.[74]

Max Reger, *Aeolsharfe*[75]

In a similar way about 17 years later the Dutch organ virtuoso and composer Gerard Bunk (1888-1958) gave the following instruction (for weighting the named three keys and producing 'Aeolian harp tones' only by drawknob): "Dieses Stimmungsbild ist folgenderweise zu spielen: Die Tasten e^2 a^2 e^3 werden *vor* dem Spiele durch drei Gegenstände, welche sie während des ganzen Stückes niederdrücken, beschwert. Erst dann werden die beiden angegebenen Register des III. (bezw. IV.) Manuals gezogen und wird in gleicher Weise nach Beendigung des Stückes das Aeoline-Register abgestossen. Das Erklingen und Verstummen der 'Aeolsharfentöne' [ppp] wird somit nur durch den *Registerzug* bewirkt."[76] This form of a 'prepared church organ'[77] that Bunk indicates for his 'atmospheric image' *Aeolsharfe* op. 12 no. 6, is predetermined in the so-called *prolongement* of some harmoniums, which makes it possible to fix single keys in the pushed down position.

VI

"Magician" (Lou Reed)

A performance of Henry Cowell's piano piece *Aeolian Harp* (1923)[78] can still be an act of 'magic' – at least (but not only) to those people who have never heard it before. They expect a composition wherein piano hammers strike some hundreds of strings (and ten fingers somewhat fewer than 88 keys) – and they find themselves experiencing the sight of only one hand touching very few (strictly selected) keys at all, while no one hears a single hammer striking a string.[79] What happens is above all a magical sound effect evoked by fingertips stroking choirs of strings, an effect corresponding to modern techniques of harp playing, which the harpist and composer Carlos Salzedo (1885-1961) had introduced in 1921[80]: Aeolian chords, Aeolian flux (furthermore: Aeolian rustling and Aeolian tremolo). Cowell's first composition for 'string piano' (as he called the new type of playing the piano by reaching for the strings and touching them with one's hand),[81] *Aeolian harp* is (concerning the harmonies) to some degree a neoclassicistic work in a tight and stringent style. The audience at the time of its origin may, having read its title, even have been in expectation of a drawingroom piece, for many of the leading salon composers had written 'Aeolian harp' opuses up to the turn of the century,[82] works which were still popular all over Europe as well as in the United States after World War I.[83] So we can suppose that those people, becoming aware of the actual (string) piano piece, were suprised between disappointment, outrage and magical amazement.

Delight rather than disillusionment had been aroused by another piano 'conjurer,' nearly a century before: Frédéric Chopin. His admirer and – for Germany – discoverer Robert Schumann invented the famous comparison[84] of Chopin's playing technique with the Aeolian harp, referring primarily to his soft, lyrical, melodious touch.

VII

"Now That the Magic Has Gone" (Joe Cocker)

The surprise was probably not far from that caused by Cowell's first 'Aeolian' performances,[85] when 54 years later a jazz fan put the LP *Dis*[86] on his turntable and switched it on for playing. Instead of the stated instruments saxophone, wood flute and guitar the listener heard a sonorous harmonic foundation of a structure not then familiar, consisting of one minute in C, and forty seconds in E, sounding (at least initially) like distant bells, before the tenor saxophone started. Together with the record cover showing a misty seashore[87] this opening suggests association with something like Vineta or Atlantis.

What first seems but mere effect, turns out to be conceptually important not only for the work of art *Dis* but also for a new lyrical and archaic view of jazz. Three takes (*Vandrere, Viddene, Dis*) on this LP are based on and extensively accompanied by wind harp sounds constituting both the musical and metaphysical background, as well as the tonal foundation for composition or carefully planned improvisation. The composer and interpreter Jan Garbarek uses Lydian scales as well as what's called *Molldur* in Germany: the scale resulting from use of the flattened submediant in major; furthermore he lays an accent on the 'pagan'[88] tones of the natural harmonic row and especially their coincidence with *blue notes*.[89]

Natural tones, minor (mode), sound stream, soft intensity and lontano effect, sound change, mediant relation, modulation – these characteristics of Garbarek's Aeolian harp compositions are, equally, some of the components constituting the 'wind harpist topos' in music;[90] moreover, several features of the other above-mentioned works belong to that *topos* (which in this case means a fluctuating combination of musical elements): slow motion, the sound remaining stationary, especially as a seventh (and ninth) chord, upper register, whole tone scale or harmony, arpeggio, open string tones, and a structurally emphasised position.

It is not frequent in the realms of jazz that a work meets such a wide reception and even more that it is used time and again as an atmospheric or illustrative musical background as has been the case with the pieces from Jan Garbarek's and Ralph Towner's long-playing record *Dis*.[91] Whether this fact should be interpreted as a sign of a common longing for some sort of magic again, and nowadays in the face of science and technology, may remain an open question; however, there is still a great yearning for the pure, natural and elementary experience, which seems to be granted by some 'unplugged' musical instruments, like – in a very special way – the Aeolian harp.

William Shakespeare, who could not have known real Aeolian harps but probably – with Spenser[92] – the relevant anticipatory and preparatory literary image, found some words[93] deserving consideration in this context: "If the true concord of well-tuned sounds, / By unisons married, do offend thine ear, / [...] /

[...] / Mark how one string, sweet husband to another, / Strikes each in each by mutual ordering; [...]."

Notes

[1] Wolfgang Caspar Printz, *Phrynis Mitilenaeus, Oder Satyrischer Componist* III 93. In the original edition the here italicised words are set in spaced type.

[2] Johann Friedrich Hugo Freiherr (baron) von Dalberg, one of those paving the way for the 'Romantic movement,' chose these Shakespearian words as a motto for the poem in prose *Die Aeolsharfe: Ein allegorischer Traum* ['An allegorical dream'] 1. Cf. Andrew Brown, *The Aeolian Harp in European Literature 1591-1892* 30, concerning Dalberg's "misquotations."

[3] Wolfgang Caspar Printz, *Historische Beschreibung der Edelen Sing- und Kling-Kunst* 200.

[4] See (for instance) Athanasius Kircher, *Ars Magna Lucis et Umbrae* 826f.

[5] See the exhibition catalogue: Vonderau Museum Fulda, *Magie des Wissens*.

[6] See Franz Maria Feldhaus, *Die Technik: Ein Lexikon: Der Vorzeit, der Geschichtlichen Zeit und der Naturvölker* 824, or various encyclopaedias.

[7] Several websites (search for instance under "athanasius kircher camera obscura") suggest, by mentioning or showing his depiction of a portable camera obscura in *Ars Magna Lucis et Umbrae* 912/913, Kircher being the inventor of the 'modern' camera obscura.

[8] Athanasius Kircher, *Neue Hall- und Thon-Kunst* 106; illustration to "Drittes Kunst=Werck. *Technasma III.* Ein anders von sich selbst klingend= und thönendes *instrument* [...]" (ibid. 105).

[9] Giovan Battista della Porta, *Magia naturalis* 1650: 559-571; 1713: 959-972. A camera obscura was described by Porta already in the first (four-book) edition – see Feldhaus, *Die Technik: Ein Lexikon* 226.

[10] Porta, *Magia naturalis* 1650: 632; 1713: 1040. See Feldhaus, *Die Technik: Ein Lexikon* 27.

[11] Athanasius Kircher, *Musurgia universalis sive Ars magna consoni et dissoni* II 352-355 (partly quoted in Friedrich Hugo von Dalberg, *Die Aeolsharfe* XII), and Kircher, *Phonurgia nova sive Conjugum mechanico-physicum artis & naturae paranympha phonosophia concinnatum* [...] 143-149 resp. Kircher, *Neue Hall- und Thon-Kunst Oder Mechanische Gehaim-Verbindung der Kunst und Natur durch Stimme und Hall-Wissenschafft gestifftet* [...] 104-108.

[12] These chapters, probably.

[13] Jean Paul, *Sämtliche Werke* I/1 677f.; 679.

[14] Ernst Theodor Amadeus Hoffmann, *Schriften* II 315f.; see Herman Meyer, "Zitierkunst in E. T. A. Hoffmanns 'Kater Murr'" 193f.

[15] E. T. A. Hoffmann, *Schriften* II 438.

[16] See Walter Windisch-Laube, *Einer luftgebornen Muse geheimnisvolles Saitenspiel: Zum Sinn-Bild der Äolsharfe in Texten und Tönen* I 30.

[17] See Stephen Bonner, *Aeolian Harp* I/1 80f.; I/2, plate 12 fig. 36; IV 113.

[18] See Bonner, ibid. I/1 80; Kircher, *Musurgia universalis* II 354, *Phonurgia nova* 147, *Neue Hall- und Thon-Kunst* 107.

[19] He did not, however, invent the term 'Aeolian harp,' which gradually appeared: "Æolium instrumentum" 1677, "Aeolus's Harp" and "The Harp of Aeolus" 1748, "Æolian Harp" 1750 – see Windisch-Laube, *Einer luftgebornen Muse geheimnisvolles Saitenspiel* I 49f.; 76; 237.

[20] Kircher, *Musurgia* II 352f., *Phonurgia* 145f., *Neue Hall- und Thon-Kunst* 105f.

[21] See Windisch-Laube, *Einer luftgebornen Muse geheimnisvolles Saitenspiel* I 67; 72.

[22] Before Kircher, too: Spenser, Balde. See for all these: Windisch-Laube, ibid. I 72f.

[23] See ibid. I 182f.

[24] See Ulf Scharlau, "Athanasius Kircher und die Musik um 1650: Versuch einer Annäherung an Kirchers Musikbegriff;" especially 56-59.

[25] See Jiri Marek, "Athanasius Kircher und die 'neue' Physik im 17. Jahrhundert."

[26] Cf. Schopenhauer's *Die Welt als Wille und Vorstellung* III § 52.

[27] This poetical idea extends within Romantic poetry from Dalberg 1801 (*Die Aeolsharfe: Ein allegorischer Traum* 6; cf. 39-45; 50-54 – concerning Thomson's *Ode on Aeolus's Harp* and Macpherson's *'Ossian'*) via E. T. A. Hoffmann 1812 (*Deux Trios* [...] *par Louis van Beethoven* – concerning the 'Ghost Trio,' *Schriften* V 127) as well as Shelley 1813 (*Queen Mab* v. 52/53) up to Wilhelm Hauff 1827 (*Der Schwester Traum*) and even Eduard Mörike 1837 (*An eine Äolsharfe*).

[28] William Wordsworth, *The Prelude* 8f. (1804: 1.104 resp. 1850: 1.96).

[29] John Keats, *Endymion* 1.786, in *The Poems* 126.

[30] See *Herzensergießungen eines kunstliebenden Klosterbruders.*

[31] Tobias Smollett's *The Adventures of Ferdinand Count Fathom* (1753), referring to baroque picaresque novels (*Don Quijote, Gil Blas*), was the first literary work mentioning heavenly choirs in connection with the Aeolian harp (231) after James Thomson's "Ethereal Race, inhabitants of Air" (the beginning of his *Ode on Aeolus's Harp*); Georg Christoph Lichtenberg's *Von der Äolus-Harfe* (1792; *Schriften und Briefe* II 297), opening with a citation of Edmund Spenser's *Ruins of time* (*Complaints I*), takes up that ethereal image (appearing again in Heinrich Christoph Koch's *Musikalischem Lexikon*, publ. in 1802). The Smollett (erotic) Aeolian harp scene later shows an effect on Hugo von Hofmannsthal's *Jedermann* in its first version (1905; *Dramen* III 78).

[32] The most popular one, "Tom Bowling," yearly performed within the 'Last Night of the Proms.'

[33] "[...] generally a mixture of narration and singing by a single person seated behind a table and facing an audience" – Roger Fiske and Irena Cholij, art. "Dibdin, Charles" 299.

[34] At the New Sans Souci Theatre (Leicester Square), London.

[35] Charles Dibdin, *The Aeolian Harp* 2.

[36] See Windisch-Laube, *Einer luftgebornen Muse geheimnisvolles Saitenspiel* I 305-308.

[37] Christian Schreiber, *16 Lieder* [1808] 25; the words (by the composer himself) were published already in 1804 – see Windisch-Laube, ibid. I 98f.

[38] While in French music there are rather few instances of Aeolian harp symbolism (about 11 counted so far by the author, whereas the vast majority of nearly 300 are to be found in German, Austrian and also British or American composers), the number of pioneer works amongst them is remarkable: the first instrumental and first orchestral work (Berlioz), the first piano piece (Friedrich Wilhelm Kalkbrenner, 1834), the first work in connection with a scholarly one (Kastner), and one of the first songs headed with 'Aeolian harp' (Adolphe Charles Adam). Although – some of the early vocal compositions with piano can also be performed purely instrumental, especially *Aeolian Harmony* by William Reeve (1806), the harp and piano score of which (from the "Romantic Drama of the White Plume," with large instrumental pre- and interludes) is suitable for performance as an individual piece; and Charles Dibdin's *The Aeolian Harp* (see above) has been published with a counter-melody, probably for violin, and an enclosed ver-

sion for two flutes.

39 *Musurgia universalis* II 352; 308; *Phonurgia nova* 145; *Neue Hall- und Thon-Kunst* 105.

40 See for instance (even the book title itself): Jan Brauers, *Von der Äolsharfe zum Digitalspieler: 2000 Jahre mechanische Musik* [...].

41 Hector Berlioz, *New Edition* IV 59f. See Peter A. Bloom, "Orpheus' Lyre Resurrected: A *Tableau Musical* by Berlioz."

42 See Windisch-Laube, *Einer luftgebornen Muse geheimnisvolles Saitenspiel* II 289f.

43 See ibid. I 256f.; II 284-288, and Peter A. Bloom (note 41).

44 First performance of this version: Weimar, 21/2/1855; earliest working on *Le Retour à la vie* from May 1831 on.

45 See also Hans Bartenstein, *Hector Berlioz' Instrumentalkunst* 84 / 185 (note); 89 / 190 (note).

46 Concerning the symphony (not the opera) orchestra; see ibid. 103 / 201f.

47 1841 in: *Revue et gazette musical de Paris* VIII 568; book edition 1843 and, modified, 1855.

48 Hector Berlioz, *Grand Traité* (1855) 144; *Instrumentationslehre* (1898) 87.

49 See Windisch-Laube, *Einer luftgebornen Muse geheimnisvolles Saitenspiel* II 293-300. For extensive depiction of *Lélio* see Wolfgang Dömling, *Hector Berlioz: Die symphonisch-dramatischen Werke* 38-54 and notes 134f. Ibid. 150-170: libretto of this work.

50 See Michael Heinemann, "Konstruktion und Programmatik: Zum fünften Satz des *Lelio* von Hector Berlioz" 77. He also gives a detailed analysis of the structure of this movement.

51 His origin was the quasi bilingual Alsace.

52 He lived from 1810 to 1869, and so he was of the same age as F. Chopin and R. Schumann. In Paris Kastner's teacher was Anton Reicha.

53 See below both under Primary Sources and Secondary Literature.

54 Besides that, he was one of the pioneers in composing for the saxophone (which had been developed by Adolphe Sax in 1841).

55 Johann Georg Kastner, *Traité Général d'Instrumentation* 24.

56 None of the musical works within them yet seems to be performed or recorded at all (cf. Philipp Spitta, *Johann Georg Kastner* 358).

57 Johann Georg Kastner, *Stephen ou La Harpe d'Éole* 31-36; 67f.; 88f.; 91f.; 116-120; 124.

58 In a narrow, limiting sense, not in that of instrumentology.

59 See within the section Primary Sources, below.

60 See Windisch-Laube, *Einer luftgebornen Muse geheimnisvolles Saitenspiel* II 673-678.

61 See his late concise piano pieces, as *Nuages gris or Am Grabe Richard Wagners* (both partly almost atonal), or his melodrama *Der traurige Mönch* (with extensive use of whole tone scales and augmented chords; *Musikalische Werke* VII/3 163-167) after Lenau; of course his great symphonic poems, too.

62 In a letter to Agnes Street-Klindworth from November 18, 1860 (cit. *Riemann Musik Lexikon* III 927).

63 See Windisch-Laube, *Einer luftgebornen Muse geheimnisvolles Saitenspiel* I 262f.

64 See ibid. also I 352.

65 Franz Liszt *Musikalische Werke* VII/3 17.

66 'The harps have been hung up in the willows' while the people mourns and cries being abroad.

67 Mediant relations are part of the 'musical-literary topos' "WindHarfner" ('wind harpist') already explained by the author of this study – see Windisch-Laube, *Einer luftgebornen Muse geheimnisvolles Sai-*

tenspiel I 327-352.

Nearly the same chords as used by Liszt in this passage had been taken by Hector Berlioz within his *La Harpe Éolienne* (see above) in the form of harp arpeggios which (almost symmetrically) structure the movement – see Michael Heinemann, "Konstruktion und Programmatik: Zum fünften Satz des *Lelio* von Hector Berlioz" 83.

[68] Peter Cornelius, *Musikalische Werke* (I) 200.

[69] Cf. the Aeolian harp itself, which in fact, from an organological point of view, is a zither, not a harp.

[70] See Windisch-Laube, *Einer luftgebornen Muse geheimnisvolles Saitenspiel* I 317f.

[71] Max Reger, *Sämtliche Werke* (XXXII) 175f.

[72] For a both extensive and critical acknowledgement see Windisch-Laube, *Einer luftgebornen Muse geheimnisvolles Saitenspiel* II 570-578; 656-664.

[73] See ibid. II 641-655.

[74] A detailed survey (the first and apparently only one to date) on Reger's remarkable composition can be found ibid. II 606-640 (the chapter on the whole, resp.: 569-664).

[75] Max Reger, *Sämtliche Werke* (XXXII) 175.

[76] Gerard Bunk, *Acht Charakterstücke für Orgel* (2) 8. The italicised words are underlined in the original.

[77] A line can be drawn from here by way of Henry Cowell (see below) up to John Cage's 'prepared piano' and further to Terry Riley (above all interesting in this context: *The Harp of New Albion*, 1985/86).

[78] For detailed analysis, explanation and interpretation see the relevant chapter in the author's core study on the Aeolian harp symbol (Windisch-Laube, *Einer luftgebornen Muse geheimisvolles Saitenspiel* II 804-847).

[79] The keys (in coordination with the pedal) are in this case only used to operate the dampers.

[80] Within his *Modern Study of the Harp*; see Windisch-Laube, ibid. I 54; II 805f.

[81] See ibid. II 834f.

[82] See ibid. II 535-537.

[83] See ibid. II 807-809.

[84] Robert Schumann, *Gesammelte Schriften über Musik und Musiker* II 103f.; see Windisch-Laube, *Einer luftgebornen Muse geheimnisvolles Saitenspiel* II 508-513; 517-520.

[85] Cowell was, temporarily (in the twenties), what today is called a composer-performer.

[86] Jan Garbarek: *Dis*, ECM 1093, Munich 1977 (recorded December 1976, Oslo); interpreters: Jan Garbarek, Saxophones and Wood Flute, Ralph Towner, Guitars.

[87] All the more (in correlation with the title *Dis*, which Norwegian means 'misty'), a jazz expert's expectation has been directed towards a 'jazzy' opening, for he may have had connotations of the famous song and piece *Misty*, first published by its composer Errol Garner in 1955.

[88] Here means: not corresponding with our tonal system.

[89] For detailed analysis of the *Dis*-pieces see Windisch-Laube, *Einer luftgebornen Muse geheimnisvolles Saitenspiel* II 848-860.

[90] See note 67, first part (especially cf. I 328).

[91] See Windisch-Laube, *Einer luftgebornen Muse geheimnisvolles Saitenspiel* II 860-864, and some more besides since then.

[92] Edmund Spenser, 1591 in *Complaints* v.603-616: the wind in Orpheus' apotheosised lyre – *The Works* (VIII) 54.

[93] Sonnet VIII, verse 5 up to verse 10.

Works Cited

1. Primary Sources

Berlioz, Hector. "De l'Instrumentation (Cinquième article)." *Revue et gazette musical de Paris* 8 (19 Dec. 1841): 567-569.

---. *Grand Traité d'Instrumentation et d'Orchestration Modernes.* Paris: Henri Lemoine, 1855 [Concerning the date of this edition see Windisch-Laube (under Secondary Literature), II 286: footnote].

---. *Instrumentationslehre.* Authorised German ed. Ed. Alfred Dörffel. 6th ed. Leipzig: Peters, 1898.

---. *New Edition of the Complete Works.* Vol. VI. Ed. David Gilbert. Kassel: Bärenreiter, 1998. Vol. VII. Ed. Peter A. Bloom. Kassel: Bärenreiter, 1992.

Bunk, Gerard. *Acht Charakterstücke für Orgel.* Issue 2. Leipzig: Leuckart, 1920.

Cornelius, Peter. *Musikalische Werke. Erste Gesamtausgabe.* Ed. Max Hasse. Vol. I: *Einstimmige Lieder und Gesänge mit Pianofortebegleitung.* Leipzig: Breitkopf and Härtel, 1905.

Cowell, Henry. *Piano Music by Henry Cowell.* Ed. Oliver Daniel. New York: Associated Music Publishers, 1960.

Dalberg, [Johann] Friedrich Hugo [Freiherr] von. *Die Aeolsharfe: Ein allegorischer Traum.* Erfurt: Beyer and Maring, 1801.

Dibdin, Charles. *The Æolian Harp, written & composed by Mr. Dibdin and Sung by him in his New Entertainment called Most Votes.* London: by the Author, 1802.

Garbarek, Jan. *Dis.* Munich: ECM Records (ECM 1093), 1977 [long-playing record; since 1986 available on CD].

Hoffmann, Ernst Theodor Amadeus. [*Schriften*] 5 vols. [II:] *Die Elixiere des Teufels: Lebens-Ansichten des Katers Murr.* Ed. Wolfgang Kron. Darmstadt: Wissenschaftliche Buchgesellschaft, 1979.

Hofmannsthal, Hugo von. *Gesammelte Werke in zehn Einzelbänden.* Ed. Bernd Schöller (and Rudolf Hirsch). Vol. III: *Dramen* III. 1893-1927. Frankfurt am Main: Fischer, 1979.

Jean Paul [Friedrich Richter]. *Werke.* Ed. Norbert Miller. Vol. I/1: *Die unsichtbare Loge: Hesperus.* Munich: Hanser, 1970. Reprint Frankfurt am Main: Zweitausendeins, 1996.

Kastner, [Jean] Georges. *Stéphen ou La Harpe d'Éole: Grand monologue lyrique avec chœurs.* Words by Francis Maillan. Music by Georges Kastner. Paris: Brandus, Renouard, 1856 (see section Secondary Literature).

---. *Traité Général d'Instrumentation.* Paris: Philipp, 1837.

Keats, John. *The Poems of John Keats.* Ed. Jack Stillinger. London: Heinemann, 1978.

Kircher, Athanasius. *Ars Magna Lucis et Umbrae.* Rome: H. Scheus, 1646.

---. *Musurgia universalis sive Ars magna consoni et dissoni.* Tomus I and Tomus II in one volume. Rome: F.Corbeletti, 1650. Repr. ed. Ulf Scharlau. Hildesheim: Olms, 1970.

---. *Neue Hall- und Thon-Kunst Oder Mechanische Gehaim-Verbindung der Kunst und Natur durch Stimme und Hall-Wissenschafft gestifftet* [...]. Ellwangen: Arnold Heyl, 1684. Repr. Hannover: Th. Schäfer, 1983.

---. *Phonurgia nova sive Conjugum mechanico-physicum artis & naturae paranympha phonosophia concinnatum* [...]. Campidonae [Kempten]: Rudolph Dreherr, 1673.

Koch, Heinrich Christoph. *Musikalisches Lexikon, welches die theoretische und praktische Tonkunst, encyclopädisch bearbeitet,* [...] *und die alten und neuen Instrumente beschrieben, enthält.* Frankfurt am Main: August Hermann d. J., 1802. Repr. Hildesheim: Olms, 1964, 1985.

Lichtenberg, Georg Christoph. *Schriften und Briefe.* Ed. Franz H. Mautner. Vol. II: *Aufsätze. Satirische Schriften.* Frankfurt am Main: Insel, 1992.

Liszt, Franz. *Musikalische Werke.* Ed. Franz-Liszt-Stiftung. 34 vols. Leipzig: Breitkopf and Härtel, 1907-1936. Repr. Farnborough: Gregg, 1967.

Porta, Giovan Battista (Giambattista) della. *Magiae Naturalis Libri Viginti.* Rothomagi [Rouen]: J. Berthelin, 1650.

---. *Magia Naturalis Oder: Hauß-Kunst- und Wunder-Buch.* Ed. Christian Peganium / Rautner. Nuremberg: Johann Ziegers Wittwe, 1713.

Printz (von Waldthurn), Wolfgang Caspar. *Historische Beschreibung der Edelen Sing- und Kling-Kunst* [...]. Dresden, 1690. Repr. ed. Othmar Wessely. Graz: Akademische Druck- und Verlagsanstalt, 1964.

---. *Phrynis Mitilenaeus, Oder Satyrischer Componist,* [...]. 3 vols. Dresden: Johann Christoph Mieth and Joh. Christoph Zimmermann, 1696.

Reeve, William. *Eolian Harmony: Sung by Mr. Hill at the Theatre Royal Covent Garden in the Romantic Drama of the White Plume: Composed & Arranged for the Harp & Piano Forte by William Reeve* [Words by Thomas Dibdin]. London: W. Turnbull, 1806.

Reger, Max. *Sämtliche Werke.* 35+3 vols. Vol. XXXII: *Sologesänge mit Klavier* II. Ed. Fritz Stein. Wiesbaden: Breitkopf and Härtel, 1958.

Schreiber, Christian. *16 Lieder.* 1. Sammlung. Leipzig: Breitkopf and Härtel, [1808].

Schreker, Franz. *Der ferne Klang.* Oper in 3 Aufzügen. Klavierauszug [...] von Alban Berg, erleichterte Ausgabe von Ferdinand Rebay. Vienna: Universal-Edition, 1911.

---. *Der ferne Klang.* Oper in 3 Aufzügen. Studienpartitur. Schliengen: Edition Argus, 1998.

Schumann, Robert. *Gesammelte Schriften über Musik und Musiker.* 4 vols. Leipzig 1854. Repr. ed. Gerd Nauhaus. Leipzig: Breitkopf and Härtel, 1985.

Smollett, Tobias. *The Adventures of Ferdinand Count Fathom.* Vol. I. Oxford: Basil Blackwell, 1925.

Spenser, Edmund. *The Works.* Vol. VIII: *The Minor Poems* II. Ed. Charles G. Osgood and Henry G. Lotspeich, assisted by Dorothy E. Mason. Baltimore: Johns Hopkins Press, 1966.

Wordsworth, William. *The Prelude or Growth of a Poet's Mind.* Ed. from the manuscripts by Ernest de Selincourt. 2nd ed. Oxford: Clarendon Press, 1959.

2. Secondary Literature

Bartenstein, Hans. *Hector Berlioz' Instrumentationskunst und ihre geschichtlichen Grundlagen: Ein Beitrag zur Geschichte des Orchesters.* Straßburg: Heitz, 1939.

Bloom, Peter A. "Orpheus' Lyre Resurrected: A *Tableau Musical* by Berlioz." *The Musical Quarterly* 41 (1975): 189-211.

Bonner, Stephen, ed. *Aeolian Harp.* Vol. II (by St.B.): *The History and Organology of the Aeolian Harp.* 2 parts. Vol. IV (by St.B.): *The Acoustics of the Aeolian Harp. Supplement to Volumes 1-3.* Duxford: Bois de Boulogne, 1970 (II), 1974 (IV).

Brauers, Jan. *Von der Äolsharfe zum Digitalspieler: 2000 Jahre mechanische Musik – 100 Jahre Schallplatte.* Munich: Klinkhardt and Biermann, 1984.

Brown, Andrew. *The Aeolian Harp in European Literature 1591-1892*, with additional material by Nicholas Boyle. *Aeolian Harp.* Vol. III. Ed. Stephen Bonner. Duxford: Bois de Boulogne, 1970.

Dömling, Wolfgang. *Hector Berlioz: Die symphonisch-dramatischen Werke.* Stuttgart: Reclam, 1979.

Erhardt-Siebold, Erika von. "Some Inventions of the Pre-Romantic Period [Camera obscura, Colour-organ, Aeolian harp] and their Influence upon Literature." *Englische Studien* 66 (1931-32): 347-363.

Feldhaus, Franz Maria. *Die Technik: Ein Lexikon: Der Vorzeit, der Geschichtlichen Zeit und der Naturvölker.* Leipzig, Berlin: W. Engelmann, 1914. Repr. Munich: Moos, 1970. Unveränderte Sonderausgabe. Wiesbaden: R. Löwit.

Fiske, Roger and Irena Cholij. Art. "Dibdin, Charles", in *The New Grove Dictionary of Music and Musicians.* Ed. Stanley Sadie. 29 vols. London: Macmillan, 2001. Vol. VII. 298-301.

Heinemann, Michael. "Konstruktion und Programmatik: Zum fünften Satz des *Lelio* von Hector Berlioz." *Musik & Ästhetik* 8 .29 (2004): 74-84.

Kastner, [Jean] Georges. *La Harpe d'Éole et la musique cosmique: Études sur les rapports des phénomènes sonores de la nature avec la science et l'art. Suivies de Stéphen ou La Harpe d'Éole* [...; see Primary Sources]. Paris: Brandus, Renouard, 1856.

Marek, Jiri. "Athanasius Kircher und die 'neue' Physik im 17. Jahrhundert," in *Athanasius Kircher und seine Beziehungen zum gelehrten Europa seiner Zeit.* Ed. John Fletcher. Wiesbaden: Otto Harrassowitz, 1988. 37-51.

Meyer, Herman. "Zitierkunst in E. T. A. Hofmanns 'Kater Murr'," in *Interpretationen.* Ed. Jost Schillemeit. Vol. IV: *Deutsche Erzählungen von Wieland bis Kafka.* Frankfurt am Main: Fischer, 1966. 179-195.

Riemann Musik Lexikon. 12. völlig neubearbeitete Auflage. 3+2 vols. Vol. III: Sachteil. Ed. Willibald Gurlitt and Hans Heinrich Eggebrecht. Mainz: Schott, 1967.

Scharlau, Ulf. "Athanasius Kircher und die Musik um 1650. Versuch einer Annäherung an Kirchers Musikbegriff," in *Athanasius Kircher und seine Beziehungen zum gelehrten Europa seiner Zeit.* Ed. John Fletcher. Wiesbaden: Otto Harrassowitz, 1988. 53-67.

Spitta, Philipp. "Johann Georg Kastner," in *Musikgeschichtliche Aufsätze.* Berlin: Gebrüder Paetel, 1894. 335-361.

Vonderau Museum Fulda, ed. *Magie des Wissens: Athanasius Kircher (1602-1680): Jesuit und Universalgelehrter.* Kurzführer zur Ausstellung 17. Januar – 16. März 2003. Petersberg: Michael Imhof, 2003.

Windisch-Laube, Walter. *Einer luftgebornen Muse geheimnisvolles Saitenspiel: Zum Sinn-Bild der Äolsharfe in Texten und Tönen.* Parts I and II in 3 vols. Mainz: Are, 2004 (Musik im Kanon der Künste. Ed. Ulrike Kienzle and Thomas Lindner. Vol. 3).

Notes on Contributors

Tobias Döring is Professor of English Literature at the Ludwig-Maximilians-Universität Munich.

Walter Göbel is Professor of English and American Studies at the Universität Stuttgart.

Brean Hammond is Professor Emeritus of Modern English Literature at the University of Nottingham.

Andreas Höfele is Professor of English Literature at the Ludwig-Maximilians-Universität Munich.

Stephan Laqué is Senior Lecturer of English Literature at the Ludwig-Maximilians-Universität Munich.

Roger Luckhurst is Professor in Modern and Contemporary Literature at Birkbeck College, University of London.

Jarmila Mildorf is Senior Lecturer of English Literature at the Universität Paderborn.

Susanne Rupp is Professor of English Literature at the Universität Hamburg.

Elmar Schenkel is Professor of English Literature at the Universität Leipzig.

Gregor Schiemann is Professor of Philosophy at the Bergische Universität Wuppertal.

Johann N. Schmidt is Professor Emeritus of English Literature at the Universität Hamburg.

Hans Ulrich Seeber is Professor Emeritus of English Literature at the Universität Stuttgart.

Robert Stockhammer is Professor of Comparative Literature at the Ludwig-Maximilians-Universität Munich.

Dirk Vanderbeke is Professor of English Literature at the Friedrich-Schiller-Universität Jena.

Martin Windisch is Senior Lecturer of English Literature at the Universität Stuttgart.

Walter Windisch-Laube is musicologist and literary scholar, works as a creative artist and writer, and is director of Alsfeld school of music.

Kultur und Technik

Schriftenreihe des Internationalen Zentrums für Kultur- und Technikforschung der Universität Stuttgart

hrsg. von Prof. Dr. Georg Maag, Prof. Dr.Ing. Helmut Bott, Prof. Dr. Gerd de Bruyn, Prof. Dr. Walter Göbel, und Prof. Dr. Ortwin Renn

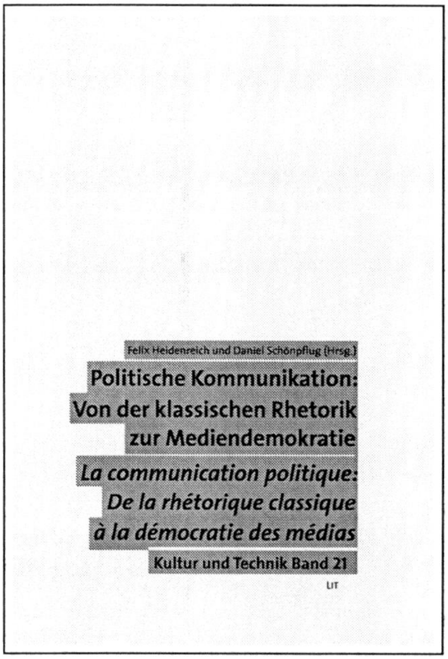

Felix Heidenreich; Daniel Schönpflug (Hg.)

Politische Kommunikation: Von der klassischen Rhetorik zur Mediendemokratie. Communication politique: De la rhétorique classique à la démocratie des médias

Jedes neue Verbreitungsmedium hat spezifische Formen politischer Kommunikation ermöglicht: Die klassische Rhetorik reflektierte die Rationalitätspotenziale einer stilistisch anspruchsvollen Rede; auch Briefe und politische Biographien ordnen politische Kommunikation durch genrespezifische Erwartungen und Ansprüche.

Die elektronischen Medien scheinen eine neue Dynamik zu entfalten, die man mit Begriffen wie Mediokratie, Boulevardisierung oder Politainment zu verstehen sucht. Denn die Massenmedien lassen sich für eine meist bildorientierte Rhetorik nutzen, die die klassischen Topoi der politischen Inszenierung zwar aufgreift, diese aber zugleich umdeutet.

Bd. 21, 2012, 192 S., 24,90 €, br.,
ISBN 978-3-643-11742-7

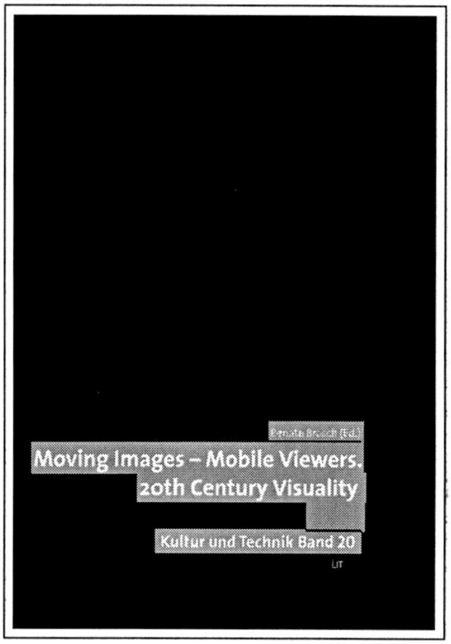

Renate Brosch (Ed.)

Moving Images – Mobile Viewers: 20th Century Visuality

Vision and movement seem to have shifted centre stage in modes of experience in the last century: as a result of their joint effect slow contemplative gazes at static images seem to be increasingly displaced by distracted, övernacularö ways of seeing. Looking out of the window of a speeding car, receiving photographs of planet earth from outer space, watching the flickering images of the TV screen, or sending images via mobile phones or webcams - all these are unique visual experiences that were impossible before various inventions in the 20th century originated completely new kinds of movement.

Bd. 20, 2011, 256 S., 24,90 €, br.,
ISBN 978-3-643-11164-7

LIT Verlag Berlin – Münster – Wien – Zürich – London

Auslieferung Deutschland / Österreich / Schweiz: siehe Impressumsseite